AS & A Level

BIOLOGY
through diagrams

W R Pickering

OXFORD

UNIVERSITY PRESS

OXFORD
UNIVERSITY PRESS

Great Clarendon Street, Oxford OX2 6DP

Oxford University Press is a department of the University of Oxford.
It furthers the University's objective of excellence in research, scholarship,
and education by publishing worldwide in

Oxford New York

Auckland Bangkok Buenos Aires Cape Town Chennai
Dar es Salaam Delhi Hong Kong Istanbul Karachi Kolkata
Kuala Lumpur Madrid Melbourne Mexico City Mumbai Nairobi
São Paulo Shanghai Taipei Tokyo Toronto

Oxford is a registered trade mark of Oxford University Press
in the UK and in certain other countries

British Library Cataloguing in Publication Data

Data available

ISBN 0 19 914197-5

10 9 8 7 6 5 4

Typeseting, artwork and design by Hardlines, Charlbury, Oxon

Printed in Great Britain

CONTENTS

Specification structures

	Edexcel	AQA Specification A	AQA Specification B
Unit 1	1 h 20 min AS 33.3% A 16.7% Structured questions	1 h 30 min AS 35% A 17.5% Structured questions	1 h 15 min AS 30% A 15% Structured questions
	Molecules and cells: cells and organelles; biological molecules; enzymes; chromosomes and the genetic code; protein synthesis and applications of gene technology	**Molecules, cells and systems:** cells and organelles; biological molecules; enzymes; tissues; heart and circulation	**Core principles:** biological molecules; cells; cell transport; exchanges with the environment; enzymes; digestion
Unit 2	1 h 20 min AS 33.3% A 16.7% Structured questions	1 h 30 min AS 35% A 17.5% Structured questions	1 h 15 min AS 30% A 15% Structured questions
	Exchange, transport and reproduction: gas exchange; transport systems; adaptation to the environment; sexual reproduction; placental development and birth	**Making use of biology:** use of enzymes; chromosomes and the genetic code; protein synthesis and applications of gene technology; forensic science; plant cultivation; manipulation of reproduction	**Genes and genetic engineering:** the genetic code and protein synthesis; the cell cycle; sexual reproduction; applications of gene technology
Unit 3	1 h written test AS 18% A 9% Longer structured questions	AS 30% A 15%	1 h 15 min AS 25% A 12.5% Structured questions
	Energy and the environment: heterotrophic nutrition; energy flow through ecosystems; recycling of nutrients; human influences on the environment	No written paper	**(a) Physiology and transport:** transport systems; the control of breathing and heartbeat; energy and exercise; plant transport
	Course work AS 15.3% A 5.7%	Coursework/practical	(b) Coursework: AS 15% A 7.5%
Unit 4	1 h 30 min A 16.7% Structured questions	1 h 30 min A 15% Short answer and structured questions	1 h 30 min A 15% Structured questions and comprehension
	Core material (includes respiration and coordination) plus one **option** from: microbiology and biotechnology; food and nutrition; human health and fitness	**A: Variation and mechanisms of inheritance and evolution:** meiosis; Mendelian genetics; selection, evolution and speciation; the five kingdom system of classification **B: The biology of ecosystems:** respiration and photosynthesis; energy flow through ecosystems; carbon and nitrogen cycles; human effects on ecosystems	**Energy, control and continuity:** energy supply; photosynthesis; respiration; survival and co-ordination; homeostasis; muscles; inheritance; variation, selection and evolution; classification
Unit 5	1 h 30 min A 16.7% Structured questions and longer 'prose' question	1 h 30 min A 15% (incl. 5% synoptic) Short answer and structured questions	1 h 15 min A 7.5% (incl. 3.5% synoptic) Longer structured questions and comprehension
	Genetics, evolution and biodiversity: autotrophic nutrition; biodiversity and classification; continuity of species and speciation; succession and stability in ecosystems	**Physiology and the environment:** Plants and water; homeostatic mechanisms; gas exchange; heterotrophic nutrition; coordination	**(a) Environment:** energy flow through ecosystems; nutrient recycling; studying ecosystems; dynamics of ecosystems; human effects on ecosystems
			(b) Coursework 7.5% (incl. 2.5% synoptic)
Unit 6	1 h 10 min written test A 8.7% Longer structured questions + essay	1 h 45 min A 10% Longer structured questions + essay	2 h Section A – A 10% (incl. 4% synoptic) Section B – A 10% (all synoptic) Long structured questions + essay
	Synoptic test: content from all modules in AS and A2	Synoptic assessment: coverage of all modules in AS and A2	A: Options: applied ecology; microbes and disease; behaviour and population B: synoptic coverage of modules 1–5
	Practical examination or coursework A 8%	Centre-assessed practical A 10% (all synoptic)	

WHAT ARE
Structured questions?

This type of question is broken up into smaller parts. There is usually only a single mark, or a few marks, for each part. Some parts will ask you to

- recall a fact – the name of a biological molecule, for example
- define a biological term – ecological succession, for example
- obtain information from graphs or tables of data

- draw labelled diagrams, or add labels/annotations to a diagram.

Other parts might

- lead you to the explanation of a complex problem by asking you to provide a series of answers in sequence – a calculation of probability, for example.

	OCR	CCEA	WJEC
Unit 1	1 h 15 min AS 30% A 15% Structured questions with few requiring extended answers	1 h 30 min AS 16.7% A 8.3% Short structured questions + one requiring a longer 'prose' answer	1 h 40 min AS 35% A 17.5% Short structured questions + one essay
	Biology foundation: biological molecules; cell structure and transport across membranes; enzymes; genetic code and protein synthesis; genetic engineering; mitosis; energy and ecosystems	**Cell biology:** biological molecules; cell structure and transport across membranes; enzymes; genetic code and protein synthesis; mitosis; genetic engineering	**Fundamental concepts and organisation:** biological molecules; cell structure and transport across membranes; enzymes; genetic code and protein synthesis; mitosis
Unit 2	1 h 15 min AS 30% A 15% Structured questions with few requiring extended answers	1 h 30 min AS 16.7% A 8.3% Short structured questions + one requiring a longer 'prose' answer	1 h 40 min AS 35% A 17.5% Short structured questions + one essay
	Human health and disease: diet and health; gas exchange and exercise; smoking and disease; infectious diseases and immunity	**Physiology and ecology:** transport systems; gas exchange; effects of exercise; energy flow through ecosystems; populations; biological control	**Adaptations and ecology:** transport systems; gas exchange; energy flow through ecosystems; populations; human effects on the environment
Unit 3	1h AS 20% A 10% Structured questions with few requiring extended answers	1 h AS 9.3% A 4.6% Structured questions + some requiring graphical and planning skills	
	Transport: heart and circulation; transport in plants	**Practical processes**	Practical contents based on Units 1 and 2
	Course work or practical examination AS 20% A 10%	Coursework AS 7.4% A 4.8%	Practical assessment 3h 45 min AS 30% A 15%
Unit 4	1 h 30 min A 15% Structured questions with more requiring extended answers	1 h 30 min A 16.7% Short structured questions + one requiring a longer 'prose' answer	1 h 40 min A 15% Longer structured questions
	Central concepts: energy and respiration; photosynthesis; populations and interactions; meiosis and inheritance; selection, evolution and classification; control, co-ordination and homeostasis	**Coordination, biochemistry and environment:** energy and respiration; photosynthesis; immunity; co-ordination and control	**Biochemistry and health:** energy and respiration; photosynthesis; digestion and absorption; microbiology and disease; pathogens and control of infection
Unit 5	1 h 30 min A 15% (incl 5% synoptic) Structured questions with more requiring extended answers	1 h 30 min A 16.7% Short structured questions + one requiring a longer 'prose' answer	2 h A 20% (incl. 6.5% synoptic) Longer structured questions + 'essay-style' question
	Options – one of: growth, development and reproduction; applications of genetics; environmental biology; microbiology and biotechnology; mammalian physiology and behaviour	**Reproduction, genetics and taxonomic diversity:** inheritance and variation; mutation; population genetics; sexual and asexual reproduction; classification	**A: Variety and control:** inheritance and variation; mutation; sexual reproduction; classification; control systems and coordination **B:** synoptic assessment of units 1–5A
Unit 6	1 h A 10% (all synoptic) Structured questions with more requiring extended answers	1 h A 9.3% Longer structured questions, including requirement for statistics – one essay	
	Content of units 1–4	Content of units 1–5	No written paper
	Coursework or practical examination A 10%	Coursework A 7.4%	Practical assessment/synoptic practical 4 h A 15% (incl. 7.5% synoptic)

Comprehension questions?

In this type of question you will be given a passage about a biological topic and then asked a series of questions to test your understanding of the topic and the scientific principles in it. The actual content of the passage may be biological material that you're not familiar with.

'Essay' questions?

These questions will test your ability to recall information and to organise it into the form of written prose. Some credit will be given for 'style' (good organisation of material) as well as for the biological content.

Synoptic questions?

These questions are only used for A2 examination papers. They will test your ability to deal with information from different parts of your course, and to use biological skills in contexts which might be unfamiliar to you. These questions may be structured in style, but will be longer than those used in the AS papers. 20% of the A level marks are allocated to synoptic questions.

Pathways

The following pathways identify the **main sections** in this book that relate to course units for each Examination Board. Note that
- you will not necessarily need all of the material given in any one section
- there may be material in other sections that you need to know
- you should identify the relevant material by referring to the specification you are following.
- If you own the book you could highlight all of the relevant information.

Units 1, 2 and 3 are AS units and Units 4, 5 and 6 are the A2 units. All units make up the full 'A' level.

	Edexcel	Book sections
Unit 1	**Molecules and cells:** cells and organelles biological molecules enzymes chromosomes and the genetic code protein synthesis	1, 2, 3, 4, 5, 6, 7, 8, 9, 10, 11, 12, 13, 14, 34, 35 15, 16, 17, 18, 19, 20, 21, 32, 33 21, 22, 23, 24 165, 166, 167, 170 167, 168, 169
Unit 2	**Exchange, transport and reproduction:** gas exchange digestion and absorption transport systems adaptation to the environment sexual reproduction placental development and birth	36, 52, 106, 107, 108, 110, 111 85, 86, 87, 88, 89 43, 44, 45, 46, 47, 48, 49, 50, 112, 113, 114, 115, 116, 117, 118 51, 58 54, 55, 56, 57, 146, 147, 148, 149, 150, 151 152, 153
Unit 3	**Energy and the environment:** heterotrophic nutrition autotrophic nutrition energy flow through ecosystems recycling of nutrients human influences on the environment	92, 93, 94 36, 37, 38, 39, 40, 41, 42 59, 63, 64, 65, 66 67, 68 69, 70, 71, 72, 74, 75
	Course work AS 15.3% A 5.7%	
Unit 4	**Core material** respiration regulation coordination plus one **option** from microbiology and biotechnology food and nutrition human health and fitness	25, 26, 27, 28, 29, 30 126, 127, 128, 129, 130 146, 147, 148, 154, 155, 156, 157, 158, 159 134, 135, 136, 144 84, 90, 91, 92, 93, 94, 95, 96, 97, 98, 99, 100 137, 138, 139, 140, 161, 162, 163, 164
Unit 5	**Genetics, evolution and biodiversity:** autotrophic nutrition biodiversity and classification continuity of species and speciation genetics applications of gene technology	36, 37, 38, 39, 40, 41, 42 53, 60, 61, 62, 66, 196, 197, 198 193, 194, 195 178, 180, 183, 184, 185, 186, 187, 188, 189, 190, 191, 192, 193 176, 177, 178, 179, 180, 181
Unit 6	Synoptic test: content from all modules in AS and A2	
	Practical examination or coursework A 8%	See introductory pages.

	AQA Specification A	Book sections
Unit 1	**Molecules, cells and systems:** cells and organelles techniques membranes biological molecules enzymes tissues heart and circulation	1, 2, 11, 12, 13 3, 4, 12 7, 9 16, 17, 18, 19, 20, 21 22, 23 34, 35 106, 107, 108, 110, 112, 113, 114, 118, 119, 120, 121, 122, 123
Unit 2	**Making use of biology:** use of enzymes chromosomes and the genetic code protein synthesis applications of gene technology plant cultivation manipulation of reproduction	24 176, 177, 181 178, 179 199, 200, 201, 204 73, 77, 78, 79, 80 152
Unit 3	No written paper	
	Coursework/practical	See introductory pages.

	AQA Specification A	Book sections
Unit 4	**A: Variation and mechanisms of inheritance and evolution:** Meiosis Mendelian genetics selection, evolution and speciation the five kingdom system of classification **B: The biology of ecosystems:** respiration photosynthesis energy flow through ecosystems carbon and nitrogen cycles human effects on ecosystems	 182 186, 187, 188, 189, 190, 191, 192 183, 193, 194, 195 196, 197 25, 26, 27, 28, 29, 30 36, 37, 38, 39, 40, 41, 42 59, 60, 61, 62, 63, 64, 65, 66 67, 68 72
Unit 5	**Physiology and the environment:** Plants and water homeostatic mechanisms gas exchange heterotrophic nutrition co-ordination	 43, 44, 49, 50, 51 101, 102, 126, 127, 128, 129, 130, 132 106, 115, 116, 117 85, 86, 87, 88, 89, 105 146, 147, 148, 154, 155, 156, 157, 158, 159
Unit 6	Synoptic assessment: coverage of all modules in AS and A2 Centre-assessed practical A 10% (all synoptic)	 See introductory pages.

	AQA Specification B	Book sections
Unit 1	**Core principles:** biological molecules cells; cell transport exchanges with the environment enzymes digestion	 5, 6, 16, 17, 18, 19, 20, 21 2, 3, 4, 10, 11, 13, 34, 35 106, 107, 108, 110 22, 23 85, 86, 88, 89
Unit 2	**Genes and genetic engineering:** the genetic code and protein synthesis the cell cycle sexual reproduction applications of gene technology	 32, 33, 176, 177, 178, 179, 184, 185 181 57, 182 199, 200, 201, 202, 203, 204
Unit 3	**(a) Physiology and transport:** transport systems energy and exercise plant transport **(b) Coursework:** AS 15% A 7.5%	 112, 113, 114, 115, 116, 117, 118, 119, 120 22, 31 43, 44, 45, 46, 47, 48, 49, 50, 51, 52
Unit 4	**Energy, control and continuity:** photosynthesis respiration survival and coordination homeostasis nervous system muscles inheritance variation, selection and evolution classification	 36, 37, 38, 39, 40, 41, 42 25, 26, 27, 28, 29, 30 151, 156 126, 127, 129, 130, 132 147, 148, 149, 155, 157, 158, 159, 160 163 182, 186, 187, 189, 190, 191, 192 183, 193, 194, 195 196, 197
Unit 5	**(a) Environment:** energy flow through ecosystems nutrient recycling studying ecosystems dynamics of ecosystems human effects on ecosystems **(b) Coursework:** 7.5% (incl. 2.5% synoptic)	 59, 63, 64, 65 67, 68 60, 61 62, 66 72, 74, 75, 80
Unit 6	A: Options: applied ecology microbes and disease behaviour and population B: synoptic coverage of modules 1–5	 59, 60, 61, 74, 76, 77, 78, 79 134, 135, 136, 137, 138, 139, 140, 141, 144 109, 124, 125, 154, 168, 169, 170, 171, 172, 173, 174, 175

	OCR	Book sections
Unit 1	**Biology foundation:** biological molecules cell structure and transport across membranes enzymes genetic code and protein synthesis genetic engineering mitosis energy and ecosystems	 5, 6, 15, 16, 17, 18, 19, 20, 21 1, 2, 3, 4, 7, 9, 10, 11, 13, 34, 35 22, 23 32, 33, 165, 166, 167, 168, 169, 170 177, 178, 179 181 59, 63, 64, 65, 68
Unit 2	**Human health and disease:** diet and health gas exchange and exercise smoking and disease infectious diseases and immunity	 84, 90, 91, 92, 100, 124, 125 107, 108, 110, 112, 164 109, 124, 125 137, 138, 139, 140, 141, 144, 145
Unit 3	**Transport:** heart and circulation transport in plants	 112, 113, 114, 115, 116, 117, 118, 119, 120, 121 43, 44, 45, 46, 47, 48, 49, 50, 51, 52
	Coursework or practical examination AS 20% A 10%	
Unit 4	**Central concepts:** energy and respiration photosynthesis populations and interactions meiosis and inheritance selection, evolution and classification control, coordination and homeostasis	 25, 26, 27, 28, 29, 30 36, 37, 38, 39, 40, 41, 42 60, 62, 66, 72, 75 180, 182, 184, 186, 187, 188, 189, 191, 192 183, 193, 194, 195, 196, 197 53, 126, 127, 128, 129, 130, 132, 146, 151, 153, 155, 156, 157, 158, 159
Unit 5	**Options** – one of: growth, development and reproduction applications of genetics environmental biology microbiology and biotechnology mammalian physiology and behaviour	 53, 54, 55, 56, 57, 165, 166, 167, 168, 169, 171, 172, 173, 174, 175 183, 184, 185, 186, 190, 199, 202, 203 59, 60, 61, 69, 70, 71, 72, 76, 188 13, 14, 24, 82, 83, 135, 136, 142, 202 84, 85, 88, 89, 101, 102, 147, 148, 153, 160, 161, 162, 163
Unit 6	Content of units 1–4	
	Coursework or practical examination A 10%	See introductory pages.

	CCEA	Book sections
Unit 1	**Cell biology:** biological molecules cell structure and transport across membranes enzymes genetic code and protein synthesis mitosis genetic engineering	 5, 6, 8, 15, 16,17, 18, 19, 20, 21, 32, 33 1, 2, 3, 4, 7, 9, 10, 11, 13, 34, 35 22, 23 177, 178, 179 181 200, 201, 203
Unit 2	**Physiology and ecology:** transport systems gas exchange energy flow through ecosystems populations biological control	 36, 43, 44, 45, 46, 47, 48, 49, 50, 51, 112, 113, 114, 115, 116, 117, 118, 119 31, 85, 88, 89, 106, 107, 108, 110, 111 59, 63, 64, 65 60, 61, 62, 103 78
Unit 3	**Practical processes** Coursework AS 7.4% A 4.8%	
Unit 4	**Coordination, biochemistry and environment:** homeostasis immunity energy and respiration photosynthesis coordination and control nutrient cycling human effects on the environment	 126, 129, 130, 151 137, 138, 139, 140, 141, 143 25, 26, 27, 28, 29, 30 36, 37, 38, 39, 40, 41, 42 53, 147, 148, 149, 155, 156, 157, 158, 159, 163 67, 68 69, 70, 71, 72, 74, 75, 80
Unit 5	**Reproduction, genetics and taxonomic diversity:** inheritance and variation mutation population genetics classification	 182, 185, 186, 187, 189, 190, 191, 192, 193, 194 183, 184 188 57, 58, 196, 197, 198
Unit 6	Content of units 1–5	
	Coursework A 7.4%	See introductory pages.

	WJEC	Book sections
Unit 1	**Fundamental concepts and organisation:** biological molecules cell structure and transport across membranes enzymes genetic code and protein synthesis gene technology mitosis	5, 6, 8, 15, 16, 17, 18, 19, 20, 21 1, 2, 3, 4, 7, 9, 10, 11, 13, 34, 35 22, 23, 24 32, 33, 177, 178, 179 199, 200, 201, 203 181
Unit 2	**Adaptations and ecology:** transport systems gas exchange energy flow through ecosystems populations human effects on the environment	36, 43, 44, 45, 46, 47, 48, 49, 50, 51, 52, 112, 113 31, 85, 88, 89, 106, 107, 108, 110, 111 59, 63, 64, 67, 68, 69, 75, 103, 104 60, 61, 62, 66 72, 76, 79, 80
Unit 3	Practical contents based on Units 1 and 2	
Unit 4	**Biochemistry and health:** energy and respiration photosynthesis digestion and absorption microbiology and disease pathogens and control of infection	25, 26, 27, 28, 29, 30 36, 37, 38, 39, 40, 41, 42 85, 86, 87, 88, 89, 101, 102 135, 136, 202 134, 137, 138, 139, 140, 141, 142, 143, 144, 145
Unit 5	**A: Variety and control:** inheritance and variation mutation sexual reproduction classification control systems and co-ordination **B:** synoptic assessment of units 1–5A	182, 183, 186, 187, 189, 191, 192, 193, 194, 195 184, 185 54, 55, 56, 165, 166, 167, 168, 169, 170, 171, 172 196, 197, 198 126, 129, 130, 150, 151, 153, 155, 156, 157, 158, 159, 160, 163
Unit 6	No written paper Practical assessment/synoptic practical 4 h A 15% (incl. 7.5% synoptic)	

There is no one method of revising which works for everyone. It is therefore important to discover the approach that suits you best. The following rules may serve as general guidelines.

GIVE YOURSELF PLENTY OF TIME

Leaving everything until the last minute reduces your chances of success. Work will become more stressful, which will reduce your concentration. There are very few people who can revise everything 'the night before' and still do well in an examination the next day.

PLAN YOUR REVISION TIMETABLE

You need to plan your revision timetable some weeks before the examination and make sure that your time is shared suitably between all your subjects.

Once you have done this, follow it – don't be side-tracked. Stick your timetable somewhere prominent where you will keep seeing it – or better still put several around your home!

RELAX

Concentrated revision is very hard work. It is as important to give yourself time to relax as it is to work. Build some leisure time into your revision timetable.

GIVE YOURSELF A BREAK

When you are working, work for about an hour and then take a short tea or coffee break for 15 to 20 minutes. Then go back to another productive revision period.

REVISION

Success in examinations

You may already have looked at the section in this book covering the brain (160). If you have you might remember that an important function of the cerebral hemispheres is to act as an *integration centre* – input of information is compared with previous experience and an appropriate action is taken. As you prepare for an examination you will be inputting factual material and skills and you will be hoping that, when you're faced with examination papers, you will be able to make the appropriate responses! The effort you make in *revision* and your willingness to *listen to advice on techniques* will greatly affect your likelihood of success, as outlined here.

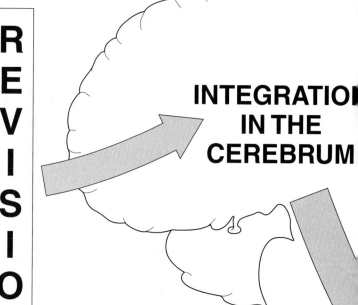

INTEGRATION IN THE CEREBRUM

KEEP TRACK

Use checklists and the relevant examination board specification to keep track of your progress. Mark off topics you have revised and feel confident with. Concentrate your revision on things you are less happy with.

MAKE SHORT NOTES, USE COLOURS

Revision is often more effective when you do something active rather than simply reading material. As you read through your notes and textbooks make brief notes on key ideas. If this book is your own property you could highlight the parts of pages that are relevant to the specification you are following. Concentrate on understanding the ideas rather than just memorizing the facts.

PRACTISE ANSWERING QUESTIONS

As you finish each topic, try answering some questions. There are some in this book to help you (see pages xxvii and xxxii). You should also use questions from past papers. At first you may need to refer to notes or textbooks. As you gain confidence you will be able to attempt questions unaided, just as you will in the exam.

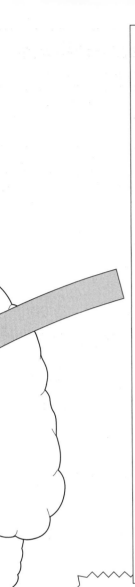

KNOW WHAT TO DO

LEARN THE KEY WORDS

Name: the answer is usually a technical term (mitochondrion, for example) consisting of no more than a few words. *State* is very similar, although the answer may be a phrase or sentence. *Name* and *state* don't need any elaboration, i.e. there's no need for explanation.

Define: the answer is a formal meaning of a particular term. *What is meant by...?* is often used instead of *define*.

List: You need to write down a number of points (each may only be a single word) with no elaboration or explanation.

Describe: your answer will simply say what is happening in a situation shown in the question, e.g. 'the temperature increased by 25 °C' – *there is no need for explanation.*

Suggest: you will need to use your knowledge and understanding of biological topics to explain an effect that may be new to you. You might use a principle of enzyme action to suggest what's happening in an industrial process, for example. There may be more than one acceptable answer to this type of question.

Explain: the answer will be in extended prose, i.e. in the form of complete sentences. You will need to use your knowledge and understanding of biological topics to expand on a statement that has been made in the question or earlier in your answer. Questions of this type often ask you to *state and explain....*

Calculate: a numerical answer is to be obtained, usually from data given in the question. Remember to
- give your answer to the correct number of significant figures, usually two or three
- give the correct unit
- show your working.

IN THE EXAMINATION

- Check that you have the correct question paper! There are many options in some specifications, so make sure that you have the paper you were expecting.
- Read through the whole paper before beginning. Select the questions you are most comfortable with – there is no rule which says you must answer the questions in the order they are printed!
- Read the question carefully – identify the key word (why not underline it?).
- Don't give up if you can't answer part of a question. The next part may be easier and may provide a clue to what you should do in the part you find difficult.
- Check the number of marks allocated to each section of the question.
- Read data from tables and graphs carefully. Take note of column headings, labels on axes, scales, and units used.
- Keep an eye on the clock – perhaps check your timing after you've finished 50% of the paper.
- Use any 'left over' time wisely. Don't just sit there and gaze around the room. Check that you haven't missed out any sections (or whole questions! Many students forget to look at the back page of an exam paper!). Repeat calculations to make sure that you haven't made an arithmetical error.

SUCCESS!

Practical assessment

Your practical skills will be assessed at both AS and A level. Make sure you know how your practical skills are going to be assessed.
You may be assessed by
- **Coursework**
- **Practical examination**

The method of assessment will depend on the specification you are following and the choice of your school/college. You may be required to take
- two practical examinations (one at AS and one at A2 level)
- two coursework assessments
- one practical examination and one coursework assessment.

PRACTISING THE SKILLS

Whichever assessment type is used, you need to learn and practise the skills during your course.

Specific skills

You will learn specific skills associated with particular topics as a natural part of your learning during the course; for example, accurate weighing when investigating water uptake by plant tissue. Make sure that you have hands-on experience of all the apparatus that is used. You need to have a good theoretical background of the topics on your course so that you can
- devise a sensible hypothesis
- identify all variables in an experiment (see p. xvii)
- control variables
- choose suitable magnitudes for variables
- select and use apparatus correctly and safely
- tackle analysis confidently
- make judgements about the outcome.

Designing experiments and making hypotheses

Remember that you can only gain marks for what you write, so take nothing for granted. Be thorough. *A description that is too long is better than one that leaves out important detail.*

Remember to
- use your knowledge of AS and A2 level biology to support your reasoning
- give quantitative reasoning wherever possible
- draw clear labelled diagrams of apparatus
- provide full details of measurements made, equipment used, and experimental procedures
- be prepared to state the obvious.

A good test of a sufficiently detailed account is to ask yourself whether it would be possible to do the experiment you describe without needing any further infomation.

PRACTICAL SKILLS
There are four basic skill areas:
- Planning (P)
- Implementing (I)
- Analysing (A)
- Evaluating (E)

The same skills are assessed in both practical examinations and coursework.

GENERAL ASSESSMENT CRITERIA
You will be assessed on your ability to   marks
- identify what is to be investigated
- devise a hypothesis or theory of the expected outcome   P [8]
- devise a suitable experiment, use appropriate resources, and plan the procedure
- carry out the experiment or research
- describe precisely what you have done   I [8]
- present your data or information in an appropriate way
- draw conclusions from your results or other data   A [8]
- evaluate the uncertainties in your experiment
- evaluate the success or otherwise of the experiment and suggest how it might have been improved.   E [6]

GENERAL SKILLS

The general skills you need to practise are
- the accurate reporting of experimental procedures
- presentation of data in tables (possibly using spreadsheets) (see p. xviii)
- graph drawing (possibly using IT software) (see p. xix)
- analysis of graphical and other data
- critical evaluation of experiments

When *analysing data* remember to
- use a large gradient triangle in graph analysis to improve accuracy
- set out your working so that it can be followed easily
- ensure that any quantitative result is quoted to an accuracy that is consistent with your data and analysis methods
- include a unit for any result you obtain.

Carrying out investigations

Keep a notebook
Record
- all your measurements
- any problems you have met
- details of your procedures
- any decisions you have made about apparatus or procedures including those considered and discarded
- relevant things you have read or thoughts you have about the problem.

Define the problem
Write down the aim of your experiment or investigation. Note the variables in the experiment. Define those that you will keep constant and those that will vary.

Suggest a hypothesis
You should be able to suggest the expected outcome of the investigation on the basis of your knowledge and understanding of science. Try to make this quantitative if you can, justifying your suggestion with equations wherever possible. For example, you might be able to make a quantitative prediction about the effect of temperature on enzyme activity, remembering that $Q_{10} = 2$.

Do rough trials
Before commencing the investigation in detail do some rough tests to help you decide on
- suitable apparatus
- suitable procedures
- the range and intervals at which you will take measurements.
- Consider carefully how you will conduct the experiment in a way that will ensure safety to persons, to equipment, and to the environment.

Ideally you would consider alternative apparatus and procedures and justify your final decision.

Carry out the experiment
Remember all the skills you have learnt during your course:
- note all readings that you make
- take repeats and average whenever possible
- use instruments that provide suitably accurate data
- consider the accuracy of the measurements you are making
- analyse data as you go along so that you can modify the approach or check doubtful data.

Presentation of data
Tabulate all your observations (see p. xviii).
- Provide a title for any table of data.

Analysing data
This may include
- the calculation of a result, e.g. percentage change in mass of potato tissue
- drawing of a graph (see p. xix); remember to provide a suitable title for your graph(s)
- statistical analysis of data, e.g. a χ^2 test
- analysis of uncertainties in the original readings, derived quantities, and results
- conclusions from the experiment.

Make sure that the stages in the processing of your data are clearly set out.

Evaluation of the investigation
The evaluation should
- identify any systematic errors in the experiment
- comment on your analysis of the uncertainties in the investigation
- suggest how results affect validity
- review the strengths and weaknesses in the way the experiment was conducted
- suggest alternative approaches that might have improved the experiment in the light of experience.

Use of information technology (IT)
You may have used data-capture techniques when making measurements or used IT in your analysis of data. In your analysis you should consider how well this has performed. You might include answers to the following questions.
- What advantages were gained by the use of IT?
- Did the data-capture equipment perform better than you could have achieved by a non-IT approach?
- How well has the data-analysis software performed in representing your data graphically, for example?

THE REPORT
Remember that your report will be read by an assessor who will not have watched you doing the experiment. For the most part the assessor will only know what you did by what you write, so do not leave out important information.

If you write a good report, it should be possible for the reader to repeat what you have done should they wish to check your work.

A word-processed report is worth considering. This makes the report much easier to revise if you discover some aspect you have omitted. It will also make it easier for the assessor to read.

Note: The report may be used as portfolio evidence for assessment of Application of Number, Communication, and IT Key Skills.

Use subheadings
These help break up the report and make it more readable. As a guide, the subheadings could be the main sections of the investigation: aims, diagram of apparatus, procedure, etc.

PRACTICAL EXAMINATION
The form of the examination varies from one examination board to another, so make sure you know what your board requires you to do. Questions generally fall into three types which fit broadly into the following categories:
You may be required to
- examine a novel situation, create a hypothesis, consider variables, and design an experiment to test the hypothesis
- examine a situation, analyse data that may be given to you, and evaluate the experiment that led to the data
- obtain and analyse data in an experiment which has been devised by the examination board.

Coping with coursework

TYPES OF COURSEWORK

Coursework takes different forms with different specifications. You may undertake:

- short experiments as a routine part of your course, e.g. use of a redox indicator to investigate respiration
- long practical tasks prescribed by your teacher/lecturer, e.g. determination of water potential of potato tissue
- a long investigation of a problem decided by you and agreed with your teacher, e.g. an ecological investigation.

A short experiment

This may take one or two laboratory sessions to complete and will usually have a specific objective that is closely linked to the topic you are studying at the time.

You may only be assessed on one or two of the skills in any one assessment.

A long investigation

This may take 5 to 10 hours of class time plus associated homework time.

You will probably be assessed on all the skills in a long investigation.

Make sure you know in detail what is expected of you in the course you are following. A summary appears on the previous pages.

STUDY THE CRITERIA

Each examination board produces criteria for the assessment of coursework. The practical skills assessed are common to all boards, but the way each skill is rewarded is different for each specification. Ensure that you have a copy of the assessment criteria so that you know what you are trying to achieve and how your work will be marked.

PLAN YOUR TIME

Meeting the deadline is often a major problem in coping with coursework.

Do not leave all the writing up to the end

Using a word processor, you can draft the report as you go along. For example, you can write out the **aim** and a **list of apparatus** very early in the sequence.

You can then go back and tidy it up at the end.

DRAW UP AN INITIAL PLAN

Include the following elements:

The aim of the project

What are you going to investigate practically?
  or
What is the topic of your research?

A list of resources

What are your first thoughts on apparatus?
  or
Where are you going to look for information?
(Books; CD ROMs; Internet)
  or
Is there some organisation you could write to for information?

Theoretical ideas

What does theory suggest will be the outcome?
  or
What are the main theoretical ideas that are linked with your investigation or research project?

Timetable

What is the deadline?

What is your timetable for
Laboratory tasks
How many lab sessions are there?
Initial thoughts on how they are to be used

Non-laboratory tasks
Initial analysis of data
Writing up or word-processing part of your final report
Making good diagrams of your apparatus
Revising your time plan
Evaluating your data or procedures

Design an experiment

An experiment is designed to **test the validity of an hypothesis** and involves the **collection of data**

e.g. light intensity affects the rate of photosynthesis

using appropriate apparatus and instruments

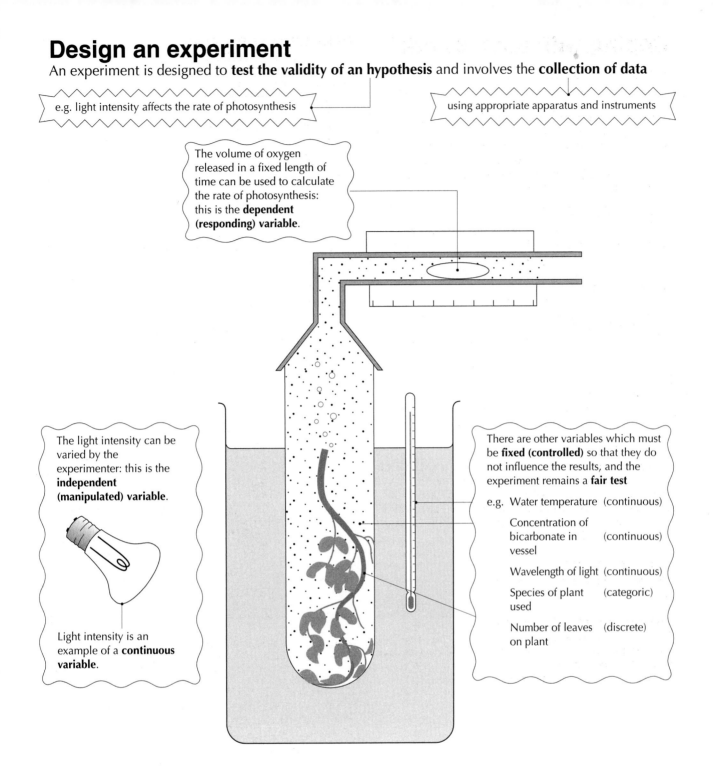

The volume of oxygen released in a fixed length of time can be used to calculate the rate of photosynthesis: this is the **dependent (responding) variable**.

The light intensity can be varied by the experimenter: this is the **independent (manipulated) variable**.

Light intensity is an example of a **continuous variable**.

There are other variables which must be **fixed (controlled)** so that they do not influence the results, and the experiment remains a **fair test**

e.g. Water temperature (continuous)

Concentration of bicarbonate in (continuous) vessel

Wavelength of light (continuous)

Species of plant (categoric) used

Number of leaves (discrete) on plant

A **control** experiment is the same in every respect **except** the manipulated variable is not changed but is kept constant.

A control allows confirmation that **no unknown variable** is responsible for any observed changes in the responding variable: it helps to make the experiment a **fair test**.

A 'repeat' is performed when the experimenter suspects that misleading data has been obtained through "operator error".

'Means' of a series of results minimise the influence of any single result, and therefore reduce the effect of any 'rogue' or anomalous data.

Dealing with data
may involve a number of steps

1 Organisation of the **raw data** (the information which you actually collect during your investigation).

2 Manipulation of the data (converting your measurements into another form).

and

3 Representation of the data in **graphical** or other form.

Steps 1 and 2 usually involve **preparation of a table of results**

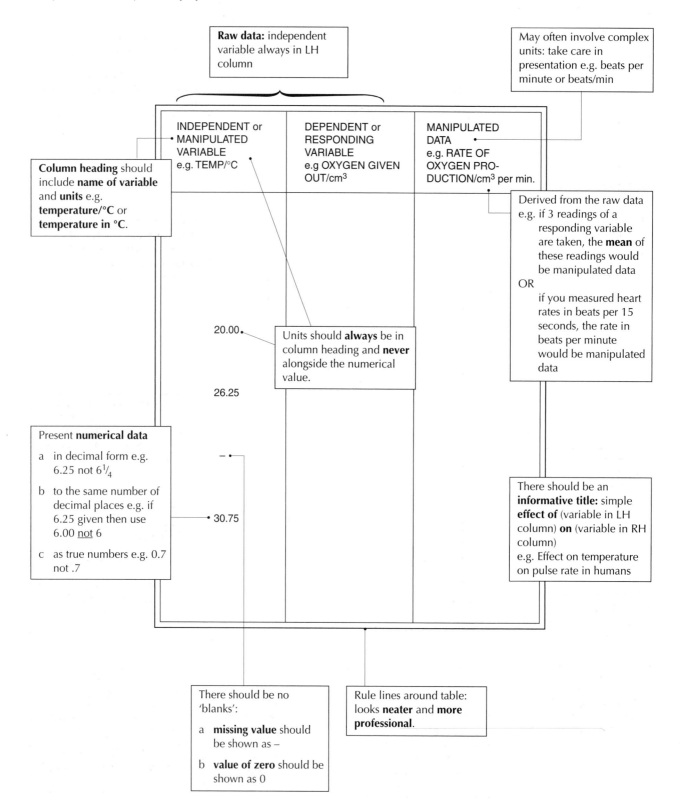

Raw data: independent variable always in LH column

May often involve complex units: take care in presentation e.g. beats per minute or beats/min

Column heading should include **name of variable** and **units** e.g. **temperature/°C** or **temperature in °C**.

INDEPENDENT or MANIPULATED VARIABLE e.g. TEMP/°C

DEPENDENT or RESPONDING VARIABLE e.g OXYGEN GIVEN OUT/cm³

MANIPULATED DATA e.g. RATE OF OXYGEN PRO-DUCTION/cm³ per min.

Derived from the raw data e.g. if 3 readings of a responding variable are taken, the **mean** of these readings would be manipulated data

OR

if you measured heart rates in beats per 15 seconds, the rate in beats per minute would be manipulated data

20.00

26.25

Units should **always** be in column heading and **never** alongside the numerical value.

Present **numerical data**

a in decimal form e.g. 6.25 not 6¼

b to the same number of decimal places e.g. if 6.25 given then use 6.00 <u>not</u> 6

c as true numbers e.g. 0.7 not .7

30.75

There should be an **informative title:** simple **effect of** (variable in LH column) **on** (variable in RH column) e.g. Effect on temperature on pulse rate in humans

There should be no 'blanks':

a **missing value** should be shown as –

b **value of zero** should be shown as 0

Rule lines around table: looks **neater** and **more professional**.

Graphical representation: A graph is a visual presentation
of data and may help to make the relationship between variables more obvious.

For example	
AIR TEMPERATURE /°C	BODY TEMPERATURE OF REPTILE /°C
20	19.4
25	25.9
30	30.4
35	35.1
40	40.1
45	44.8

isn't as helpful as

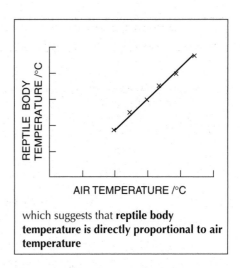

which suggests that **reptile body temperature is directly proportional to air temperature**

A graph may be produced from a table of data <u>**following certain rules**</u>

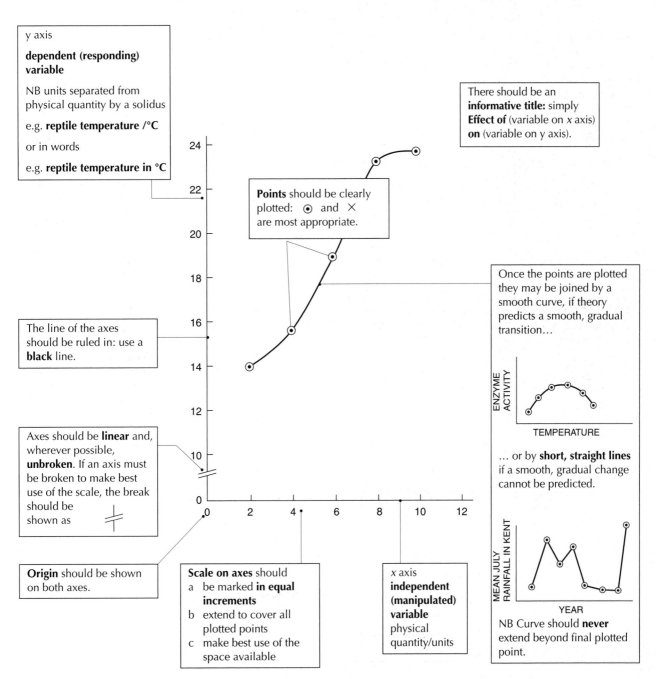

y axis

dependent (responding) variable

NB units separated from physical quantity by a solidus

e.g. **reptile temperature /°C**

or in words

e.g. **reptile temperature in °C**

There should be an **informative title:** simply **Effect of** (variable on *x* axis) **on** (variable on y axis).

Points should be clearly plotted: ⊙ and ✕ are most appropriate.

Once the points are plotted they may be joined by a smooth curve, if theory predicts a smooth, gradual transition…

The line of the axes should be ruled in: use a **black** line.

… or by **short, straight lines** if a smooth, gradual change cannot be predicted.

Axes should be **linear** and, wherever possible, **unbroken**. If an axis must be broken to make best use of the scale, the break should be shown as

Origin should be shown on both axes.

Scale on axes should
a be marked **in equal increments**
b extend to cover all plotted points
c make best use of the space available

x axis
independent (manipulated) variable
physical quantity/units

NB Curve should **never** extend beyond final plotted point.

Key Skills

What are Key Skills?

These are skills that are not specific to any subject but are general skills that enable you to operate competently and flexibly in your chosen career.

Achieving qualifications in the Key Skills will be looked on favourably by employers and by admissions tutors when considering you for a place in Higher Education.

While studying your AS or A level courses you should be able to gather evidence to demonstrate that you have achieved competence in the Key Skills areas of

- *Communication*
- *Application of Number*
- *Information Technology.*

You may also be able to prove competence in three other Key Skills areas:

- *Working with Others*
- *Improving Your own Learning*
- *Problem Solving.*

Only the first three will be considered here and only an outline of what you must do is included. You should obtain details of what you need to know and be able to do. You should be able to obtain these from your examination centre.

If you wish you can purchase a booklet that contains details about Key Skills (Ref. QCA/99/342, price £2.50 at time of writing) from QCA Publications, PO Box 99, Sudbury, Suffolk CO10 6SN.

Communication

You must be able to

- create opportunities for others to contribute to group discussions about complex subjects
- make a presentation using a range of techniques to engage the audience
- read and synthesise information from extended documents about a complex subject
- organise information coherently, selecting a form and style of writing appropriate to complex subject matter.

Application of Number

You must be able to plan and carry through a substantial and complex activity that requires you to

- plan your approach to obtaining and using information, choose appropriate methods for obtaining the results you need, and justify your choice
- carry out multistage calculations including use of a large data set (over 50 items) and re-arrangement of formulae
- justify the choice of presentation methods and explain the results of your calculations.

Information Technology

You must be able to plan and carry through a substantial activity that requires you to

- plan and use different sources and appropriate techniques to search for and select information based on judgement of relevance and quality
- use automated routines to enter and bring together information, and create and use appropriate methods to explore, develop, and exchange information
- develop the structure and content of your presentation, using others' views to guide refinements, and information from different sources.

A *complex subject* is one in which there are a number of ideas, some of which may be abstract and very detailed. Lines of reasoning may not be immediately clear. There is a requirement to come to terms with specialised vocabulary.

A *substantial activity* is one that includes a number of related tasks. The result of one task will affect the carrying out of others. You will need to obtain and interpret information and use this to perform calculations and draw conclusions.

What standard should you aim for?

Key Skills are awarded at four levels (1–4). In your A level courses you will have opportunities to show that you have reached level 3, but you could produce evidence that demonstrates that you are competent at a higher level. You may achieve a different level in each Key Skill area.

What do you have to do?

You need to show that you have the necessary underpinning knowledge in the Key Skills area and produce evidence that you are able to apply this in your day-to-day work.

You do this by producing a portfolio that contains

- evidence in the form of reports when it is possible to provide written evidence
- evidence in the form of assessments made by your teacher when evidence is gained by observation of your performance in the classroom or laboratory.

The evidence may come from only one subject that you are studying, but it is more likely that you will use evidence from all of your subjects.

It is up to you to produce the best evidence that you can.

The specifications you are working with in your AS or A level studies will include some ideas about the activities that form part of your course and can be used to provide this evidence. Some general ideas are summarised below, but refer to the specification for more detail.

Communication: in Biology/Human Biology you could achieve this by:

- undertaking a long practical or research investigation on a complex topic (e.g. use of nuclear radiation in medicine)
- writing a report based on your experimentation or research using a variety of sources (books, magazines, CD-ROMs, internet, newspapers)
- making a presentation to your fellow students
- using a presentation style that promotes discussion or criticism of your findings, enabling others to contribute to a discussion that you lead.

Application of Number: in Biology/Human Biology you could achieve this by

- undertaking a long investigation or research project that requires detailed planning of methodology
- considering alternative approaches to the work and justifying the chosen approach
- gathering sufficient data to enable analysis by statistical and graphical methods
- explaining why you analysed the data as you did
- drawing the conclusions reached as a result of your investigation.

Information Technology: in Biology/Human Biology you could achieve this by

- using CD-ROMs and the internet to research a topic
- identifying those sources which are relevant
- identifying where there is contradictory information and identifying which is most probably correct
- using a word processor to present your report, drawing in relevant quotes from the information you have gathered
- using a spreadsheet to analyse data that you have collected
- using data capture techniques to gather information and mathematics software to analyse the data.

Answering the question

There are many different styles of examination question. Some will simply expect you to recall facts, others will expect you to apply your knowledge or to demonstrate your ability to use certain skills.

In this section you will encounter a number of different question types and receive guidance on how to go about answering them. The different types are identified as
1. factual recall
2. experimental design and practical technique
3. data analysis
4. genetics and statistical analysis
5. prose and comprehension
6. essay-style and free response.

Whichever question style you are working on, it is important that you remember the following guidelines:
1. **read the question** – in particular look out for 'instructional' words such as 'explain' or 'comment on'.
2. **check the marks** – the number allocated for each section of a question will give you a good idea of how much you need to write. A rough guide is '1 appropriate point = 1 mark'.

Factual recall often involves **filling in gaps** and **very short (often one word) answers**, for example:

Example 1

Identify the structures labelled A, B, C, D, and E.   *(5 marks)*

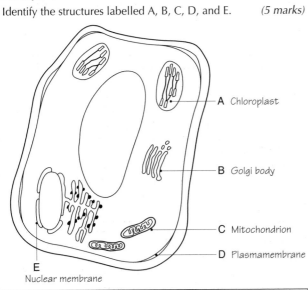

A Chloroplast

B Golgi body

C Mitochondrion

D Plasmamembrane

E
Nuclear membrane

There's little alternative but to **know your work** – there's not much scope for alternative answers here!

Example 2

Read through the following description of DNA and protein synthesis and then write on the dotted lines the most appropriate word or words to complete the passage:

The DNA molecule is made up of four types of ..nucleotide.. molecule, each of which contains ..deoxyribose.. sugar, a ..nitrogen-containing/organic.. base, and a phosphate group. Within the DNA molecule the bases are held together in pairs by ..hydrogen.. bonds – for example, guanine is always paired with ..cytosine..

DNA controls protein synthesis by the formation of a template called ..messenger RNA.. This molecule is single stranded and, compared with DNA, the sugar is ..ribose.. and the base ..uracil.. replaces the base ..thymine.. This template is made in the nucleus by the process of ..transcription.. and then passes out into the cytoplasm where it becomes attached to organelles called ..ribosomes.. A third type of nucleic acid called ..transfer RNA.. brings amino acids to these organelles where they are lined up on the template according to ..complementary.. base-pairing rules. The amino acids are joined by covalent links called ..peptide bonds.. to form a polypeptide molecule.

(14 marks)

Always **read through the whole question before writing in any answers** to make certain that the completed section makes good sense.

Examiners give some leeway – for example either 'nitrogen-containing' or 'organic' base would be acceptable – but note that they ask for the **most appropriate** term. For example, 'deoxyribose' is better than 'pentose', and 'thymine' is not the same molecule as 'thiamine'!

Other gap fillers might involve **completion of a table**:

Example 3

The table below refers to three components of human blood. If the statement is correct for the component place a tick (✔) in the appropriate box and if the statement is incorrect place a cross (✘) in the appropriate box.

Function	Red blood cell	Thrombocyte (platelet)	Plasma
Transports oxygen	✔	✘	✔
Transports hormones	✘	✘	✔
Contains enzymes involved in blood clotting	✘	✔	✔
Carries out phagocytosis	✘	✘	✘
Transports carbon dioxide	✔	✘	✔

(5 marks)

Don't forget to write in the tick or cross – a blank space would be regarded as incorrect, and so would writing 'yes' or 'no'. Many candidates don't obey this simple instruction and so lose straightforward marks.

Experimental/practical questions can also be thought of as recall, but often recall of a **general process** rather than of a particular example:

Example 4

Describe how a sample of ribosomes could be obtained from animal cells.

The technique of differential centrifugation relies on differences in mass/density of small structures to separate them from a mixture. So:
1. Break open the animal cells (homogeniser or by osmosis) to produce a mixture of parts (called an homogenate).
2. Keep mixture cool and at neutral pH, to avoid damage to organelles.
3. Spin in an ultracentrifuge.
4. Discard nuclei (first fraction) and mitochondria (second fraction) but keep pellet containing ribosomes.
5. Test for ribosomes by checking that this pellet can carry out protein synthesis.

(4 marks)

Note that by writing this in a list form you will (a) show a clear sequence of steps and (b) convince yourself that you've followed the rule of 'one sensible point = 1 mark'. The fifth point written here is an extra one – well done if you thought of it!

Many **experimental/practical questions** test whether you understand the **principles** of experimental design rather than your **recall** of particular procedures. Remember that the principle is:

> an experiment involves measuring the effect of a **manipulated (independent) variable** on the value of a **responding (dependent) variable** with all other variables **fixed** or **controlled**.

For example:

Example 5
The diagram below shows the apparatus used to compare the carbon dioxide production of different strains of yeast.
The yeast population to be investigated is suspended in sucrose solution in tube A. Nitrogen gas is bubbled through the apparatus during the experiment to ensure that respiration of the yeast is anaerobic.
Tube C contains hydrogencarbonate indicator solution through which carbon dioxide has been bubbled. This allows the colour in the tube to develop and this tube is then used as a standard. Hydrogencarbonate indicator is red when neutral, purple in alkaline and yellow in acid conditions.
The time taken for the colour to develop in tube B to match the colour in tube C is recorded.

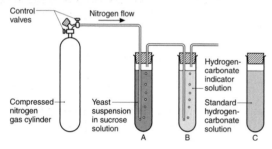

(a) (i) State how yeast suspension in tube A can be maintained at a constant temperature during the experiment.
Constant temperature/thermostatically controlled water bath.
(1 mark)

   (ii) State the colour change which would occur in the hydrogencarbonate indicator solution in tube B during the course of the experiment.
From red (neutral) to yellow (acidic).
(1 mark)

   (iii) Suggest why matching the colour by eye might not be a reliable method of determining the end-point.
Too subjective/could be colour-blind.
Reference solution may change colour.
(1 mark)

(b) Describe, giving experimental details, how you would use the apparatus to compare the carbon dioxide production in two strains of yeast.

> CO_2 is responding variable – how would you measure it? The yeast is the manipulated variable. What are the fixed variables and how would you fix them?

(5 marks)

(c) An experiment was carried out to investigate the effect of light intensity on the rate of photosynthesis of an aquatic plant, using the apparatus shown in the diagram below.

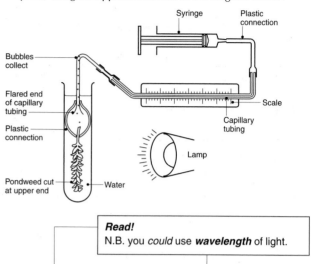

> **Read!**
> N.B. you *could* use **wavelength** of light.

State *two* environmental conditions, *other* than light intensity, which would need to be controlled. For each condition, describe how control could be achieved.

> i.e. name the **apparatus** you would use.

> 2 conditions + 2 methods.
(4 marks)

Data for analysis may be presented to you in the form of a **graph**:

Example 6
An experiment was carried out with cells of parsnip tissue to investigate the effect of temperature on the absorption of sodium ions. Cubes of the tissue were bathed in a sodium chloride solution of known concentration, and changes in the concentration of sodium ions in the solution were measured over a period of six hours. The solutions were aerated continuously, and the experiment was carried out at 2 °C and at 20 °C.

The results are shown in the graph below:

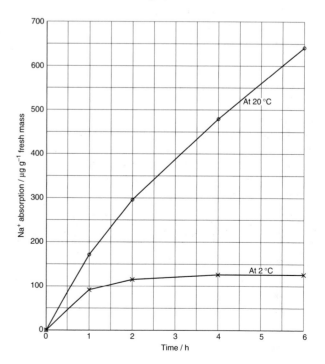

(i) Calculate the mean rate of absorption of sodium ions at 20 °C between 2 and 6 hours. Show your working.

> Forget this and lose marks!!

Abs. at 6 h = 640
Abs. at 2 h = 295

Mean rate between 2 and 6 = $\dfrac{640 - 295}{4}$

= 86.25 μg g⁻¹ h⁻¹

> 1 mark for number

> 1 mark for units

> Number of hours

(3 marks)

(ii) Compare the rates of absorption of sodium ions at 2 °C and at 20 °C during this experiment.

> You must present **both** rates.

> i.e. between 0 and 6 h

> **Be quantitative!**

At 2 °C, max. rate of absorption (135 units) is achieved by 4 h and there's very little increase from 2 to 4 h but at 20 °C, rate is higher (480 units at 4 h, for example) and continues to increase for the whole experimental period.

> Good word for comparison

> N.B. **No** request to explain these observations here

(3 marks)

(iii) Suggest an explanation for the differences in the rates of absorption of sodium ions at the two temperatures.

> i.e. try to relate these results to your biological/physical background knowledge.

Higher rate at 20 °C suggests
 – ions move more quickly at higher temperature
 – active transport involved: higher temperature releases more energy by respiration.

Some uptake at 2 °C: suggests some uptake without active transport (diffusion).

Plateau at 2 °C suggests diffusion uptake system can be saturated (max. diffusion rate achieved by 4 h).

(3 marks)

Data analysis could also involve dealing with data presented in a **table**:

Example 7

The numbers of stomata on each surface of an iris leaf and an oat leaf were estimated by counting the stomata in several 4 mm² areas of epidermis. From these measurements it was possible to calculate the mean number of stomata per cm². These are presented in the table below:

Species	Mean number of stomata per cm²	
	Upper epidermis	**Lower epidermis**
Iris	3550	1850
Oat	5100	6300

The rate of water loss from these leaves was also measured. The results of these measurements are presented in the table below:

Species	Rate of water loss / arbitrary units
Iris, lower epidermis	1.5
Iris, upper epidermis	2.2
Oat, lower epidermis	4.0
Oat, upper epidermis	5.0

Explain the relationship between the rate of water loss and the number of stomata.

> Don't just **state** the relationship.

Stomata regulate water loss, so more stomata offer greater risk of water loss: oat total (11 400 stomata per square centimetre – 9.0 units water loss) is greater than iris total (5400 stomata per square centimetre – 3.7 units water loss).

> Examiners reward you for being **quantitative.**

(2 marks)

State two other structural features of a leaf, apart from the number of stomata, which might influence the rate of water vapour loss. In each case give a reason for your answer.

> i.e. **name** them

> Don't forget – half of the marks will be for your reasons.

Feature 1

Reason

Feature 2   Leaf rolling

Reason   Traps humid atmosphere close to leaf surface to reduce water potential gradient.

(4 marks)

Genetics problems often involve **simple statistics**. Don't be afraid of this – you will only be asked to put numbers into a formula and interpret the result. **You won't be asked to recall the formula or to explain anything about it**.

Example 8

Wild type individuals of the fruit fly *Drosophila* have red eyes and straw-coloured bodies. A recessive allele of a single gene in *Drosophila* causes grey eye (g), and a recessive allele of a different gene causes black body (b).

A student carried out an investigation to test the hypothesis that the genes causing grey eye and black body show autosomal linkage. When she crossed pure-breeding wild type flies with pure breeding flies having grey eye and black body, the F₁ flies all showed the wild type phenotype for both features. On crossing the F₁ flies among themselves, the student obtained the following results for the F₂ generation.

Eye	Body	Number of flies observed in F₂ generation (O)
Wild	Wild	312
Wild	Black	64
Grey	Wild	52
Grey	Black	107
Total		535

(a) Using appropriate symbols, write down the genotypes of the F_1 flies and the grey-eyed, black-bodied F_2 flies.

F_1 flies   *Gg Bb*

GB from one parent, gb from the other.

Grey-eyed, black-bodied F_2 flies   *gg bb*

Must be homozygous to express recessive characteristics.

(2 marks)

(b) In order to generate expected numbers of F_2 flies for use in a χ^2 test, the student used the *null hypothesis* that the genes concerned were *not* linked.

(i) State the ratio of F_2 flies expected using the null hypothesis.

i.e. genes are segregating independently.

| Red, straw | •9 : 3 : 3 : 1• | Grey, black |
| Red, black | | Grey, straw |

(1 mark)

(ii) Complete the table below to give the numbers of F_2 flies expected (E) using the null hypothesis, and the differences between observed and expected numbers ($O - E$).

Eye	Body	Number of flies observed (O)	Number of flies expected (E)	$O - E$
Wild	Wild	312	301	$312 - 301 =$ 11
Wild	Black	64	100	$64 - 100 = -36$
Grey	Wild	52	100	$52 - 100 = -48$
Grey	Black	107	33	$107 - 33 =$ 74

(2 marks)

= Total (i.e. 535) multiplied by fraction

e.g. $535 \times \frac{9}{16} = 301$ red, straw flies

Nearest whole numbers

(c) (i) Use the formula

$$\chi^2 = \Sigma \frac{(O - E)^2}{E}$$

You will **not need to recall any formulae** – simply put numbers in and understand **what the answer means**.

to calculate the value of χ^2. Show your working.

Note this

Add two columns to table:

$(O - E)^2$	$\dfrac{(O - E)^2}{E}$
121	0.40
1296	12.96
2304	23.04
5476	165.94

$\chi^2 =$ 202.34

(2 marks)

(ii) How many degrees of freedom does this test involve? Explain your answer.

3: *number of d.o.f.*
= (*number of possible classes (4, in this case)* − 1) *i.e.* 4 − 1 = 3.

(2 marks)

(iii) For this number of degrees of freedom, χ^2 values corresponding to important values of P are as follows.

Value of P	0.99	0.95	0.05	0.01	0.001
Value of χ^2	0.115	0.352	7.815	11.34	16.27

What conclusions can be drawn concerning linkage of the g and b alleles? Explain your answer.

$\chi^2 = 193.21$

From this table, P must be less than 0.001, i.e. probability that genes are not linked is <0.001. Thus it is highly probable (more than 99.9% probable) that the g and b alleles are linked.

(3 marks)
(Total 12 marks)

If you are **confident** and **careful** you can **score full marks**, **very quickly** on this type of question.

xxiv Answering the question

Prose passages are used as *tests of comprehension* and to *build on factual recall*:

Example 9

Read the passage, and answer the questions which follow it.

Haematologists in Britain have been investigating the use of a blood substitute. The substitute might be used in blood transfusions and might allow the use of blood donations that have passed their 'use-by' date.

The substitute is effectively a haemoglobin solution – red cell membranes have been removed so that there are no problems with incompatible blood transfusions. The product has been developed in the United States and has overcome the problem that pure haemoglobin tends to break into two molecules and to be rapidly lost from the body. Scientists at the US-based company have developed a way of locking the two sub-units together so that the haemoglobin remains an efficient oxygen-deliverer and permits heat treatment to destroy viruses.

The artificial blood was first developed for use in accident, shock, and injury cases where oxygen delivery and fluid volume were required, but a new and exciting possibility is that it may be useful in the treatment of strokes. In tests on animals the haemoglobin solution has been able to by-pass blood clots and to reach the parts of the brain being oxygen-starved by the clot. The effects of stroke damage to the brain, including paralysis and partial loss of speech, have been shown to be reduced if the solution is infused soon after the stroke has occurred.

Even if the trials of the blood substitute prove successful there will still be a need for blood donors because the haemoglobin solution is derived from human blood. The benefits, however, are that there will be less waste of out-dated blood and artificial blood can be stored for longer periods.

Triggers recall

Ability to use 'given information for explanation

Understanding the passage

Hint: it is often worth reading the questions before the passage – you may then know what to look out for!

Note three sections to question… and 6 marks. Remember that 6 = 3 × 2!

(a) Donated blood has a 'use-by' date because red blood cells have a shorter life span than most other cells in the body. Give one reason, connected with their structure, why red blood cells have such a short life span.

Red cells have no nucleus – no replacement of mRNA for protein synthesis so limited cell repair.

(1 mark)

(b) Explain why the blood substitute can 'reach the parts of the brain being oxygen-starved by the clot'.

Has no cells so less viscous/can pass obstructions more easily.

(2 marks)

(c) Apart from its use in stroke victims, give *two* advantages of using the blood substitute rather than natural whole blood.

(i) Reduce waste of 'out-of-date' blood.

(ii) No cross-matching for blood type is necessary.

(2 marks)

(d) Explain the role of haemoglobin in the loading, transport, and unloading of oxygen.

• Uptake where pO_2 is high/co-operative binding.

• 'Plateau' on association curve means O_2 retained as oxyhaemoglobin at pO_2 arteries.

• Steepness of curve suggests almost complete unloading for small change in pO_2 at respiring tissues.

• Uptake/release can be modified by local conditions of temperature and pH (Bohr effect).

(6 marks)

Why not use a simple *diagram* of oxygen–haemoglobin association curve?

Essay style (*free response*) **questions** may gain from 10–25 marks. The key to scoring highly is to **break down the question into smaller units**.

Example 10

Write an essay on water pollution.

The plan for your essay might look like this.

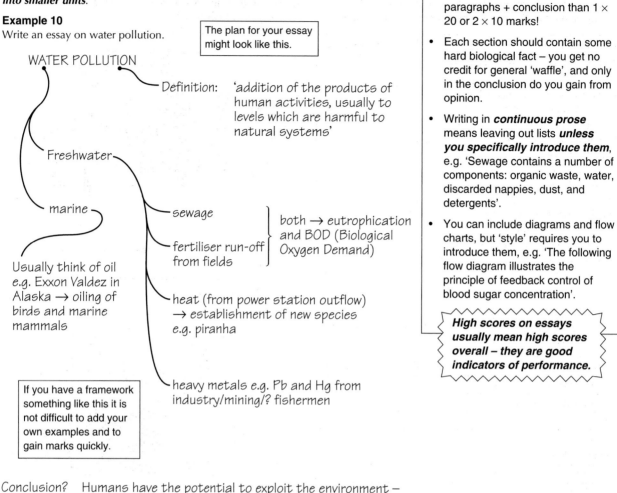

WATER POLLUTION

Definition: 'addition of the products of human activities, usually to levels which are harmful to natural systems'

Freshwater

marine

Usually think of oil e.g. Exxon Valdez in Alaska → oiling of birds and marine mammals

sewage
fertiliser run-off from fields

both → eutrophication and BOD (Biological Oxygen Demand)

heat (from power station outflow) → establishment of new species e.g. piranha

heavy metals e.g. Pb and Hg from industry/mining/? fishermen

If you have a framework something like this it is not difficult to add your own examples and to gain marks quickly.

- It is much easier to gain 2 + (4 × 4) + 2 marks for definition + 4 paragraphs + conclusion than 1 × 20 or 2 × 10 marks!

- Each section should contain some hard biological fact – you get no credit for general 'waffle', and only in the conclusion do you gain from opinion.

- Writing in **continuous prose** means leaving out lists **unless you specifically introduce them**, e.g. 'Sewage contains a number of components: organic waste, water, discarded nappies, dust, and detergents'.

- You can include diagrams and flow charts, but 'style' requires you to introduce them, e.g. 'The following flow diagram illustrates the principle of feedback control of blood sugar concentration'.

High scores on essays usually mean high scores overall – they are good indicators of performance.

Conclusion? Humans have the potential to exploit the environment – we must try to ensure that we do so in a responsible manner so that it is not made unsuitable for other organisms.

Self assessment questions: AS topics

1. The following table contains descriptions of a number of organelles. Identify the organelles from their descriptions.

Description	Name of organelle
A system of membranes that packages proteins in a cell	A
Contains the genetic material of a cell: surrounded by double membrane	B
Usually rod-shaped, and surrounded by a double membrane; inner membrane folded to increase its surface area	C
Approximately spherical, and often responsible for a 'rough' appearance of cell membranes; sometimes isolated attached to messenger RNA	D
Disc-shaped structure – surrounded by a double membrane and containing a series of grana	E

2. Complete the following table to compare the features of a eukaryotic and a prokaryotic cell. Use a + if the feature is present and a – if it is absent.

Feature	Prokaryotic cell	Eukaryotic cell
Nuclear envelope		
Cell surface membrane		
DNA		
Mesosome		
Mitochondria		
Ribosomes		
Microtubules		

3. The diagram shows a typical plant cell.
   (a) Identify the structures labelled A–F.
   (b) Suggest a label for the scale line – give your answer to the nearest 5μ.

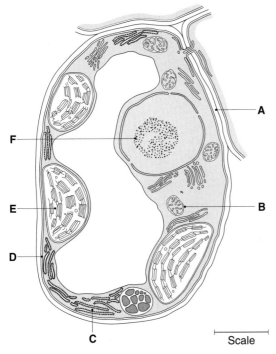

4. The diagram shows a section of cell membrane. Identify the components labelled A–F.
   What is the length of the line XY?

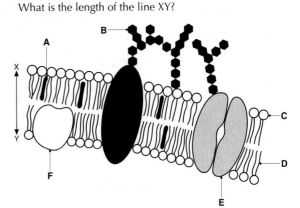

5. The diagram shows some of the structures present in an animal cell:

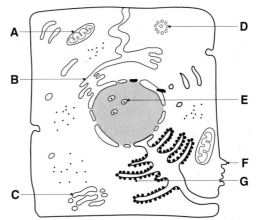

Which of these structures is responsible for
(a) manufacture of lipids and steroids
(b) release of energy
(c) manufacture of hormones and digestive enzymes
(d) production of spindle fibres in cell division
(e) endo- and exocytosis?

6. The following diagrams show some molecules found in cells.
   (a) Match the diagrams to the labels supplied:

   triglyceride; α-glucose; β-glucose; amino acid; purine; steroid; ribose

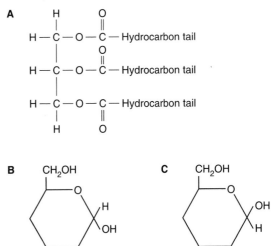

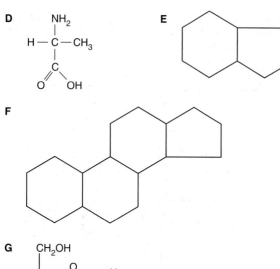

D

$$H-\underset{\underset{\underset{OH}{\overset{\|}{O}}}{\overset{NH_2}{\overset{|}{C}}}}{\overset{|}{C}}-CH_3$$

E

F

G

CH₂OH

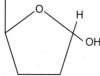

(b) Which of these molecules would be found in
  (i) glycogen
  (ii) insulin
  (iii) DNA
  (iv) cellulose
  (v) amylase?

7. The diagram represents one possible mechanism for enzyme action.
Match the letters to the following labels:

  catalase
  oxygen
  hydrogen peroxide
  water
  E–S complex

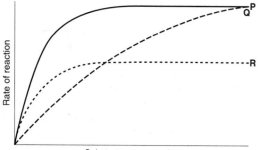

8. The following graph represents the effects of some compounds on the action of an enzyme:

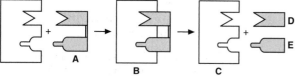

(a) Which of the following correctly identifies the three lines on the graph?

	Uninhibited enzyme	Enzyme + competitive inhibitor	Enzyme + non-competitive inhibitor
A	P	Q	R
B	P	R	Q
C	Q	R	P
D	R	P	Q
E	R	Q	P

(b) Name **(i)** an enzyme, its normal substrate, and its non-competitive inhibitor
  (ii) an enzyme, its normal substrate, and its competitive inhibitor.

9. The following diagram represents a section of an important biological molecule:

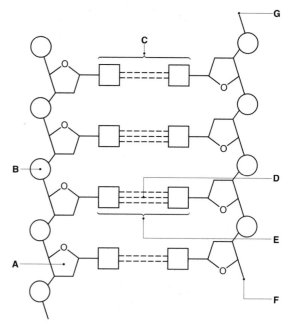

(a) Identify the molecule.
(b) Match the following labels to the letters on the diagram: A–T base pair; C–G base pair; hydrogen bond; 5' end; 3' end; deoxyribose; phosphate/pentose

10. The diagram represents a structure normally found on the underside of leaves. Match the letters on the diagram to appropriate labels from the list below:

  chloroplast; epidermal cell; nucleus; site of K⁺ pump; stomatal pore; mitochondrion; guard cell; cellulose cell wall

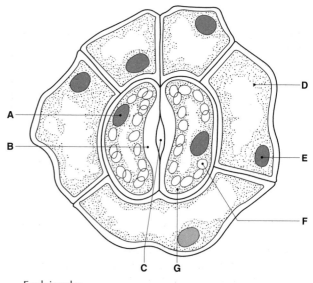

Explain why
(a) these structures are typically found on the *underside* of the leaf
(b) these structures may be sunk into pits in Xerophytes
(c) the pores usually open if internal leaf concentration of carbon dioxide falls
(d) the pores usually close as light intensity falls.

11. The following diagram represents cells from a plant tissue.
  (a) Choose appropriate labels from the following list to match with the letters on the diagram:

  sieve plate; companion cell; sieve tube; cytoplasmic strand; plasmodesma; mitochondrion; lignin

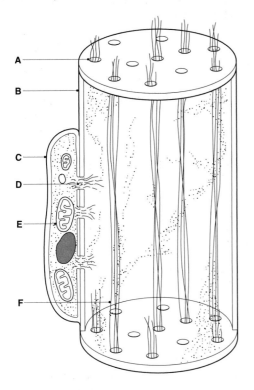

  (b) Name the tissue of which these cells are a part. What is the function of this tissue?
  (c) Choose an example from the human alimentary canal to illustrate the sequence
  ORGANELLE–CELL–TISSUE–ORGAN–SYSTEM.

12. The diagram shows the results of an experiment on the digestion of carbohydrates:

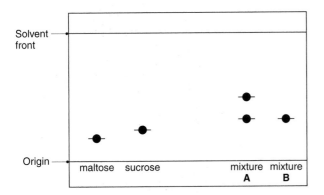

  (a) Name the technique which would have produced this result.
  (b) Name the two components of the mixture **A**.
  (c) Calculate the R_f value for fructose in this solvent.
  (d) State the site of secretion of the enzymes responsible for these digestive reactions.
  (e) What is the general name given to reactions which 'break down large molecules to smaller ones by the addition of water'?

13. The graph shows the results of an experiment in which two groups of young rats were fed on a basic diet which could be supplemented with milk.
  (a) Calculate the mean rate of growth of group A rats between 0 and 20 days. Show your working.
  (b) Suggest why group B rats were able to grow from 0 to 10 days, even in the absence of milk.
  (c) At day 30, calculate group A body mass as a percentage of group B body mass. Show your working.

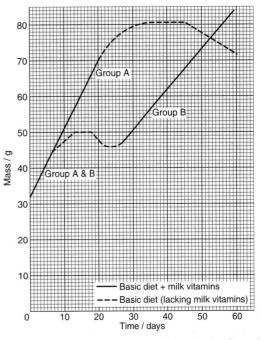

14. This diagram represents two possibilities for the flow of water over the gills of a fish:

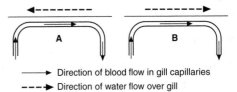

  (a) Which of the two possibilities is the one you would expect to find in a living fish? Explain why this is the more efficient method for gas exchange.
  (b) List three features of gas exchange surfaces which contribute to their efficiency.

15. The diagram represents the breathing movements of a human. Match the letters on the diagram to the appropriate labels from the following list:
  inspiration; expiration; trachea; contraction of external intercostal muscles; relaxation of diaphragm; contraction of diaphragm; air exhaled down pressure gradient; relaxation of external intercostal muscles; air inhaled down pressure gradient

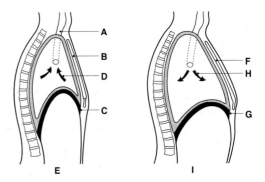

16. Choose words from the following list to complete the passage about blood and its functions. You can use each word once, more than once, or not at all.

white cells; glucose; urea; bone marrow; albumin; hydrogencarbonate; sodium; haemopoiesis; calcium; platelets; plasma; serum; solute potential; red blood cells; stem cells; homeostasis

Blood consists of a liquid called ...... in which are suspended several types of 'cell'. These include ...... (or erythrocytes), ... (including neutrophils), and ...... , which are really fragments of cells and are involved in blood clotting. All of these cells are produced from ...... by a process called ......, which occurs in the ....

Water is the main component of blood and may carry several dissolved ions including ...... (the most abundant cation), ... (another factor involved in blood clotting), and ...... (mainly formed from the solution of carbon dioxide in water). There are also plasma proteins present, including fibrinogen and ...... – as well as having individual specific functions these plasma proteins also affect the physical properties of the blood such as its viscosity and .... Blood with cells and fibrinogen removed is called ......

The blood is the major transport system of the body. For example, ... is transported from the liver to the kidney for excretion and ...... is distributed to the cells as a source of energy.

17. The following diagram shows some properties of different parts of the mammalian circulatory system.

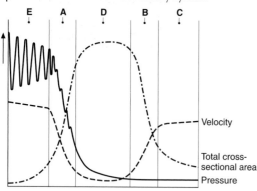

(a) Match the following structures with letters on the diagram:

arteriole; capillary; vein; artery; venule

(b) Use the letters to identify the structure in which
 (i) the pulse can be felt most strongly
 (ii) valves are most likely to be present
 (iii) the walls contain most elastic and muscular tissue
 (iv) exchange of soluble substances is most likely to take place.

18. The diagram shows a section through the mammalian heart:

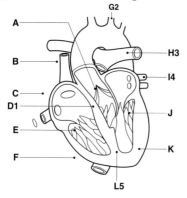

(a) Match the *letters* on the diagram with appropriate labels from the following list:

pulmonary artery; superior vena cava; wall of right atrium; pulmonary vein; chordae tendinae; tricuspid valve; aortic valve; wall of right ventricle; interventricular septum; carotid artery; wall of left ventricle; A–V node

(b) Use the *numbers* on the diagram to identify
 (i) the location of the bundle of His
 (ii) the site of receptors sensitive to blood pressure changes
 (iii) vessels carrying deoxygenated blood to the lungs
 (iv) a structure sensitive to electrical impulses arriving from the atria
 (v) vessels transporting oxygenated blood.

19. The graph represents the changes in pressure measured in the left side of the heart during the cardiac cycle:

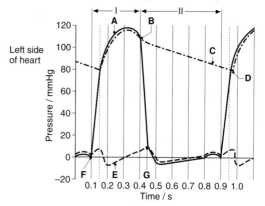

(a) Match the letters on the diagram with the appropriate label from the following list:

opening of aortic valve; pressure change in left ventricle; pressure change in left atrium; bicuspid valve opens; bicuspid valve closes; pressure changes in aorta; aortic valve closes

(b) If a doctor measured the blood pressure of this patient what would the result be?
(c) Calculate the pulse (rate of heartbeat) of this individual. Show your working.
(d) The peak systolic pressure in the left ventricle is approximately five times that in the right ventricle. Calculate the peak systolic pressure in the right ventricle. Explain why this value is different from the value measured in the left ventricle.

20. Use words from the following list to complete the passage about cell division:

pole; cell wall; interphase; cytokinesis; chromatids; nucleolus; nucleus; telophase; metaphase; centromere; spindle

During the cell cycle DNA replication takes place during ...... – sometimes mistakenly called 'resting phase'. At the beginning of prophase the chromosomes shorten and thicken so that they become visible – they are seen to consist of two identical ..... joined at the .... The ...... and the nuclear membrane are broken down, and a ..... develops in the cell. During ...... the chromosomes line up at the equator, and each one becomes attached to the ...... by its .... During anaphase one chromatid from each chromosome is pulled towards each ..... of the cell and during the final phase two new cells are formed as a result of the 'pinching' of the cytoplasm, called ....

21. Use words from the following list to identify the lettered structures on the diagram:

spindle organiser; chromatid; centromere; nuclear membrane; spindle fibre; chromosome; cell wall

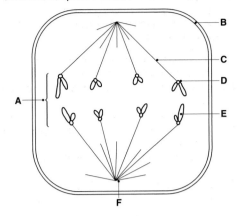

22. Use words from the following list to identify the lettered structures on the diagram:

anticodon; hydrogen bonds; unpaired folds; amino acid attachment site

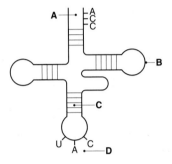

23. (a) The diagram shows the DNA content of a cell during meiosis. Use words from the following list to identify the lettered stages on the diagram:

DNA replication; separation of homologous chromosomes; separation of chromatids; haploid; cytokinesis

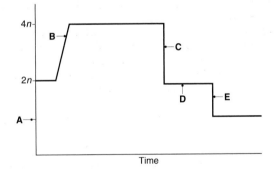

(b) Give two reasons why this process is important in the life cycle of living organisms.

24. Use words from the list to complete the passage which follows:

polyploidy; aneuploidy; X; Y; gene; crossing over; fusion; Down's; mutation; assortment

A ...... is any change in the structure or amount of DNA in an organism. Changes at a single locus on a chromosome are called ...... mutations – examples include cystic fibrosis and sickle cell anaemia. The loss or gain of a whole chromosome is called ...... – important examples include ...... syndrome (an extra 21st chromosome) and Klinefelter's syndrome (an extra ...... chromosome in males). The presence of additional whole sets of chromosomes, or ...... , has many important plant examples, including modern wheat.
All of these changes in DNA may be 'reshuffled' by free ...... during gamete formation, ... during meiosis, and random ...... during zygote formation.

25. The following flow diagram represents the stages involved in the formation of a section of recombinant DNA.
Match labels from the following list to complete the flow diagram:

DNA polymerase; plasmid; plasmid with 'sticky ends'; DNA ligase; restriction endonuclease; recombinant DNA; reverse transcriptase

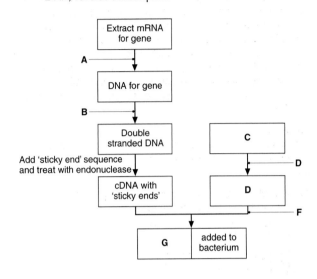

Self assessment questions: A2 topics

26. The diagram shows a single mitochondrion:

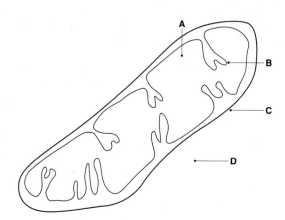

(a) Match the labels in the following list to the letters on the diagram:

glycolysis; TCA cycle; electron transport; pyruvate transport

(b) In the absence of molecular oxygen the following reactions occur. Match labels in the list to the letters on the flow chart:

NAD; ATP; ADP; alcohol; carbon dioxide; lactate; $NADH_2$

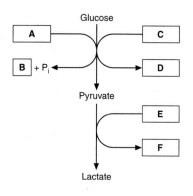

(c) What is the name given to the volume of oxygen required to reoxidise the lactate following anaerobic respiration?

27. The graph represents the effect of light intensity on the rate of photosynthesis of a culture of algae:

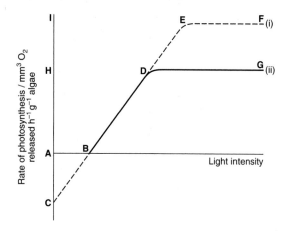

(a) Which region of the graph (use the letters to give your answer) corresponds to
   (i) the rate of respiration at zero light intensity
   (ii) a period when rate of photosynthesis is directly proportional to light intensity
   (iii) the effect of raising the value of a second limiting factor, temperature
   (iv) the maximum rate of photosynthesis possible at this light intensity and temperature?

(b) Similar results were obtained by plants in a woodland as light intensity increased during the day. What name is given to the point marked **B** on this graph?

28. The diagram represents a possible system for the light-dependent stages of photosynthesis.

(a) Match labels from the list to the letters on the diagram. A label may be used more than once.

light energy; photosystem I; photosystem II; excited electrons; ATP synthesis; protons; water; $NADPH_2$

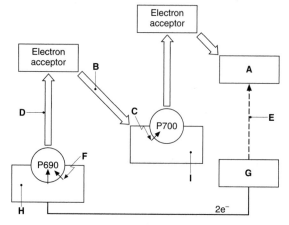

(b) These reactions occur in the structure drawn below:

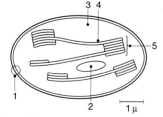

   (i) Name this structure.
   (ii) Which of the numbered regions carries out the biochemical processes above?
   (iii) Use the scale to estimate the length of this structure.
   (iv) Name an enzyme found in region 3, and state its function.

29. The diagram shows a single ovule just prior to fertilisation:

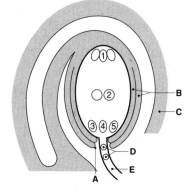

(a) Match the labels from this list to the letters on the diagram:

pollen tube; male nuclei; integuments; ovary wall; micropyle

(b) Which of the numbered cells
- **(i)** will take part in fertilisation
- **(ii)** will become part of the zygote
- **(iii)** will form part of a triploid structure
- **(iv)** are haploid?

30. Complete the following table describing the effects of some plant growth substances. Use + to indicate YES and – to indicate NO.

Effect	Auxins	Gibberellins	Abscisic acid	Ethene
Stimulate cell elongation				
Stimulate root formation in cuttings				
Stimulate fruit ripening				
Inhibit development of lateral buds				
Stimulate breaking of dormancy in seeds				
Stimulate leaf fall in deciduous trees				

31. The following is a list of some important ecological terms. Match the terms to the list of definitions:

TERMS   **A**   Biomass
             B   Community
             C   Abiotic
             D   Population
             E   Quadrat
             F   Niche
             G   Transect
             H   Ecosystem

DEFINITIONS   **1**   a piece of apparatus used to sample an area of the environment
                 2   the role of an organism in its environment
                 3   all of the living organisms within a defined area
                 4   the amount of organic matter present per unit area
                 5   the total number of any one species present within a defined area
                 6   non-living factors, such as temperature, which affect the distribution of living organisms
                 7   living organisms together with their non-living environment
                 8   a means of sampling a change in the living or non-living environment across a defined area.

32. Choose the most appropriate terms from the following list to complete the passage below:

climax; herbivore; population; ecology; habitat; ecosystem; succession; evolution; community

...... is the study of the interactions of living organisms with each other and with their environment. The living component together with the non-living part of the environment make up an ....

A field is an example of a ...... , and will have a ...... of several species of herbivorous insects. The insects may be preyed on by a ...... of one species of spider. As conditions in the field change, different plants and their herbivores may become established, and the herbivores may be preyed on by different carnivores. This gradual process of change is called ...... and the end point is known as a ...... community.

33. Use terms from the following list to complete the table listing the processes involved in the nitrogen cycle:

glucose; nitrate; nitrogen; carbon dioxide; amination; maltose; photosynthesis; respiration; nitrite; carbon dioxide

Process	Substrate	Product
Nitrification		Nitrate
Nitrogen fixation	Nitrogen	
Denitrification		
	Carbon dioxide	
	Glucose	
	Amino groups	Amino acids
Putrefaction	Starch	

34. Use words from the following list to complete the paragraph below. Each word may be used once, more than once, or not at all.

histamine; B-lymphocyte; T-lymphocyte; cytokines; T-helper; antibodies; cell-mediated; macrophage; antigen; humoral; T-cytotoxic

The body of a mammal is protected by an immune response. This has two parts – a ..... response which depends on the release of protein molecules called .... from plasma cells, and a ...... response controlled by a number of .... The plasma cells are one type of ...... , activated when another cell called a ..... 'presents' a piece of the invading organism to it. The plasma cells secrete ...... which may remove the invading ...... by several methods, including agglutination and precipitation. The cell-mediated response involves a number of interactions, controlled by chemicals called ...... These chemicals are released by ...... cells which may, for example, 'instruct' ...... cells to attack body cells infected with a virus or bacterium.

35. The diagram represents a single nephron from a mammalian kidney:

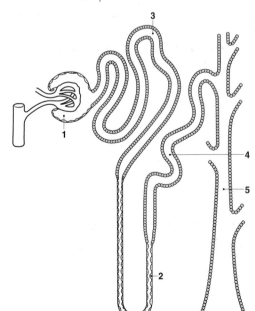

(a) Identify which of the numbered regions is
  (i) the site of ultrafiltration
  (ii) particularly sensitive to ADH
  (iii) the main site for the reabsorption of glucose and amino acids
  (iv) largely responsible for adjustment of blood pH.
(b) Which of the numbered regions would be particularly lengthy in a desert mammal? Explain your answer.

36. Use words or phrases from the following list to complete the paragraph below. Each word or phrase may be used once, more than once, or not at all.

  ammonia; soluble; by diffusion; in solution; insoluble; toxic; water conservation; urea; uric acid; as a precipitate

The nitrogenous waste of most aquatic animals is ...... which is extremely ...... in water. It is also extremely ...... so much water is consumed in diluting it. Mammals excrete ...... which is quite soluble and so can be passed out ..... in the urine. Most insects have problems of ...... and so excrete ...... which is ...... and so consumes very little water.

37. The diagram illustrates structures involved in the iris reflex. Match labels from the following list with the letters on the diagram:

  sensory neurone; retina; inhibitory neurone; excitatory neurone; circular muscle in iris; visual centre; radial muscle in iris

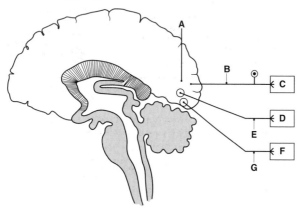

(a) What is the value of this reflex?
(b) Explain how a conditioned reflex differs from this example.

38. The following graphs were obtained during glucose tolerance tests on two hospital patients. Match the following labels with the letters on the graphs:

  diabetic patient; glucose injection; increased insulin secretion; non-diabetic patient

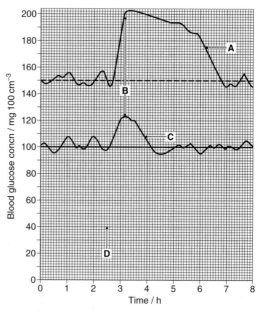

(a) Give *two* differences between the response shown by the diabetic and non-diabetic patients.
(b) Calculate the maximum percentage increase in blood glucose concentration following glucose injection in the diabetic patient. Show your working.
(c) Describe the differences in *cause* and in *method of treatment* for Type I and Type II diabetes.

39. The diagram shows light absorption by three cell types:

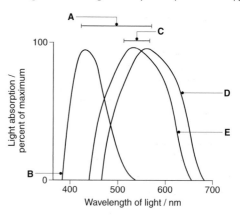

(a) Where, exactly, are these cell types found?
(b) Match labels from this list with the letters on the diagram:

  orange colour perceived; red cone; blue cone; white colour perceived; green cone

(c) One defect in colour vision results from inheritance of a sex-linked mutant allele. What is meant by *sex-linkage*?

40. The diagram represents changes in the membrane potential of a nerve cell during an action potential.

Use the numbers on the diagram to identify

(a) the point at which sodium gates are opened
(b) the point at which potassium gates are opened
(c) the main period of sodium ion influx
(d) the period during which the Na–K pump begins to work
(e) the main period of potassium ion outflow.

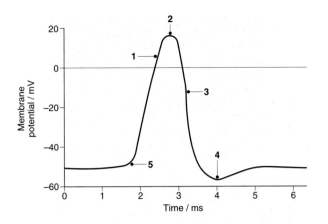

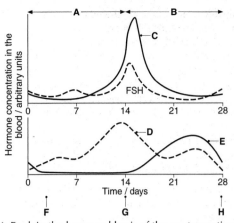

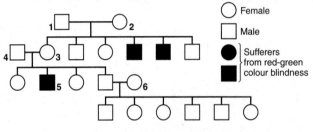

41. The table below describes some features of different skeletal systems. Complete the table using terms from the following list. Each term may be used once, more than once, or not at all.

ladybird; contains chitin; earthworm; fluid-filled muscular sac; annelid; cat; exoskeleton; endoskeleton; mollusc; chordate

Type of skeleton	Brief description	Phylum in which this type of skeleton typically occurs	Example organism
hydrostatic			
	internal framework of bones and/or cartilage		
		arthropod	

42. Use labels from this list to identify the regions on the structure shown in the diagram:

I band; H zone; M line; Z line; sarcomere; A band

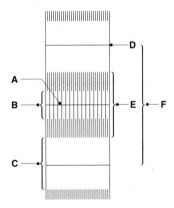

43. The following diagram shows some of the changes in blood hormone concentrations which occur during the menstrual cycle.

(a) Complete the diagram using labels from the following list:

oestrogen; ovulation; repair of endometrium; luteinising hormone; menstruation; luteal phase; progesterone; ovarian phase

(b) Explain the hormonal basis of the contraceptive pill.
(c) State and explain *two* differences between male and female gametes.

44. The diagram shows a family tree of individuals with red-green colour blindness. This condition is X-linked – there are five possible genotypes:

X^RY, X^rY, X^rY^r, X^rY^R, X^RY^R

R is dominant and normal-sighted, r is recessive and colour blind.

There are three possible phenotypes: carrier, normal, colour-blind.

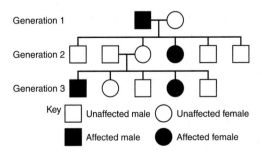

Use the family tree to work out the genotypes and phenotypes of the individuals labelled **1–6**. Present your answers in the table below:

	Genotype	Phenotype
1		
2		
3		
4		
5		
6		

45. The diagram shows the inheritance of the condition cystic fibrosis.
(a) Is the condition sex-linked? Explain your answer.
(b) Is the condition dominant or recessive? Explain your answer.
(c) Why are antibiotics often prescribed for individuals with this condition?
(d) Suggest why individuals with cystic fibrosis often have difficulty with digestion of fats.

Answers to self assessment questions

1. A = Golgi body; B = nucleus; C = mitochondrion; D = ribosome; E = chloroplast

2. Prokaryotic: − + + + − + − Eukaryotic: + + + − + + +

3. (a) A = cellulose cell wall; B = mitochondrion; C = rough endoplasmic reticulum; D = cell surface membrane/plasmamembrane; E = chloroplast; F = nucleolus
   (b) 10 μm

4. A = cholesterol; B = glycocalyx; C = hydrophilic head of phospholipid; D = hydrophobic tail of phospholipid; E = transmembrane protein; F = intrinsic protein
   XY = 8 nm

5. (a) C (b) A (c) G (d) D (e) F

6. (a) A = triglyceride; B = α-glucose; C = β-glucose; D = amino acid; E = purine; F = steroid
   (b) (i) B (ii) D (iii) E (iv) C (v) D

7. A = hydrogen peroxide; B = E–S complex; C = catalase; D/E = oxygen/water

8. (a) A
   (b) (i) Succinate dehydrogenase/succinate/fumarate
       (ii) Cytochrome oxidase/hydrogen ions – oxygen/cyanide ion

9. (a) DNA
   (b) A = deoxyribose; B = phosphate; C = A–T base pair; D = hydrogen bond; E = C–G base pair; F = 5′ end; G = 3′ end

10. A = site of K^+ pump; B = cellulose cell wall; C = stomatal pore; D = epidermal cell; E = nucleus; F = chloroplast; G = guard cell
    (a) This reduces water loss as stomata are protected from direct sunlight/heat
    (b) Provides a local humid atmosphere and so reduces water potential gradient between internal surfaces of leaf and drying atmosphere
    (c) Increased carbon dioxide necessary as carbon dioxide concentration is a limiting factor in photosynthesis
    (d) Light becomes the limiting factor in photosynthesis so that there is no requirement for stomata to remain open for carbon dioxide uptake – stomata close to limit water loss

11. (a) A = sieve plate; B = sieve tube; C = companion cell; D = plasmodesma; E = mitochondrion; F = cytoplasmic strand
    (b) Phloem – transport of organic solutes/products of photosynthesis
    (c) Mitochondrion–epithelial cell–epithelium–ileum–digestive system

12. (a) Chromatography
    (b) Glucose and fructose
    (c) 0.5
    (d) Surface of gut epithelia cells
    (e) Hydrolysis

13. (a) (70 − 32)/20 = 1.9 g per day
    (b) Vitamins stored in body (liver) following transfer across placenta
    (c) Difference in mass = 80 − 50.5 = 29.5
        Percentage difference = 29.5/50.5 × 100 = 58%

14. (a) A; the countercurrent system provides the greater surface with a concentration gradient which favours oxygen transfer from water to blood
    (b) *Three from* large surface area/thin/moist/close to blood vascular system

15. A = trachea; B = relaxation of external intercostal muscles; C = relaxation of diaphragm; D = air exhaled down pressure gradient; E = expiration; F = contraction of external intercostal muscles; G = contraction of diaphragm; H = air inhaled down pressure gradient; I = inspiration

16. Plasma; red blood cells; white cells; platelets; stem cells; haemopoiesis; bone marrow; sodium; calcium; hydrogencarbonate; albumin; solute potential; serum; urea; glucose

17. (a) A = arteriole; B = venule; C = vein; D = capillary; E = artery
    (b) (i) E (ii) C (iii) E (iv) D

18. (a) A = aortic valve; B = superior vena cava; C = wall of right atrium; D = A–V node; E = tricuspid valve; F = wall of right ventricle; G = carotid artery; H = pulmonary artery; I = pulmonary vein; J = chordae tendinae; K = wall of left ventricle; L = interventricular septum
    (b) (i) 5 (ii) 2 (iii) 3 (iv) 1 (v) 4

19. (a) A = pressure change in left ventricle; B = aortic valve closes; C = pressure changes in aorta; D = opening of aortic valve; E = pressure change in left atrium; F = bicuspid valve opens; G = bicuspid valve closes
    (b) 118/80 mmHg
    (c) 1 complete beat takes (0.9 − 0.1) s = 0.8 s
        Number of beats per minute = 60/0.8 = 72
    (d) 118/5 = 23.6 mmHg; right ventricle needs to generate only sufficient pressure to propel blood to lungs, left ventricle must generate enough pressure to propel blood around body in the systemic circulation

20. Interphase; chromatids; centromere; nucleolus; spindle; metaphase; spindle; centromere; pole; cytokinesis

21. A = chromosome; B = cell wall; C = spindle fibre; D = centromere; E = chromatid; F = spindle organiser

22. A = amino acid attachment site; B = unpaired folds; C = hydrogen bonds; D = anticodon

23. (a) A = haploid; B = DNA replication; C = separation of homologous chromosomes; D = separation of chromatids; E = cytokinesis
    (b) Restores diploid number following fertilisation; variation during gamete formation

24. Mutation; gene; aneuploidy; Down's; X; polyploidy; assortment; crossing over; fusion

25. A = reverse transcriptase; B = DNA polymerase; C = plasmid; D = restriction endonuclease; E = plasmid with 'sticky ends'; F = DNA ligase; G = recombinant DNA

26. **(a)** A = TCA cycle; B = electron transport; C = pyruvate transport; D = glycolysis
(b) A = ATP; B = ADP; C = NAD; D = $NADH_2$; E = $NADH_2$; F = NAD
(c) The oxygen debt

27. **(a)** (i) AC (ii) BD (iii) DE (iv) EF
(b) Compensation point

28. **(a)** A = $NADPH_2$; B = ATP synthesis; C = light energy; D = excited electrons; E = protons; F = light energy; G = water; H = photosystem II; I = photosystem I
(b) (i) Chloroplast (ii) 5 (iii) 5 μm (iv) carbonic anhydrase – combination of carbon dioxide and ribulose bisphosphate

29. **(a)** A = micropyle; B = integuments; C = ovary wall; D = male nuclei; E = pollen tube
(b) (i) 2 and 4 (ii) 4 (iii) 2 (iv) 1, 2, 3, 4, and 5

30. Auxins: + + – + – – gibberellins: + – – – + – abscisic acid: – – – – – + ethene: – – + – – –

31. A – 4; B – 3; C – 6; D – 5; E – 1; F – 2; G – 8; H – 7

32. Ecology; ecosystem; habitat; community; population; succession; climax

33. nitrite; nitrate; nitrate; nitrogen; photosynthesis; glucose; respiration; carbon dioxide; amination; glucose

34. humoral; antibodies; cell-mediated; T-lymphocytes; B-lymphocyte; macrophage; antibodies; antigen; cytokines; T-helper; T-cytotoxic

35. **(a)** (i) 1 (ii) 5 (iii) 3 (iv) 4
(b) 2 (loop of Henle) – very lengthy to create high solute concentration in medulla so that water potential gradient favours reabsorption of water from collecting duct. This is vital in an animal living in an area of water shortage

36. Ammonia; soluble; toxic; urea; in solution; water conservation; uric acid; insoluble

37. A = visual centre; B = sensory neurone; C = retina; D = circular muscle; E = excitatory neurone; F = radial muscle; G = inhibitory neurone (D/E and F/G can be exchanged)
(a) Prevents bleaching of retina and consequent loss of vision
(b) Conditioned reflex replaces normal stimulus with another one not directly related to survival

38. A = diabetic patient: B = increased insulin secretion; C = non-diabetic patient; D = glucose injection
(a) Difference in height and duration of glucose peak
(b) $(202 – 146) = 56$, $56/146 \times 100 = 38\%$
(c)

	Cause	Treatment
Type I	Deficient insulin secretion	Insulin injection
Type II	Liver insensitivity to insulin secretion	Control dietary intake of glucose/fat

39. **(a)** In the retina
(b) A = white colour perceived; B = blue cone; C = orange colour perceived; D = green cone; E = red cone
(c) A condition caused by a gene carried on one of the sex chromosomes (usually the X-chromosome)

40. **(a)** 1 **(b)** 2 **(c)** 1 **(d)** 4 **(e)** 3

41. Fluid-filled muscular sac; annelid; earthworm; endoskeleton; chordate; cat; exoskeleton; contains chitin; ladybird

42. A = M line; B = H zone; C = I band; D = Z line; E = A band

43. **(a)** A = ovarian phase; B = luteal phase; C = luteinising hormone; D = oestrogen; E = progesterone; F = repair of endometrium; G = ovulation; H = menstruation
(b) Progesterone acts as feedback inhibitor of FSH and so prevents ovulation – no ovulation means no conception
(c)

Difference	Reason
Male gamete much smaller	More cytoplasm in female gamete provides more food reserves for early embryonic development
Male gamete has flagellum	Male gamete must swim to female gamete

44.

	Genotype	Phenotype
1	$X^R Y$	Normal (male)
2	$X^R X^r$	Carrier (female)
3	$X^R X^r$	Carrier (female)
4	$X^R Y$	Normal (male)
5	$X^r Y$	Colour blind (male)
6	$X^R X^r$ or $X^R X^R$	Normal/carrier (female)

45. **(a)** Unlikely – there are similar numbers of males and females with the condition
(b) Recessive – two 'normal' parents in generation 2 produce a CF child in generation 3
(c) Thick mucus in lungs becomes infected with bacteria – antibiotics control bacterial multiplication
(d) Thick mucus blocks pancreatic duct and so limits release of lipase in pancreatic juice

Section A Molecules and cells

Use of the light microscope

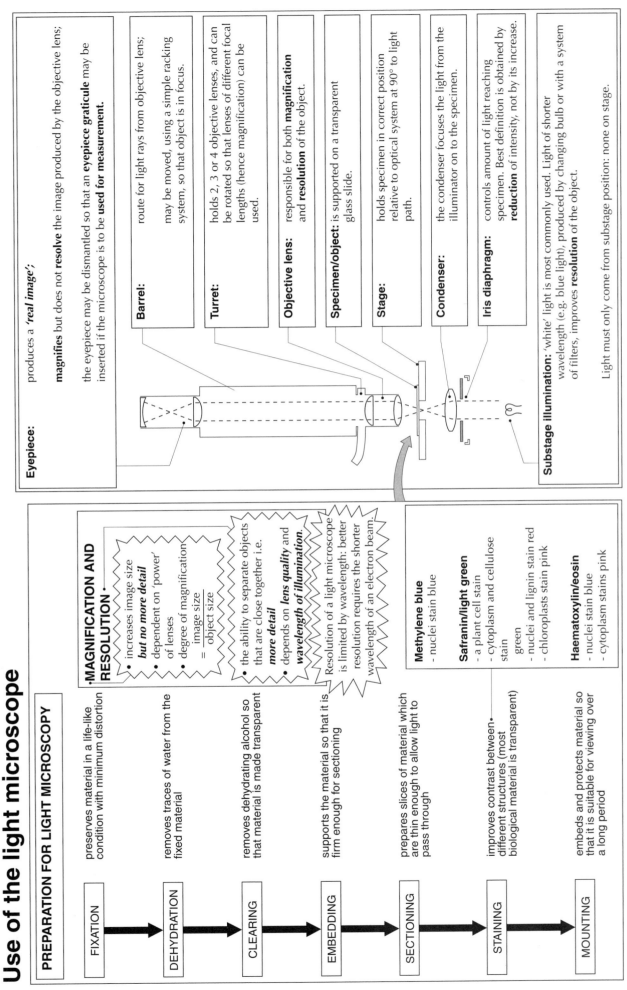

Eyepiece: produces a *'real image'*;

magnifies but does not **resolve** the image produced by the objective lens;

the eyepiece may be dismantled so that an **eyepiece graticule** may be inserted if the microscope is to be **used for measurement.**

Barrel: route for light rays from objective lens;

may be moved, using a simple racking system, so that object is in focus.

Turret: holds 2, 3 or 4 objective lenses, and can be rotated so that lenses of different focal lengths (hence magnification) can be used.

Objective lens: responsible for both **magnification** and **resolution** of the object.

Specimen/object: is supported on a transparent glass slide.

Stage: holds specimen in correct position relative to optical system at 90° to light path.

Condenser: the condenser focuses the light from the illuminator on to the specimen.

Iris diaphragm: controls amount of light reaching specimen. Best definition is obtained by **reduction** of intensity, not its increase.

Substage illumination: 'white' light is most commonly used. Light of shorter wavelength (e.g. blue light), produced by changing bulb or with a system of filters, improves **resolution** of the object.

Light must only come from substage position: none on stage.

PREPARATION FOR LIGHT MICROSCOPY

FIXATION → preserves material in a life-like condition with minimum distortion

DEHYDRATION → removes traces of water from the fixed material

CLEARING → removes dehydrating alcohol so that material is made transparent

EMBEDDING → supports the material so that it is firm enough for sectioning

SECTIONING → prepares slices of material which are thin enough to allow light to pass through

STAINING → improves contrast between different structures (most biological material is transparent)

MOUNTING → embeds and protects material so that it is suitable for viewing over a long period

MAGNIFICATION AND RESOLUTION

- increases image size **but no more detail**
- dependent on 'power' of lenses
- degree of magnification

$$= \frac{\text{image size}}{\text{object size}}$$

- the ability to separate objects that are close together i.e. **more detail**
- depends on **lens quality** and **wavelength of illumination.**

Resolution of a light microscope is limited by wavelength: better resolution requires the shorter wavelength of an electron beam.

Methylene blue
- nuclei stain blue

Safranin/light green
- a plant cell stain
- cytoplasm and cellulose stain green
- nuclei and lignin stain red
- chloroplasts stain pink

Haematoxylin/eosin
- nuclei stain blue
- cytoplasm stains pink

Plant and animal cells

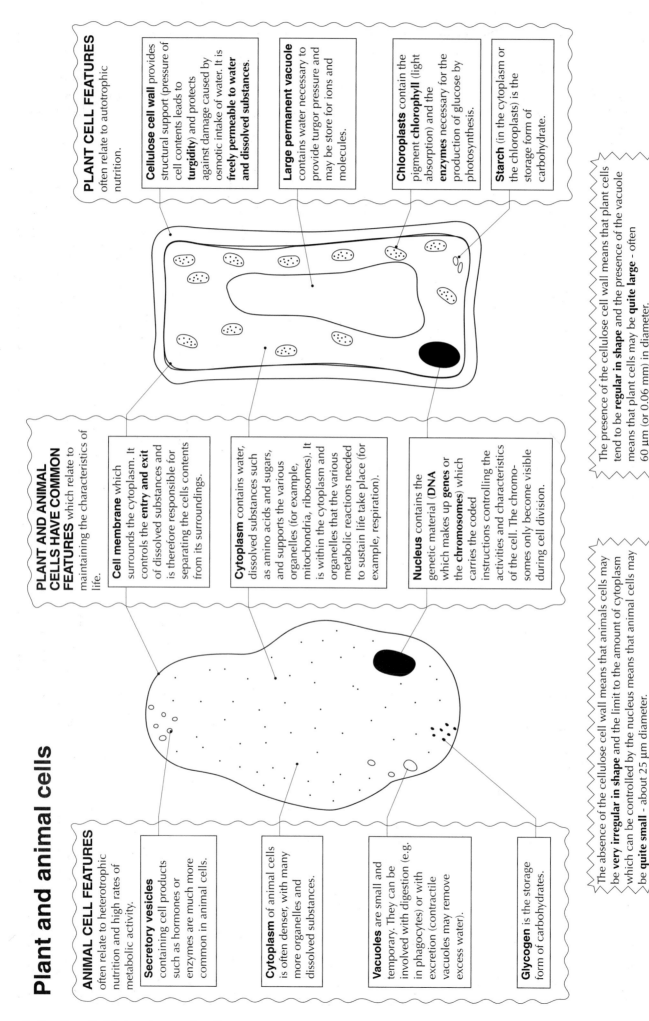

PLANT CELL FEATURES often relate to autotrophic nutrition.

Cellulose cell wall provides structural support (pressure of cell contents leads to **turgidity**) and protects against damage caused by osmotic intake of water. It is **freely permeable to water and dissolved substances.**

Large permanent vacuole contains water necessary to provide turgor pressure and may be store for ions and molecules.

Chloroplasts contain the pigment **chlorophyll** (light absorption) and the **enzymes** necessary for the production of glucose by photosynthesis.

Starch (in the cytoplasm or the chloroplasts) is the storage form of carbohydrate.

PLANT AND ANIMAL CELLS HAVE COMMON FEATURES which relate to maintaining the characteristics of life.

Cell membrane which surrounds the cytoplasm. It controls the **entry and exit** of dissolved substances and is therefore responsible for separating the cells contents from its surroundings.

Cytoplasm contains water, dissolved substances such as amino acids and sugars, and supports the various organelles (for example, mitochondria, ribosomes). It is within the cytoplasm and organelles that the various metabolic reactions needed to sustain life take place (for example, respiration).

Nucleus contains the genetic material (**DNA** which makes up **genes** or the **chromosomes**) which carries the coded instructions controlling the activities and characteristics of the cell. The chromosomes only become visible during cell division.

ANIMAL CELL FEATURES often relate to heterotrophic nutrition and high rates of metabolic activity.

Secretory vesicles containing cell products such as hormones or enzymes are much more common in animal cells.

Cytoplasm of animal cells is often denser, with many more organelles and dissolved substances.

Vacuoles are small and temporary. They can be involved with digestion (e.g. in phagocytes) or with excretion (contractile vacuoles may remove excess water).

Glycogen is the storage form of carbohydrates.

The presence of the cellulose cell wall means that plant cells tend to be **regular in shape** and the presence of the vacuole means that plant cells may be **quite large** - often 60 μm (or 0.06 mm) in diameter.

The absence of the cellulose cell wall means that animals cells may be **very irregular in shape** and the limit to the amount of cytoplasm which can be controlled by the nucleus means that animal cells may be **quite small** - about 25 μm diameter.

Transmission electron microscope

Cathode: metal electrode (commonly platinum) which emits high velocity electron beam. Electrons are negatively charged particles (e⁻).

Anode: positively charged electrode at potential of 50 kV with respect to cathode - accelerates the electron beam.

Condenser: electromagnetic lens which focuses the electron beam on to the specimen.

Air lock/specimen port: allows the introduction of the specimen into the microscope without the loss of vacuum.

Objective: electromagnetic lens which focuses and magnifies (depending on applied voltage) the first image.

Projector: further magnification by selection of region of image to be viewed.

To vacuum pump: creation of vacuum to minimize electron scattering and any heating due to electron/air molecule collision.

Fluorescent, swing out, screen: coated with electron sensitive compounds - necessary since deflected electron beam (the image) cannot be viewed directly.

Photographic plate: allows a black and white permanent record of the image to be made. Printing may offer further magnification.

Concrete base: stable support which minimizes vibration and thus eliminates unwanted deflection of electron beam.

SPECIMEN

VIEWING/ CAMERA PORT

Sample is

Fixed: to avoid deformation of all cell components. Use small sample (rapid penetration) and immerse in **glutaraldehyde** or **glutaraldehyde/osmic acid.**

Dehydrated: to prepare material for infiltration by embedding or infiltration medium which is not miscible with water. Dehydration should be gradual to preserve fine detail, using a series of progressively increasing concentrations of **ethanol** or **propanone.**

Cleared: alcohol or propanone may be immiscible with embedding agents and so is replaced with a clearing agent (commonly **xylol**) which is miscible and also makes the material transparent.

Embedded: **plastic** or **resin** is used to support the material so that it is not distorted during sectioning.
ready for

Sectioning
The material must be cut into **ultrathin sections** (20-100 nm thick) since the electron beam has very low penetrating power.

ULTRA MICROTOME moves embedded sample forward in 20 nm steps

EMBEDDED SAMPLE

DIAMOND OR GLASS KNIFE CUTS ULTRATHIN SECTIONS

RIBBON OF SECTIONS COLLECTS ON WATER

Staining: biological structures are transparent, or nearly so, to electrons. To increase electron beam deflection (i.e. contrast between different structures) sections are treated with **solutions of heavy metal salts** such as **uranyl** or **lead acetate.**

Mounting: sections are supported on a small copper grid (~3 mm diameter). The electron beam may pass through the gaps in the grid (a glass slide would not permit transmission of electron beam).

IMAGE INTERPRETATION

A — A' A number of ultrathin sections, e.g.
B — B' A-A', B-B', must be examined to
C — C' provide a true three-dimensional
D — D' representation of the sample.

High magnification means that several photographs may be necessary to give a composite image of the specimen.

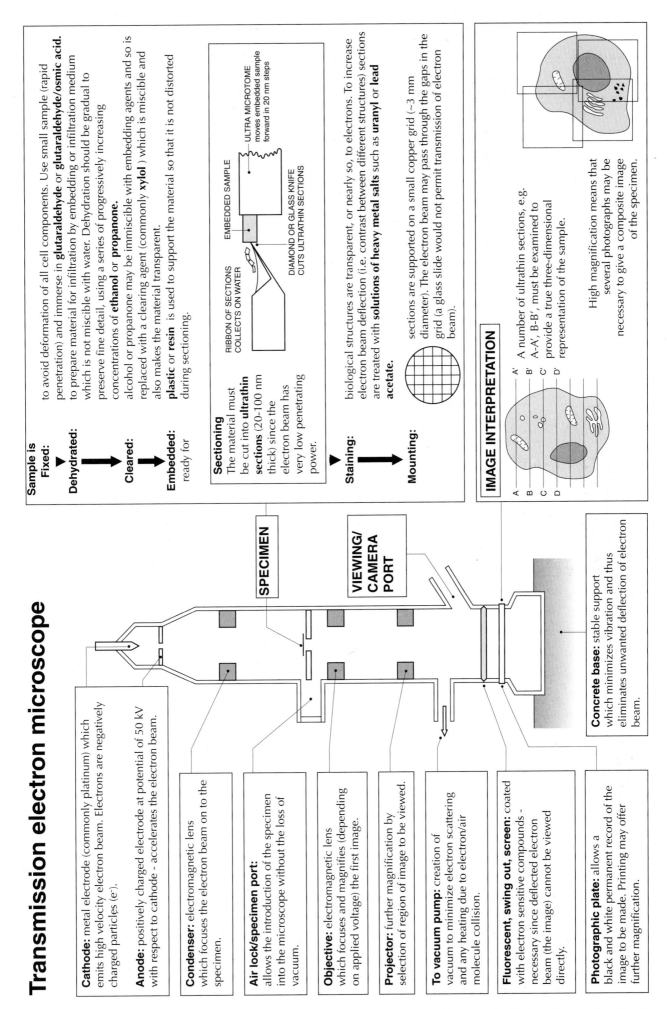

Scanning electron microscope

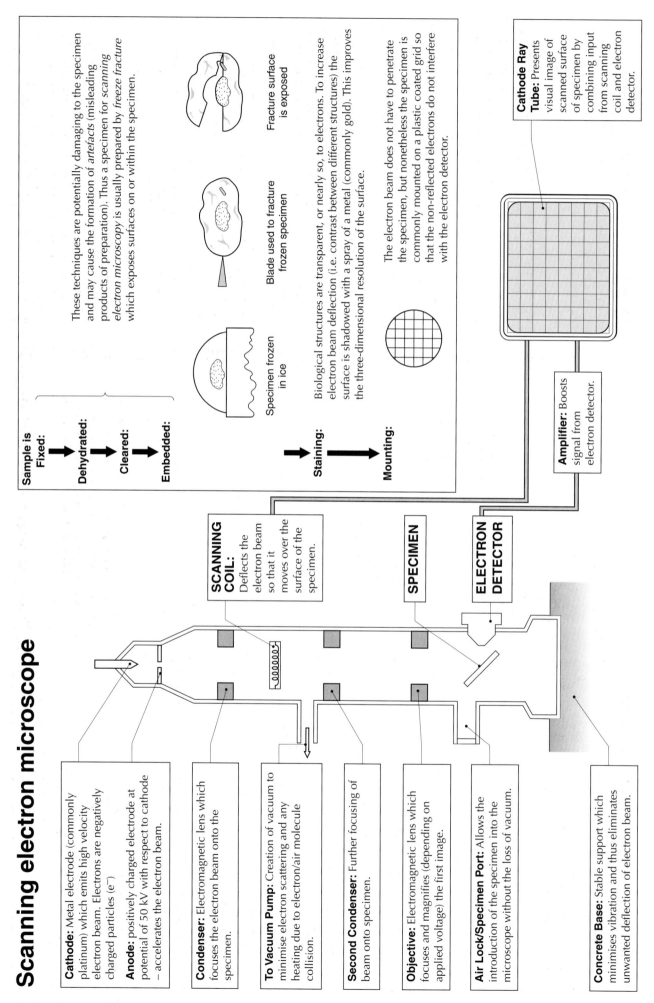

Cathode: Metal electrode (commonly platinum) which emits high velocity electron beam. Electrons are negatively charged particles (e⁻)

Anode: positively charged electrode at potential of 50 kV with respect to cathode – accelerates the electron beam.

Condenser: Electromagnetic lens which focuses the electron beam onto the specimen.

To Vacuum Pump: Creation of vacuum to minimise electron scattering and any heating due to electron/air molecule collision.

Second Condenser: Further focusing of beam onto specimen.

Objective: Electromagnetic lens which focuses and magnifies (depending on applied voltage) the first image.

Air Lock/Specimen Port: Allows the introduction of the specimen into the microscope without the loss of vacuum.

Concrete Base: Stable support which minimises vibration and thus eliminates unwanted deflection of electron beam.

SCANNING COIL: Deflects the electron beam so that it moves over the surface of the specimen.

SPECIMEN

ELECTRON DETECTOR

Amplifier: Boosts signal from electron detector.

Cathode Ray Tube: Presents visual image of scanned surface of specimen by combining input from scanning coil and electron detector.

Sample is
Fixed:
Dehydrated:
Cleared:
Embedded:

These techniques are potentially damaging to the specimen and may cause the formation of *artefacts* (misleading products of preparation). Thus a specimen for *scanning electron microscopy* is usually prepared by *freeze fracture* which exposes surfaces on or within the specimen.

Specimen frozen in ice

Blade used to fracture frozen specimen

Fracture surface is exposed

Staining:

Biological structures are transparent, or nearly so, to electrons. To increase electron beam deflection (i.e. contrast between different structures) the surface is shadowed with a spray of a metal (commonly gold). This improves the three-dimensional resolution of the surface.

Mounting:

The electron beam does not have to penetrate the specimen, but nonetheless the specimen is commonly mounted on a plastic coated grid so that the non-reflected electrons do not interfere with the electron detector.

Physical properties of water

are explained by hydrogen bonding between the individual molecules

High specific heat capacity The specific heat capacity of water (the amount of heat, measured in joules, required to raise 1 kg of water through 1°C) is very high: much of the heat absorbed is used to break the hydrogen bonds which hold the water molecules together.

High latent heat of vaporization Hydrogen bonds attract molecules of liquid water to one another and make it difficult for the molecules to escape as vapour: thus a relatively high energy input is necessary to vaporize water and water has a much higher boiling point than other molecules of the same size.

Molecular mobility The weakness of individual hydrogen bonds means that individual water molecules continually jostle one another when in the liquid phase.

Cohesion and surface tension Hydrogen bonding causes water molecules to 'stick together', and also to stick to other molecules - the phenomenon of *cohesion*. At the surface of a liquid the inwardly-acting cohesive forces produce a 'surface tension' as the molecules are particularly attracted to one another.

Density and freezing properties As water cools towards its freezing point the individual molecules slow down sufficiently for each one to form its maximum number of hydrogen bonds. To do this the water molecules in liquid water must move further apart to give enough space for all four hydrogen bonds to fit into. As a result water expands as it freezes, so that ice is less dense than liquid water and therefore floats upon its surface.

Colloid formation Some molecules have strong intramolecular forces which prevent their solution in water, but have charged surfaces which attract a covering of water molecules. This covering ensures that the molecules remain dispersed throughout the water, rather than forming large aggregates which could settle out. The dispersed particles and the liquid around them collectively form a *colloid*.

Because hydrogen and oxygen atoms are different in **size** and **electronegativity** the water molecule (H_2O) is **non-linear** and **polar**.

OXYGEN ATOM

105°

HYDROGEN ATOM

HYDROGEN ATOM

$H^{\delta+}$

$O^{\delta-}$

$H^{\delta+}$

$H^{\delta+}$

$O^{\delta-}$

$O^{\delta-}$

$H^{\delta+}$

$H^{\delta+}$

$H^{\delta+}$

$O^{\delta-}$

$H^{\delta+}$

$H^{\delta+}$

Hydrogen bond - one water molecule may form hydrogen bonds with up to *four* other water molecules.

This polarity means that individual water molecules can form **hydrogen bonds** with other water molecules. Although these individual hydrogen bonds are weak, collectively **they make water a much more stable substance than would otherwise be the case.**

Solvent properties The polarity of water makes it an excellent solvent for other polar molecules ....

The electrostatic attractions between polar water molecules and ions are greater than those between the anion and cation.

Ions become **hydrated in aqueous solution.**

ANION –

CATION +

δ^- δ^-

+

δ^- δ^-

δ^+ δ^+

δ^+ δ^+

–

δ^+ δ^+

δ^+ δ^+

Such polar substances, which dissolve in water, are said to be **hydrophilic** ('water-loving').

... but means that non-polar (**hydrophobic** or 'water-hating') substances do not readily dissolve in water.

ADD TO WATER

Non-polar molecules arrange themselves to expose the minimum possible surface to the water molecules.

The biological importance
of water depends on its physical properties

Solvent properties:
allow water to act as a transport medium for polar solutes. For example,
- movements of minerals to lakes and seas;
- transport via blood and lymph in multicellular animals;
- removal of metabolic wastes such as urea and ammonia in urine.

Transpiration stream: the continuous column of water is able to move up the xylem because of cohesion between water molecules and adhesion between water and the walls of the xylem vessels.

Molecular mobility: the rather weak nature of individual hydrogen bonds means that water molecules can move easily relative to one another – this allows *osmosis* (vital for uptake and movement of water) to take place.

Expansion on freezing: since ice floats it forms at the surface of ponds and lakes – it therefore insulates organisms in the water below it, and allows the ice to thaw rapidly when temperatures rise. Changes in density also maintain circulation in large bodies of water, thus helping nutrient cycling. Floating ice also means that penguins and polar bears have somewhere to stand!

Phew!

Metabolic functions
Water is used directly …
1. as a reagent (source of reducing power) in photosynthesis
2. to hydrolyse macromolecules to their subunits, in digestion for example.
… and is also the medium in which all biochemical reactions take place.

Lubricant properties: water's cohesive and adhesive properties mean that it is viscous, making it a useful lubricant in biological systems. For example,
- **synovial fluid** - lubricates many vertebrate joints;
- **pleural fluid** - minimizes friction between lungs and thoracic cage (ribs) during breathing;
- *mucus* - permits easy passage of faeces down the colon, and lubricates the penis and vagina during intercourse.

Thermoregulation: the high specific heat capacity of water means that bodies composed largely of water (cells are typically 70-80% water) are very thermostable, and thus less prone to heat damage by changes in environmental temperatures.
The high latent heat of vaporization of water means that a body can be considerably cooled with a minimal loss of water - this phenomenon is used extensively by mammals (sweating) and reptiles (gaping) and may be important in cooling transpiring leaves.

Volatility/stability: is balanced at Earth's temperatures so that a water cycle of evaporation, transpiration and precipitation is maintained.

Supporting role: the cohesive forces between water molecules mean that it is not easily compressed, and thus it is an excellent medium for support. Important biological examples include the *hydrostatic skeleton* (e.g. earthworm), *turgor pressure* (in herbaceous parts of plants), *amniotic fluid* (which supports and protects the mammalian foetus) and as a *general supporting medium* (particularly for large aquatic mammals such as whales).

Transparency: water permits the passage of visible light. This means that photosynthesis (and associated food chains) is possible in relatively shallow aquatic environments.

Osmosis

Water molecules, like other molecules, are mobile. In pure water, or in solutions containing very few solute molecules, the water molecules can move very freely (they have a high **free kinetic energy**). As a result, many of the water molecules may cross the membrane, which is freely permeable to water.

Partially permeable membrane allows the free passage of some particles but is not freely permeable to others. Biological membranes are **freely permeable to water** but have **restricted permeability to solutes** such as sodium ions and glucose molecules, i.e. they are **selectively permeable.**

In a solution with many solute molecules the movement of the water molecules is restricted because of solute-water interactions. Fewer of the water molecules have a **free kinetic energy** which is great enough to enable them to cross the membrane.

Solute molecules cannot cross the membrane as freely or as rapidly as water molecules can.

MANY WATER MOLECULES CAN MOVE IN THIS DIRECTION

FEW WATER MOLECULES CAN MOVE IN THIS DIRECTION

THERE IS A NET MOVEMENT OF WATER MOLECULES IN THIS DIRECTION * .

This movement of water depends on how many water molecules have sufficient free kinetic energy to 'escape from' the system

so that any system in which the water molecules have a **high** average kinetic energy will have a greater tendency to lose water than will a system in which the water molecules have a **low** average kinetic energy

and when describing water movements scientists replace the term **free kinetic energy** with the term **water potential**, so that

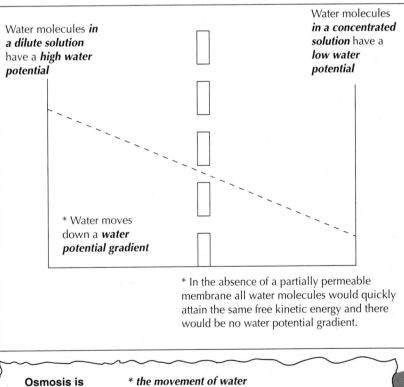

Water molecules **in a dilute solution** have a **high water potential**

Water molecules **in a concentrated solution** have a **low water potential**

* Water moves down a **water potential gradient**

* In the absence of a partially permeable membrane all water molecules would quickly attain the same free kinetic energy and there would be no water potential gradient.

Osmosis is
* **the movement of water**
* **down a water potential gradient**
* **across a partially permeable membrane**
* **to a solution with a more negative water potential.**

pH

Hydrogen ion concentrations in aqueous solutions are measured on the pH scale.

Water dissociates to a very slight extent, thus

$$H_2O \rightleftharpoons H^+ + OH^-$$

In pure water the hydrogen ion concentration $[H^+]$ is 0.000 0001 $mol\,dm^{-3}$ (that is, 1 in every 10^7 water molecules is dissociated into H^+ and OH^-). This is an awkward number to handle and so is better expresssed in a logarithmic form, i.e. in pure water

$$[H^+] = 10^{-7}\ mol\,dm^{-3}$$

This is more commonly written as

Hydrogen ions available in solution

pH = 7.0

Power – the negative power to which 10 is raised

pH	Examples
1	stomach acid
2	gastric juice (pH 1.0–3.0), lemon juice
3	vinegar, beer, soft drinks, orange juice
4	tomato juice, grapes
5	black coffee, rainwater
6	urine (5–7), milk (6.5), saliva (6.5–7.4)
7	pure water, blood (7.3–7.5), egg white
8	duodenal contents (7.4–8.0), seawater (7.8–8.3)
9	baking soda, phosphate detergents
10	soap solutions, Milk of Magnesia (antacid)
11	non-phosphate detergents
12	washing soda (Na_2CO_3)
13	
14	

increasingly acidic

neutral

increasingly basic

The pH range is from 1 to 14, because in 'neutral' water at pH 7.0, $[H^+] = [OH^-] = 10^{-7}$. The product of the concentrations of these two ions is thus 10^{-14} and, by the law of chemical equilibrium, as the concentration of one of them increases the other must decrease to maintain this product at 10^{-14}. Thus

$[H^+]$	1	10^{-1}	10^{-2}	10^{-3}	10^{-4}	10^{-5}	10^{-6}	10^{-7}	10^{-8}	10^{-9}	10^{-10}	10^{-11}	10^{-12}	10^{-13}	10^{-14}
pH	1	2	3	4	5	6	7	8	9	10	11	12	13	14	

BUFFER SOLUTIONS

Buffers are able to accept or release hydrogen ions, and thus are able to resist minor changes in pH. Typically a buffer is a mixture of a weak acid and one of its salts. Thus

$$H.A \rightleftharpoons H^+ + A^-$$

Weak acid can act as a proton donor if $[H^+]$ falls (i.e. pH rises)

Weak base can act as proton acceptor if $[H^+]$ rises (i.e. pH falls)

Note the pH scale is logarithmic, so that a change in pH of 1 unit corresponds to a change in hydrogen ion concentration of 10. For example, gastric juice is 10 times more concentrated than lemon juice and 100 times more concentrated than a soft drink and the hydrogen ion concentration in human urine may fluctuate over a hundred-fold range.

Important buffers in biological fluids are

Hydrogencarbonate: H_2CO_3/HCO_3^-

Hydrogenphosphate: $H_2PO_4^-/HPO_4^{2-}$

Haemoglobin: histidine residues in side chain accept H^+

These buffers, and the action of the distal convuluted tubule in the nephron of the kidney, are able to maintain blood pH at between 7.3 and 7.5. This is the optimum for maintenance of structure and function of blood proteins.

Structural components of membranes

membranes permit fluidity, selective transport and recognition, integrity and compartmentalization.

Because of the different solubility properties of the two ends of phospholipid molecules ...

polar, so very soluble in water

non-polar, so very insoluble in water

... such molecules form a layer at a water surface

and a *phospholipid bilayer* can act as a barrier between two aqueous environments.

Hydrophilic heads point outwards: form hydrogen bonds with water

Hydrophobic tails point towards one another: this maximizes hydrophobic attractions and excludes water

WATER

WATER

Diffusion through aqueous channels in pore proteins:

transmembrane proteins may have aqueous channels through which charged molecules may pass and thus avoid the hydrophobic tails of the phospholipid molecules.

Na$^+$

Some channels are open all of the time, but others are *gated* (they open and close only in response to a stimulus, such as a change in the membrane's electrical potential).

Such *gated channels* are vital to the operation of nerve and muscle, where movements of Na$^+$, K$^+$ and Ca^{2+} initiate information transfer.

Diffusion across the lipid bilayer is responsible for the movement of *small, uncharged molecules.*

Thus O$_2$, H$_2$O, CO$_2$, urea and ethanol cross rapidly (they 'squeeze between') the polar phospholipid heads then dissolve in the lipid on one side of the membrane and emerge on the other.

Large or *charged molecules* cannot cross the lipid bilayer.

Thus Na$^+$, K$^+$, Cl$^-$, HCO$_3^-$ and glucose do not cross in this way.

Active transport
uses a *carrier protein* to transport a solute across a membrane but *energy is required* since transport may be *against a concentration gradient.* Typically ATP is hydrolysed and the binding of the phosphate group to the carrier changes the protein's conformation in such a way that the solute molecule is moved across the membrane.

Facilitated diffusion
uses a *carrier protein* to transfer a molecule across a membrane *along* its electrochemical gradient. The binding of the solute alters the conformation of the carrier so that its position in the membrane changes and the solute molecule is discharged on the other side of the membrane. Glucose uptake by erythrocytes occurs in this way.

N.B. There is *no requirement for ATP*, as there is *no energy consumption.*

SOLUTE BINDING

CARRIER INVERSION

SOLUTE RELEASE AND CARRIER RETURN

Surface carbohydrates
(collectively the *glycocalyx)* are usually oligosaccharides which are positioned to aid in cell recognition functions.

Lipid composition influences membrane fluidity: unsaturated fatty acid tails are 'kinked', limit close packing of the hydrophobic tails and so *increase* fluidity, but cholesterol may interfere with lateral movement of hydrophobic tails and thus *reduce* membrane fluidity.

Cell adhesion proteins firmly attach adjacent cells to one another, this is particularly important in epithelia. These proteins also serve as internal anchorage points for protein tubules of the cytoskeleton.

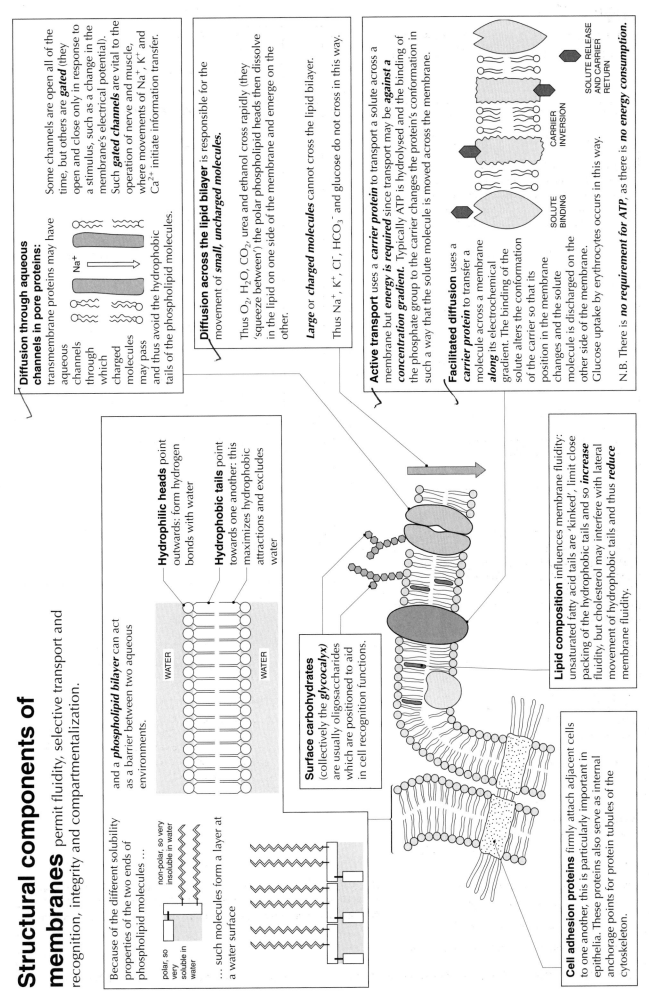

Animal cell ultrastructure

Centrioles are a pair of structures, held at right angles to one another, which act as organizers of the nuclear spindle in preparation for the separation of chromosomes or chromatids during nuclear division.

Secretory vesicle undergoing exocytosis. May be carrying a synthetic product of the cell (such as a protein packaged at the Golgi body) or the products of degradation by lysosomes. Secretory vesicles are abundant in cells with a high synthetic activity, such as the cells of the *Islets of Langerhans*.

Smooth endoplasmic reticulum is a series of flattened sacs and sheets that are the sites of synthesis of steroids and lipids.

Rough endoplasmic reticulum is so-called because of the many ribosomes attached to its surface. This intracellular membrane system aids cell compartmentalization and transports proteins synthesized at the ribosomes towards the Golgi bodies for secretory packaging.

Golgi apparatus consists of a stack of sacs called *cisternae*. It modifies a number of cell products delivered to it, often enclosing them in vesicles to be secreted. Such products include trypsinogen (from *pancreatic acinar cells*), insulin (from *beta-cells of the Islets of Langerhans*) and mucin (from *goblet cells in the trachea*). The Golgi is also involved in lipid modification in cells of the ileum, and plays a part in the formation of lysosomes.

Plasmalemma (plasmamembrane) is the surface of the cell and represents its contact with its environment. It is differentially permeable and regulates the movement of solutes between the cell and its environment. There are many specializations of the membrane, often concerning its protein content.

Microfilaments are threads of the protein *actin*. They are usually situated in bundles just beneath the cell surface and play a role in endo- and exocytosis, and possibly in cell motility.

Microvilli are extensions of the plasmamembrane which increase the cell surface area. They are commonly abundant in cells with a high absorptive capacity, such as *hepatocytes* or cells of the *first coiled tubule of the nephron*. Collectively the microvilli represent a *brush border* to the cell.

Free ribosomes are the sites of protein synthesis, principally for proteins destined for intracellular use. There may be 50 000 or more in a typical eukaryote cell.

Endocytic vesicle may contain molecules or structures too large to cross the membrane by active transport or diffusion.

Nucleus is the centre of the regulation of cell activities since it contains the hereditary material, DNA, carrying the information for protein synthesis. The DNA is bound up with histone protein to form chromatin. The nucleus contains one or more nucleoli in which ribosome subunits, ribosomal RNA, and transfer RNA are manufactured. The nucleus is surrounded by a double nuclear membrane, crossed by a number of nuclear pores. The nucleus is continuous with the endoplasmic reticulum. There is usually only one nucleus per cell, although there may be many in very large cells such as those of striated (skeletal) muscle. Such multinucleate cells are called coenocytes.

Cytoplasm is principally water, with many solutes including glucose, proteins and ions. It is permeated by the *cytoskeleton*, which is the main architectural support of the cell.

Mitochondrion (pl. mitochondria) is the site of aerobic respiration. Mitochondria have a highly folded inner membrane which supports the proteins of the electron transport chain responsible for the synthesis of ATP by oxidative phosphorylation. The mitochondrial matrix contains the enzymes of the TCA cycle, an important metabolic 'hub'. These organelles are abundant in cells which are physically (*skeletal muscle*) and metabolically (*hepatocytes*) active.

Typical plant cell

contains chloroplasts and a permanent vacuole, and is surrounded by a cellulose cell wall.

Plasmodesmata are minute strands of cytoplasm which pass through pores in the cell wall and connect the protoplasts of adjacent cells. This represents the **symplast** pathway for the movement of water and solutes throughout the plant body. These cell-cell cytoplasm connections are important in cell survival during periods of drought. The E.R. of adjacent cells is also in contact through these strands.

Cell wall is composed of long cellulose molecules grouped in bundles called **microfibrils** which, in turn, are twisted into rope-like **macrofibrils.** The macrofibrils are embedded in a matrix of **pectins** (which are very adhesive) and **hemicelluloses** (which are quite fluid). There may be a **secondary cell wall**, in which case the outer covering of the cell is arranged as:

Plasmalemma

Secondary cell wall: laid down on inside of primary wall. Often impregnated with **lignin** (gives mechanical strength to xylem) or **suberin** (waterproofs endodermis).

Primary cell wall: laid down first, by plasmamembrane.

Middle lamella: contains gums and calcium pectate to cement cells together.

The function of the cell wall is a mechanical one - pressure from the cell protoplast maintains cell turgidity. The wall is freely permeable to water and most solutes so that the cell wall represents an important transport route - the **apoplast system** - throughout the plant body.

Rough endoplasmic reticulum is the site of protein synthesis (on the attached ribosomes), storage and preparation for secretion. The endoplasmic reticulum (E.R.) also plays a part in the compartmentalization of the cell.

Nucleus is surrounded by the nuclear envelope and contains the genetic material, DNA, associated with histone protein to form chromatin. The nucleus thus controls the activity of the cell through its regulation of protein synthesis. The nucleolus is the site of synthesis of transfer RNA, ribosomal RNA, and ribosomal subunits.

Mitochondrion contains the enzyme systems for ATP synthesis by oxidative phosphorylation. May be abundant in sieve tube companion cells, root epidermal cells and dividing meristematic cells.

Golgi body (dictyosome) synthesizes polysaccharides and packages them in vesicles which migrate to the plasmamembrane for eventual incorporation in the cell wall.

Chloroplast is the site of photosynthesis. It is one of a number of plastids, all of which develop from **proplastids** which are small, pale green or colourless organelles.

Other typical plastids of complex cells are **chromoplasts** which may develop from chloroplasts by internal rearrangements. Chromoplasts are coloured due to the presence of carotenoid pigments and are most abundant in cells of flower petals or fruit skins.

Leucoplasts are a third type of plastid common in cells of higher plants - they include **amyloplasts** which synthesize and store starches and **elaioplasts** which synthesize oils.

Vacuole may occupy 90% of the volume of a mature plant cell. It is filled with cell sap (a solution of salts, sugars and organic acids) and helps to maintain turgor pressure inside the cell. The vacuole also contains anthocyanins, pigments responsible for many of the red, blue and purple colours of flowers. Vacuoles also contains enzymes involved in recycling of cell components such as chloroplasts. The vacuolar membrane is called the **tonoplast.**

Microtubules are hollow structures (about 25 nm in diameter) composed of the protein tubulin. They occur just below the plasmamembrane where they may aid the addition of cellulose to the cell wall. They are also involved in the cytoplasmic streaming of organelles such as Golgi bodies and chloroplasts, and they form the spindles and cell plates of dividing cells.

Plasmamembrane (plasmalemma, cell surface membrane) is the differentially-permeable cell surface, responsible for the control of solute movements between the cell and its environment. It is flexible enough to move close to or away from the cell wall as the water content of the cytoplasm changes. The membrane is also responsible for the synthesis and assembly of cell wall components.

Smooth endoplasmic reticulum is the site of lipid synthesis and secretion.

Differential centrifugation may be used to isolate cell components.

Isotonic Solution prevents osmotic damage to cells and organelles

Buffer Solution prevents pH changes which might otherwise denature proteins, especially enzymes and membrane proteins.

Motor-driven Homogeniser forces cells between wall of tube and rotating pestle. Shearing forces developed are just sufficient to rupture the cells but not damage the organelles

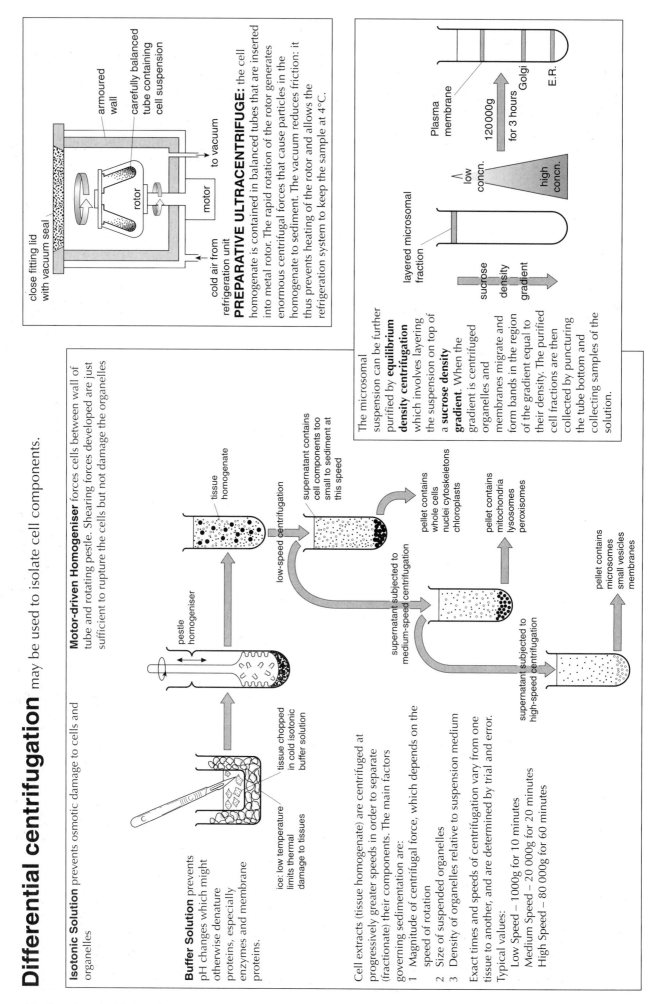

tissue chopped in cold isotonic buffer solution

ice: low temperature limits thermal damage to tissues

pestle homogeniser

tissue homogenate

low-speed centrifugation

supernatant contains cell components too small to sediment at this speed

pellet contains whole cells nuclei cytoskeletons chloroplasts

supernatant subjected to medium-speed centrifugation

pellet contains mitochondria lysosomes peroxisomes

supernatant subjected to high-speed centrifugation

pellet contains microsomes small vesicles membranes

Cell extracts (tissue homogenate) are centrifuged at progressively greater speeds in order to separate (fractionate) their components. The main factors governing sedimentation are:
1 Magnitude of centrifugal force, which depends on the speed of rotation
2 Size of suspended organelles
3 Density of organelles relative to suspension medium

Exact times and speeds of centrifugation vary from one tissue to another, and are determined by trial and error.
Typical values:
Low Speed – 1000g for 10 minutes
Medium Speed – 20 000g for 20 minutes
High Speed – 80 000g for 60 minutes

close fitting lid with vacuum seal

armoured wall

carefully balanced tube containing cell suspension

rotor

motor

to vacuum

cold air from refrigeration unit

PREPARATIVE ULTRACENTRIFUGE: the cell homogenate is contained in balanced tubes that are inserted into metal rotor. The rapid rotation of the rotor generates enormous centrifugal forces that cause particles in the homogenate to sediment. The vacuum reduces friction: it thus prevents heating of the rotor and allows the refrigeration system to keep the sample at 4°C.

The microsomal suspension can be further purified by **equilibrium density centrifugation** which involves layering the suspension on top of a **sucrose density gradient**. When the gradient is centrifuged organelles and membranes migrate and form bands in the region of the gradient equal to their density. The purified cell fractions are then collected by puncturing the tube bottom and collecting samples of the solution.

layered microsomal fraction

sucrose density gradient

low concn.

high concn.

Plasma membrane

120000g

for 3 hours

Golgi

E.R.

A prokaryotic cell

(e.g. a bacterium) has no true organelles.

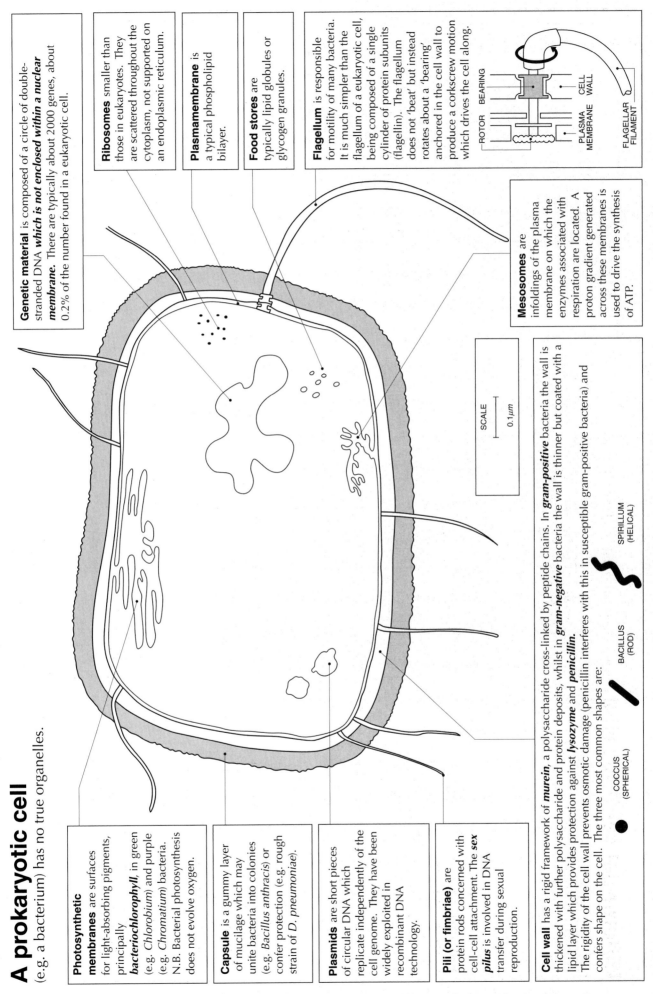

Genetic material is composed of a circle of double-stranded DNA *which is not enclosed within a nuclear membrane.* There are typically about 2000 genes, about 0.2% of the number found in a eukaryotic cell.

Ribosomes smaller than those in eukaryotes. They are scattered throughout the cytoplasm, not supported on an endoplasmic reticulum.

Plasmamembrane is a typical phospholipid bilayer.

Food stores are typically lipid globules or glycogen granules.

Flagellum is responsible for motility of many bacteria. It is much simpler than the flagellum of a eukaryotic cell, being composed of a single cylinder of protein subunits (flagellin). The flagellum does not 'beat' but instead rotates about a 'bearing' anchored in the cell wall to produce a corkscrew motion which drives the cell along.

ROTOR   BEARING

PLASMA MEMBRANE

CELL WALL

FLAGELLAR FILAMENT

Mesosomes are infoldings of the plasma membrane on which the enzymes associated with respiration are located. A proton gradient generated across these membranes is used to drive the synthesis of ATP.

SCALE

0.1μm

Photosynthetic membranes are surfaces for light-absorbing pigments, principally *bacteriochlorophyll*, in green (e.g. *Chlorobium*) and purple (e.g. *Chromatium*) bacteria. N.B. Bacterial photosynthesis does not evolve oxygen.

Capsule is a gummy layer of mucilage which may unite bacteria into colonies (e.g. *Bacillus anthracis*) or confer protection (e.g. rough strain of *D. pneumoniae*).

Plasmids are short pieces of circular DNA which replicate independently of the cell genome. They have been widely exploited in recombinant DNA technology.

Pili (or fimbriae) are protein rods concerned with cell-cell attachment. The *sex pilus* is involved in DNA transfer during sexual reproduction.

Cell wall has a rigid framework of *murein*, a polysaccharide cross-linked by peptide chains. In *gram-positive* bacteria the wall is thickened with further polysaccharide and protein deposits, whilst in *gram-negative* bacteria the wall is thinner but coated with a lipid layer which provides protection against *lysozyme* and *penicillin*. The rigidity of the cell wall prevents osmotic damage (penicillin interferes with this in susceptible gram-positive bacteria) and confers shape on the cell. The three most common shapes are:

COCCUS (SPHERICAL)

BACILLUS (ROD)

SPIRILLUM (HELICAL)

Viruses are acellular and parasitic

A typical virus is a simple structure composed of a nucleic acid core completely enclosed within a protein coat.

Viruses are the smallest structures which can show signs of life – they range in length from 20 to 300 nm. They are not visible with the light microscope, and they pass easily through filters which retain bacteria.

Viruses are classified on the border between living and non-living. They are considered living since they contain genetic material and are capable of reproduction, but non-living since they have no cellular organisation. They certainly show no signs of life outside the cells of their host.

Viruses and disease Since they are obligate parasites viruses necessarily damage the cells which are their hosts, and are therefore responsible for some serious diseases. Some examples are outlined below:

Disease	Causative agent	Effects	Control method
Influenza	A number of DNA viruses (myxovirus)	Aches, symptoms of common cold, respiratory distress	Vaccination with 'killed' virus of the correct strain
Rabies	RNA virus	Hyperexcitability, paralysis hydrophobia, death	Elimination of animal carriers. Vaccination of humans
Tobacco mosaic	TMV (an RNA virus)	Mottling of tobacco leaves causing death/reduced yield	Destruction of infected plants
German measles	Rubella virus	Swollen lymph nodes, inflamed respiratory passages, skin rash	Living attenuated virus – more important for girls because of pregnancy complications
Foot and mouth	RNA	Lesions in mouth of cattle, pigs	Vaccination. Slaughter of infected animals

Acquired immune deficiency syndrome (AIDS) is caused by a depression of the immune response.

May be due to HIV infection of dendritic cells (display antigen) or T$_{HELPER}$ lymphocytes (co-ordinate lymphocyte action).

Allows attack by normally harmless organisms, e.g.
Pneumocystis carinii → pneumonia
Candida albicans → thrush
Cytomegalovirus → retinal disease and characteristic skin lesions of Kaposi's sarcoma.

Viral diseases are often difficult to treat *since antibiotics cannot be used* (there are no metabolic processes for them to inhibit), vaccines may be ineffective since the viruses may mutate the antigenic factors on their capsids, and chemotherapy may damage host cells as well as inhibiting viral replication. The most common forms of control are vaccination and removal of the source of infection.

Some retroviruses may actually play a part in disease treatment – they may be used to insert DNA into host cells which have a defective form of the particular gene (PKU – phenylketonuria – may be treated in this way).

The *core* is the region within the capsid. It is not cytoplasm, and contains no organelles.

The *genetic material* may be either DNA or RNA, but never both. The number of genes is very small – these genes contain only the information necessary for replication of viral subunits and their assembly into a complete virus or *virion*, once within the host cell.

The *protein coat* is composed of a large number of identical units called *capsomeres* which self-assemble into a highly symmetrical *capsid*. The form of the assembled capsid may be used in the classification of viruses.

Surface extensions of the capsid are responsible for the antigenic properties of the virus.

A *lipoprotein envelope* surrounds the capsid in some viruses, usually the larger ones. The envelope is frequently derived from the host cell and encloses the virion as it is dispersed from the host. This envelope may be important in the virus's ability to penetrate the host cell's defences. The human immunodeficiency virus (HIV) is surrounded by such a lipoprotein envelope.

Origins of viruses: the fact that viruses cannot replicate outside of a parasitised host cell suggests that they may originate from 'escaped' DNA rather than represent primitive ancestors of more typical cells. *Prions* are also infective particles, but are proteins with no nucleic acid component. They are thought to be involved in V CJD and BSE.

Elements in the human body

Name	Chemical symbol	Proportion of total mass (%)	Role or function
BULK ELEMENTS			
Oxygen	O	65	Required for oxidation reactions, principally those of respiration. Present in most organic compounds and in water.
Carbon	C	18	Forms backbone of organic compounds.
Hydrogen	H	10	Present in most organic compounds and in water.
Nitrogen	N	3	Component of all proteins, nucleic acids, and many other organic compounds.
Calcium	Ca	1.5	Structural component of bones and teeth. Important in conduction of nerve impulses across synapses, blood clotting, muscle contraction, fertilisation.
Phosphorus	P	1.0	Component of nucleic acids, nucleotides involved in energy transfer, phospholipids. Structural component of bone.
Potassium	K	0.4	Principal intracellular cation. Important in conduction of nerve impulses.
Sulphur	S	0.3	Component of most proteins.
Sodium	Na	0.2	Principal extracellular cation. Important in controlling fluid movement between body compartments. Conduction of nerve impulses.
TRACE ELEMENTS			
Magnesium	Mg	0.1	Enzyme cofactor (kinases).
Chlorine	Cl	0.1	Principal anion of interstitial fluid. Important in fluid balance. Chloride shift in oxygen transport.
Iron	Fe	trace	Component of haemoglobin and myoglobin. Electron transfer molecules. Enzyme cofactor (catalase).
Iodine	I	trace	Component of thyroid hormones.
Cobalt	Co	trace	Component of vitamin B_{12}.

Other elements found in trace amounts include manganese (Mn), copper (Cu), zinc (Zn), fluorine (F), molybdenum (Mo), and selenium (Se).

BIOTIP

Note that on this page there are different ways of expressing data: one set of figures is expressed as 'percentage of total atoms present', and another as 'percentage of total mass'. The two sets of data should not, therefore, be directly compared. It is very important to always consider the *units* as well as the *number* when looking at a quantity.

RELATIVE PERCENTAGES OF ELEMENTS FOUND IN LIVING ORGANISMS COMPARED WITH THE EARTH'S CRUST

(These data expressed as a percentage of the total atoms present)

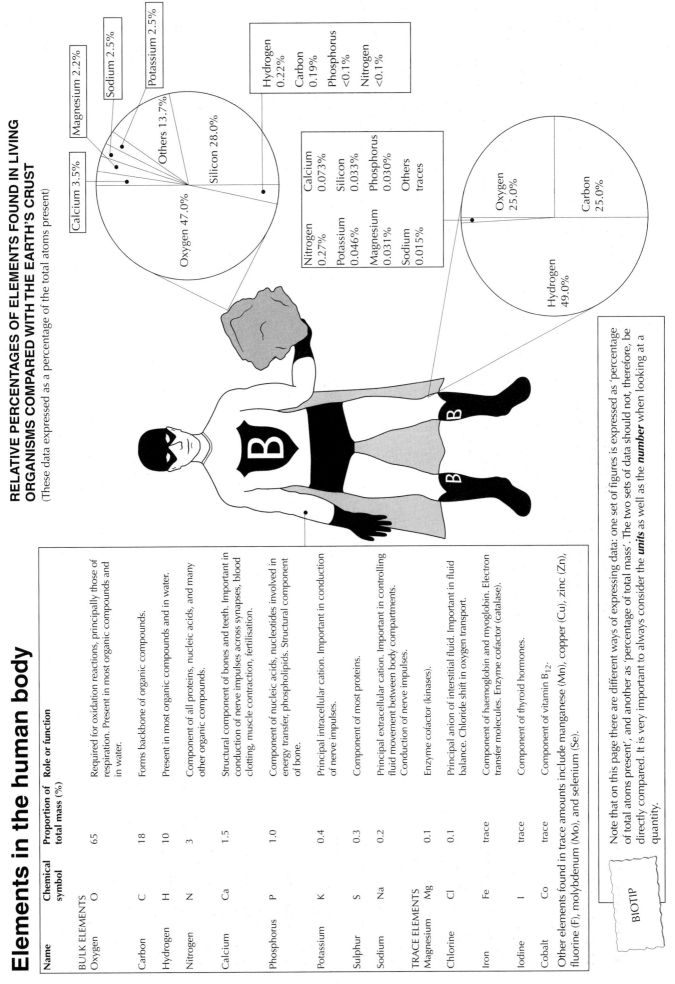

Earth's crust:
- Oxygen 47.0%
- Silicon 28.0%
- Others 13.7%
- Calcium 3.5%
- Magnesium 2.2%
- Sodium 2.5%
- Potassium 2.5%

Others:
- Hydrogen 0.22%
- Carbon 0.19%
- Phosphorus <0.1%
- Nitrogen <0.1%

Human body:
- Hydrogen 49.0%
- Oxygen 25.0%
- Carbon 25.0%
- Nitrogen 0.27%
- Calcium 0.073%
- Potassium 0.046%
- Silicon 0.033%
- Magnesium 0.031%
- Phosphorus 0.030%
- Sodium 0.015%
- Others traces

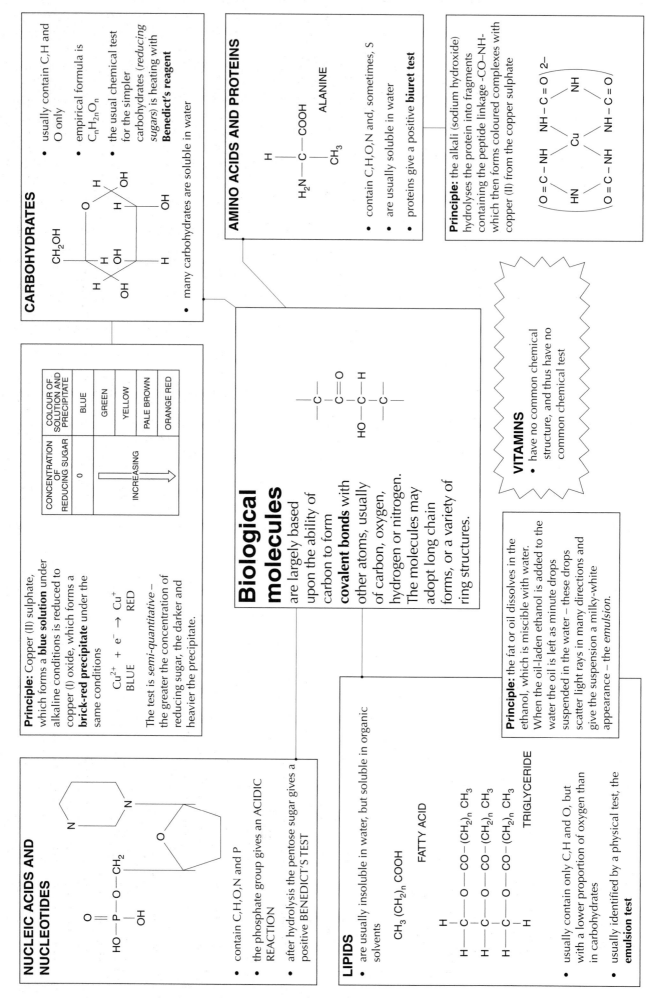

Biological molecules

are largely based upon the ability of carbon to form **covalent bonds** with other atoms, usually of carbon, oxygen, hydrogen or nitrogen. The molecules may adopt long chain forms, or a variety of ring structures.

CARBOHYDRATES

- usually contain C, H and O only
- empirical formula is $C_nH_{2n}O_n$
- the usual chemical test for the simpler carbohydrates (*reducing sugars*) is heating with **Benedict's reagent**
- many carbohydrates are soluble in water

Principle: Copper (II) sulphate, which forms a **blue solution** under alkaline conditions is reduced to copper (I) oxide, which forms a **brick-red precipitate** under the same conditions

$$Cu^{2+} + e^- \rightarrow Cu^+$$
BLUE   RED

The test is *semi-quantitative* – the greater the concentration of reducing sugar, the darker and heavier the precipitate.

CONCENTRATION OF REDUCING SUGAR	COLOUR OF SOLUTION AND PRECIPITATE
0	BLUE
INCREASING	GREEN
	YELLOW
	PALE BROWN
	ORANGE RED

AMINO ACIDS AND PROTEINS

ALANINE

- contain C,H,O,N and, sometimes, S
- are usually soluble in water
- proteins give a positive **biuret test**

Principle: the alkali (sodium hydroxide) hydrolyses the protein into fragments containing the peptide linkage -CO-NH- which then forms coloured complexes with copper (II) from the copper sulphate

NUCLEIC ACIDS AND NUCLEOTIDES

- contain C,H,O,N and P
- the phosphate group gives an ACIDIC REACTION
- after hydrolysis the pentose sugar gives a positive BENEDICT'S TEST

LIPIDS

- are usually insoluble in water, but soluble in organic solvents

$CH_3 (CH_2)_n$ COOH
FATTY ACID

TRIGLYCERIDE

- usually contain only C,H and O, but with a lower proportion of oxygen than in carbohydrates
- usually identified by a physical test, the **emulsion test**

Principle: the fat or oil dissolves in the ethanol, which is miscible with water. When the oil-laden ethanol is added to the water the oil is left as minute drops suspended in the water – these drops scatter light rays in many directions and give the suspension a milky-white appearance – the *emulsion*.

VITAMINS

- have no common chemical structure, and thus have no common chemical test

Lipid structure and function

TRUE LIPIDS are esters of fatty acids and alcohols, formed by condensation reactions. Many of their properties result from their insolubility in water.

GLYCEROL   3 FATTY ACIDS

TRIGLYCERIDE $3H_2O$

Since the hydrocarbon chains are long (19 C in arachidonic acid) most of the weight of the triglyceride is fatty acid.

Water-repellent properties: oily secretions of the sebaceous glands help to waterproof the fur and skin. The preen gland of birds produces a secretion which performs a similar function on the feathers.

Cell membranes: phospholipids (phosphatides) are found in all cell membranes. These molecules have a polar 'phosphate-base' group substituted for one of the fatty acids in a triglyceride.

This part of the molecule is very *soluble* in water

ORGANIC BASE — PHOSPHATE

This part of the molecule is very *insoluble* in water

Electrical insulation: myelin is secreted by Schwann cells and insulates some neurones in such a way that impulse transmission is made much more rapid.

Hormones: an important group of hormones, including cortisone, testosterone and oestrogen, are *steroids*. Steroids are not true esters but have the same solubility properties as them.

BASIC STEROID NUCLEUS

Physical protection: the shock-absorbing ability of subcutaneous fat stores protects delicate organs such as the kidneys from mechanical damage.

Thermal insulation: fats conduct heat very poorly - subcutaneous fat stores help heat retention in endothermic animals. Incompressible blubber is an important insulator in diving mammals.

Attraction: plant scents are derivatives of fatty acids. They are attractive to insects and thus aid pollination.

Waterproofing: the waxy cuticle of insects and plants reduces water losses by evaporation since water cannot cross the insoluble lipid layer. Waxes are esters of higher fatty acids with long chain alcohols (i.e. *not* with glycerol).

Storage: high energy yield per unit mass and insolubility in water make fats and oils ideal energy storage compounds, particularly where dispersal or locomotion requires mass to be kept to a minimum, as in some seeds and fruits.

Honeycomb: bees use wax in constructing their larval chambers.

Nutrition: both bile acids and vitamin D (involved in fat digestion and Ca^{2+} absorption respectively) are manufactured from steroids.

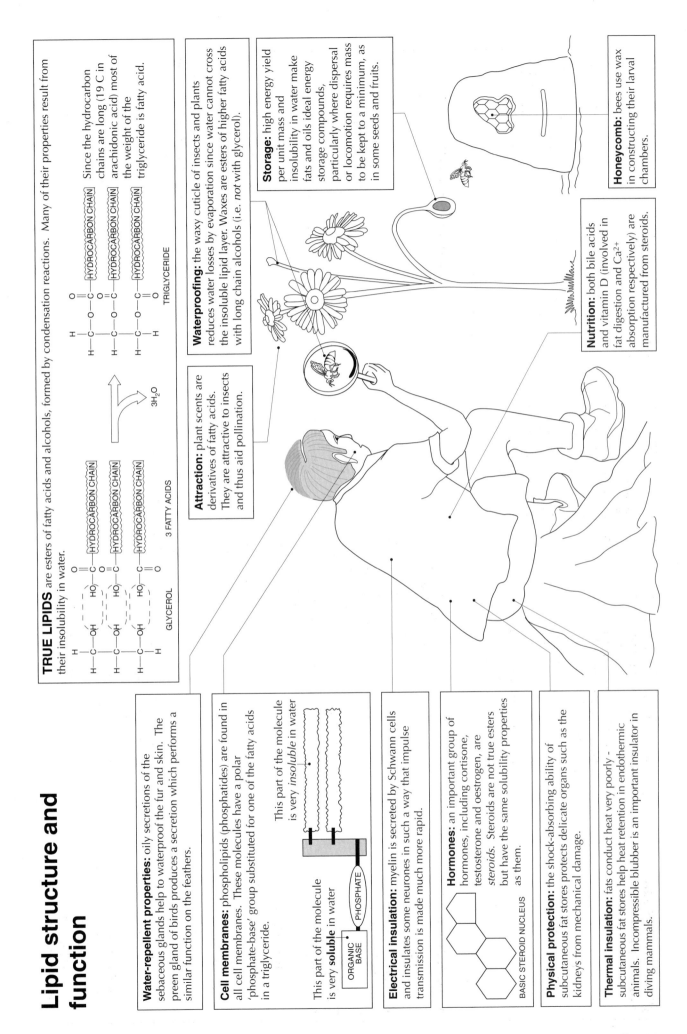

Functions of soluble carbohydrates include transport, protection, recognition and energy release.

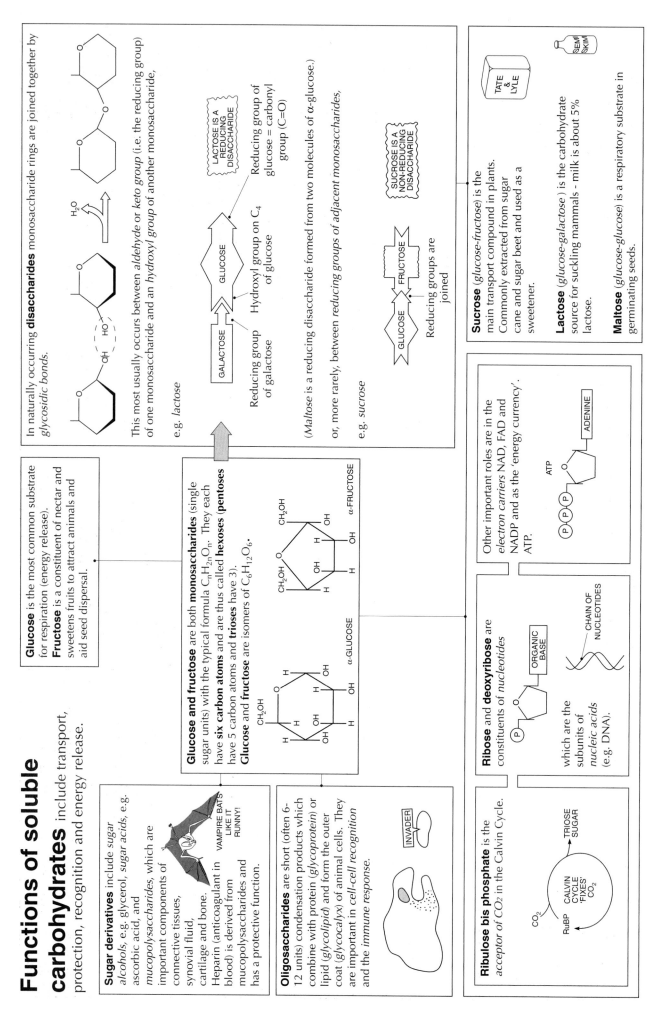

In naturally occurring **disaccharides** monosaccharide rings are joined together by *glycosidic bonds*.

This most usually occurs between *aldehyde or keto group* (i.e. the reducing group) of one monosaccharide and an *hydroxyl group* of another monosaccharide,

e.g. *lactose*

LACTOSE IS A REDUCING DISACCHARIDE

Reducing group of glucose = carbonyl group (C=O)

Hydroxyl group on C_4 of glucose

GALACTOSE

GLUCOSE

Reducing group of galactose

(*Maltose* is a reducing disaccharide formed from two molecules of α-glucose.)

or, more rarely, between *reducing groups of adjacent monosaccharides*,

e.g. *sucrose*

SUCROSE IS A NON-REDUCING DISACCHARIDE

GLUCOSE

FRUCTOSE

Reducing groups are joined

Sucrose (*glucose-fructose*) is the main transport compound in plants. Commonly extracted from sugar cane and sugar beet and used as a sweetener.

TATE & LYLE

Lactose (*glucose-galactose*) is the carbohydrate source for suckling mammals - milk is about 5% lactose.

SEMI SKIM

Maltose (*glucose-glucose*) is a respiratory substrate in germinating seeds.

Glucose is the most common substrate for respiration (energy release).
Fructose is a constituent of nectar and sweetens fruits to attract animals and aid seed dispersal.

Glucose and fructose are both **monosaccharides** (single sugar units) with the typical formula $C_nH_{2n}O_n$. They each have **six carbon atoms** and are thus called **hexoses** (**pentoses** have 5 carbon atoms and **trioses** have 3).
Glucose and **fructose** are isomers of $C_6H_{12}O_6$.

CH₂OH

α-GLUCOSE

CH₂OH

CH₂OH

α-FRUCTOSE

Sugar derivatives include *sugar alcohols*, e.g. glycerol, *sugar acids*, e.g. ascorbic acid, and *mucopolysaccharides*, which are important components of connective tissues, synovial fluid, cartilage and bone. Heparin (anticoagulant in blood) is derived from mucopolysaccharides and has a protective function.

VAMPIRE BATS LIKE IT RUNNY!

Oligosaccharides are short (often 6-12 units) condensation products which combine with protein (*glycoprotein*) or lipid (*glycolipid*) and form the outer coat (*glycocalyx*) of animal cells. They are important in *cell-cell recognition* and the *immune response*.

INVADER

Other important roles are in the *electron carriers* NAD, FAD and NADP and as the 'energy currency'. ATP.

ATP

P P P

ADENINE

Ribose and **deoxyribose** are constituents of *nucleotides*

P

ORGANIC BASE

CHAIN OF NUCLEOTIDES

which are the subunits of *nucleic acids* (e.g. DNA).

Ribulose bis phosphate is the *acceptor of CO_2 in the Calvin Cycle.*

CO₂

RuBP

CALVIN CYCLE 'FIXES' CO₂

TRIOSE SUGAR

Polysaccharides are polymers

formed by glycosidic bonding of monosaccharide subunits

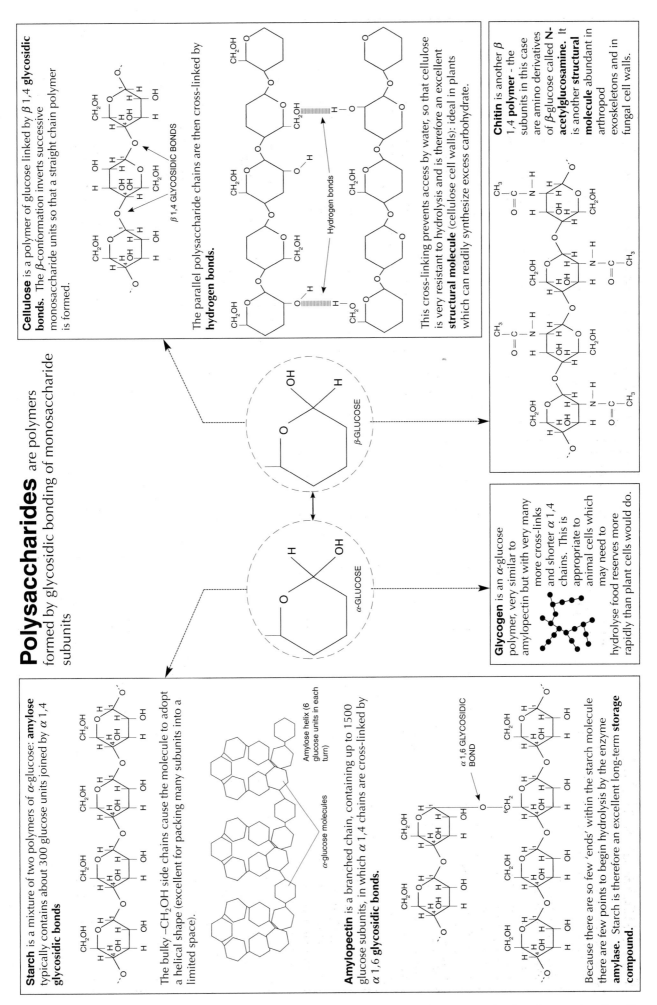

Cellulose is a polymer of glucose linked by β 1,4 glycosidic bonds. The β-conformation inverts successive monosaccharide units so that a straight chain polymer is formed.

β 1,4 GLYCOSIDIC BONDS

The parallel polysaccharide chains are then cross-linked by **hydrogen bonds.**

Hydrogen bonds

This cross-linking prevents access by water, so that cellulose is very resistant to hydrolysis and is therefore an excellent **structural molecule** (cellulose cell walls): ideal in plants which can readily synthesize excess carbohydrate.

Chitin is another β 1,4 **polymer** - the subunits in this case are amino derivatives of β-glucose called **N-acetylglucosamine.** It is another **structural molecule** abundant in arthropod exoskeletons and in fungal cell walls.

β-GLUCOSE

α-GLUCOSE

Glycogen is an α-glucose polymer, very similar to amylopectin but with very many more cross-links and shorter α 1,4 chains. This is appropriate to animal cells which may need to hydrolyse food reserves more rapidly than plant cells would do.

Starch is a mixture of two polymers of α-glucose: **amylose** typically contains about 300 glucose units joined by α 1,4 **glycosidic bonds**

The bulky –CH₂OH side chains cause the molecule to adopt a helical shape (excellent for packing many subunits into a limited space).

Amylose helix (6 glucose units in each turn)

α-glucose molecules

Amylopectin is a branched chain, containing up to 1500 glucose subunits, in which α 1,4 chains are cross-linked by α 1,6 **glycosidic bonds.**

α 1,6 GLYCOSIDIC BOND

Because there are so few 'ends' within the starch molecule there are few points to begin hydrolysis by the enzyme amylase. Starch is therefore an excellent long-term **storage compound.**

Levels of protein structure

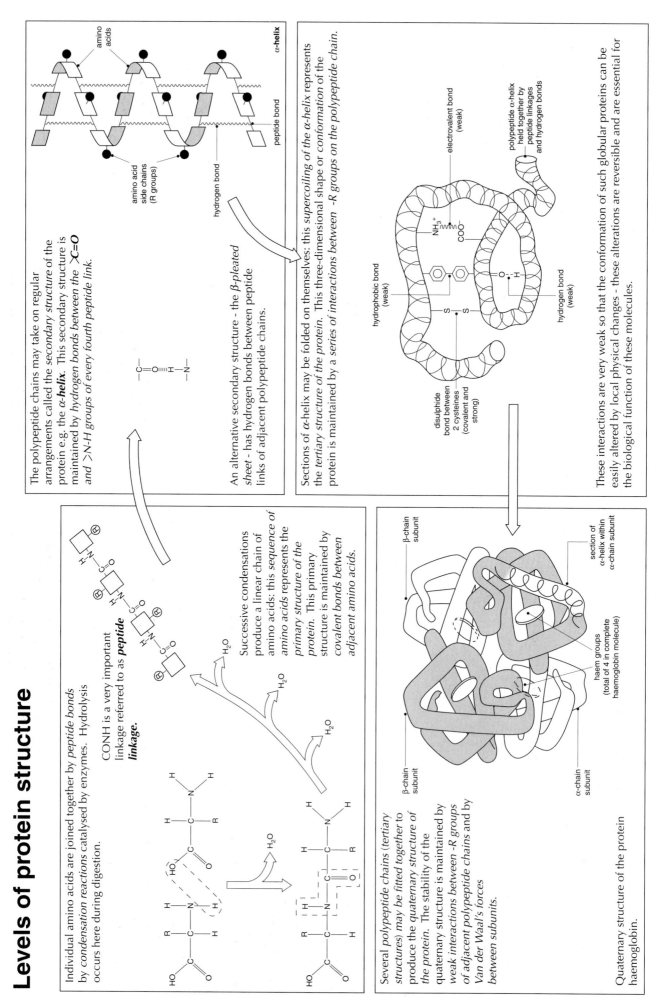

The polypeptide chains may take on regular arrangements called the *secondary structure* of the protein e.g. the α-**helix**. This secondary structure is maintained by *hydrogen bonds between the* ⟩C=O *and* ⟩N-H *groups of every fourth peptide link.*

amino acids

amino acid side chains (R groups)

hydrogen bond

peptide bond

α-helix

An alternative secondary structure - the *β-pleated sheet* - has hydrogen bonds between peptide links of adjacent polypeptide chains.

Sections of α-helix may be folded on themselves: this *supercoiling* of the α-*helix* represents the *tertiary structure of the protein*. This three-dimensional shape or *conformation* of the protein is maintained by *a series of interactions between -R groups on the polypeptide chain.*

electrovalent bond (weak)

polypeptide α-helix held together by peptide linkages and hydrogen bonds

hydrophobic bond (weak)

hydrogen bond (weak)

disulphide bond between 2 cysteines (covalent and strong)

These interactions are very weak so that the conformation of such globular proteins can be easily altered by local physical changes - these alterations are reversible and are essential for the biological function of these molecules.

Individual amino acids are joined together by peptide bonds by condensation reactions catalysed by enzymes. Hydrolysis occurs here during digestion.

CONH is a very important linkage referred to as **peptide linkage.**

Successive condensations produce a linear chain of amino acids: this *sequence of amino acids* represents the *primary structure of the protein.* This primary structure is maintained by covalent bonds between adjacent amino acids.

H_2O

Several polypeptide chains (tertiary structures) may be fitted together to produce the *quaternary structure of the protein.* The stability of the quaternary structure is maintained by weak interactions between -R groups of adjacent polypeptide chains and by Van der Waal's forces between subunits.

β-chain subunit

section of α-helix within α-chain subunit

haem groups (total of 4 in complete haemoglobin molecule)

α-chain subunit

Quaternary structure of the protein haemoglobin.

Functions of proteins include

transport, catalysis, protection, storage, sensitivity, structure and co-ordination.

Collagen is a *fibrous structural protein* of connective tissue in skin, tendons and ligaments.

Cytochrome c is a *transport protein* which plays a part in electron transfer of the respiratory chain.

Pepsin and **lipase** are hydrolytic *digestive enzymes* of the gut.

Keratin is a fibrous **structural protein** found in scales, horns, hooves, hair, and nails.

Ferritin is a *storage protein* which holds iron in egg yolk, spleen and liver.

Albumin may act as a *water storage protein* in egg 'white'.

Opsin is a part of the light-sensitive pigment *rhodopsin* found in rod cells of the retina.

Cobra venom is a *toxic protein* which kills by blocking nerve function.

Insulin is a *hormone* which *controls blood glucose concentration*.

Antibodies are *protective proteins* which as part of the immune response reduce the harmful effects of foreign proteins (**antigens**).

Proteins play an important part in *transport across membranes*, e.g. Na/K pumps transport ions across nerve cell membranes in preparation for transmission of an action potential.

Haemoglobin *transports oxygen in blood*, and *myoglobin stores* oxygen in muscles.

Fibrinogen and **prothrombin** are *protective proteins* essential for the clotting of blood.

Many diseases are caused by *viruses* which are enclosed in *structural viral coat proteins*.

Collagen, found in connective tissue of skin, tendons, and ligaments, is the most abundant of all animal proteins. Changes in collagen structure are partially responsible for the wrinkling of skin and the hardening of arteries.

Locomotion depends on the activity of the two *contractile proteins*, **actin** and **myosin** in muscle tissue.

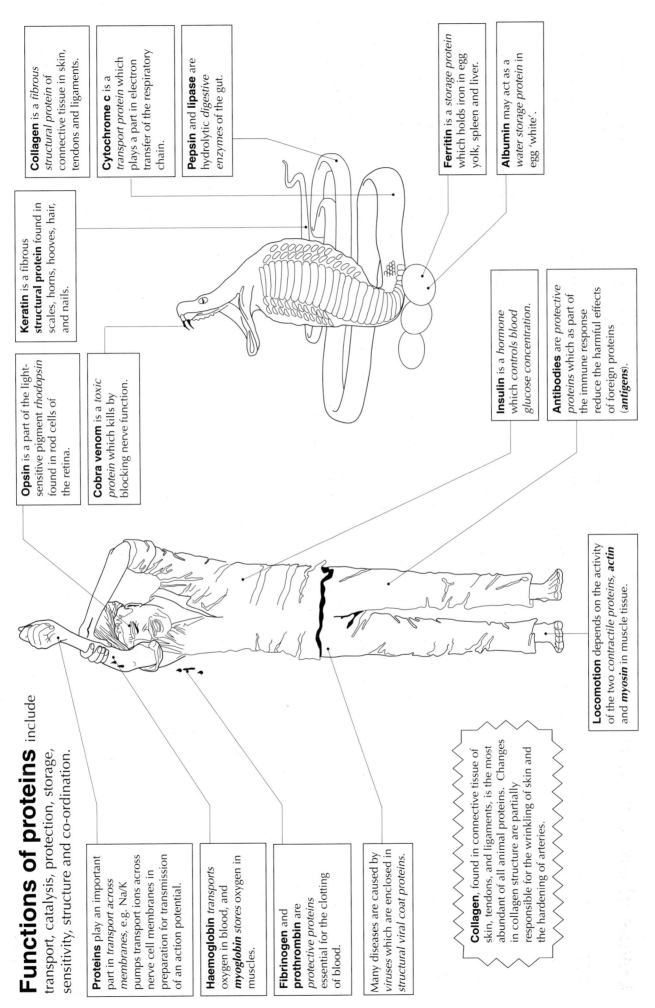

Catalysis by enzymes

An important step in enzyme catalysis is substrate binding to the active sites.

Enzymes form **enzyme-substrate complexes** which reduce the activation energy for reactions which they catalyse.

Consider the reaction: SUBSTRATE (S) $\longrightarrow$ PRODUCT (P)

which can be illustrated by a **reaction profile.**

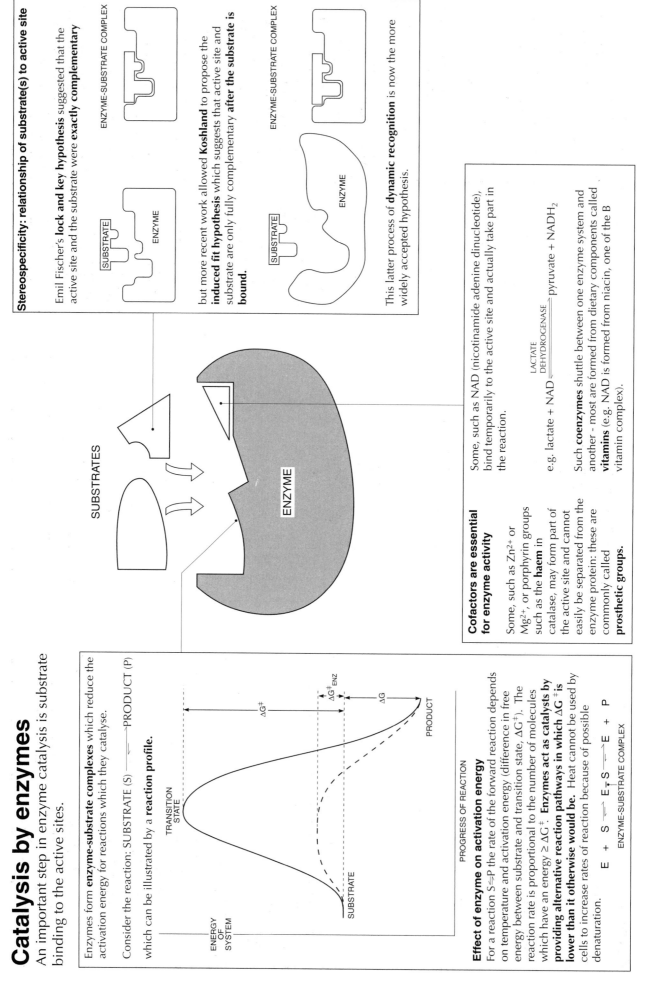

SUBSTRATES

ENZYME

Stereospecificity: relationship of substrate(s) to active site

Emil Fischer's **lock and key hypothesis** suggested that the active site and the substrate were **exactly complementary**

SUBSTRATE

ENZYME

ENZYME-SUBSTRATE COMPLEX

but more recent work allowed **Koshland** to propose the **induced fit hypothesis** which suggests that active site and substrate are only fully complementary **after the substrate is bound.**

SUBSTRATE

ENZYME

ENZYME-SUBSTRATE COMPLEX

This latter process of **dynamic recognition** is now the more widely accepted hypothesis.

Effect of enzyme on activation energy

For a reaction S⇌P the rate of the forward reaction depends on temperature and activation energy (difference in free energy between substrate and transition state, $\Delta G^{\ddagger}$). The reaction rate is proportional to the number of molecules which have an energy $\geq \Delta G^{\ddagger}$. **Enzymes act as catalysts by providing alternative reaction pathways in which $\Delta G^{\ddagger}$ is lower than it otherwise would be.** Heat cannot be used by cells to increase rates of reaction because of possible denaturation.

$$E + S \rightleftharpoons E \cdot S \rightleftharpoons E + P$$
ENZYME-SUBSTRATE COMPLEX

ENERGY OF SYSTEM

TRANSITION STATE

$\Delta G^{\ddagger}$

$\Delta G^{\ddagger}_{ENZ}$

ΔG

SUBSTRATE

PRODUCT

PROGRESS OF REACTION

Cofactors are essential for enzyme activity

Some, such as Zn^{2+} or Mg^{2+}, or porphyrin groups such as the **haem** in catalase, may form part of the active site and cannot easily be separated from the enzyme protein: these are commonly called **prosthetic groups.**

Some, such as NAD (nicotinamide adenine dinucleotide), bind temporarily to the active site and actually take part in the reaction.

e.g. lactate + NAD $\underset{\text{DEHYDROGENASE}}{\overset{\text{LACTATE}}{\rightleftharpoons}}$ pyruvate + $NADH_2$

Such **coenzymes** shuttle between one enzyme system and another - most are formed from dietary components called **vitamins** (e.g. NAD is formed from niacin, one of the B vitamin complex).

Factors affecting enzyme activity

exert their effects by altering the ease with which an enzyme-substrate complex is formed.

Any factor which alters the conformation (dependent on tertiary structure) of the enzyme will alter the shape of the active site, affect the frequency of enzyme-substrate complex formation and thus influence the rate of the enzyme-catalysed reaction.

Competitive inhibitors compete for the active site with the normal substrate. These inhibitors therefore must have a similar structure to the natural substrate.
The success of the binding of I to the active site depends on the relative concentrations of I and S, and such inhibition is therefore **reversible by an increase in substrate concentration,** e.g. malonate competes with succinate for the active site on the enzyme **succinate dehydrogenase.**

Irreversible inhibition occurs if the enzyme-inhibitor binding is covalent and the distortion of the active site may be permanent, e.g. cyanide (CN⁻) binds irreversibly to the active site of the enzyme **cytochrome oxidase.**

Non-competitive inhibitors reduce enzyme activity by distortion of enzyme conformation caused by binding to some site **other than the active site.** If the binding is non-covalent the inhibition may be **reversible if the inhibitor concentration is diminished.** Many such inhibitors are natural **allosteric regulators** of metabolism, e.g. ATP controls the rate of respiration by inhibition of the enzyme **phosphofructokinase.**

Activators may be necessary to complete the structural relationship between active site and substrate, e.g. chloride ions (Cl⁻) are required for activity of the enzyme **salivary amylase.**

There are also **allosteric activators** which enhance enzyme-substrate binding by alteration of enzyme conformation when binding to another ('**allosteric**') site on the enzyme.

EFFECT OF TEMPERATURE

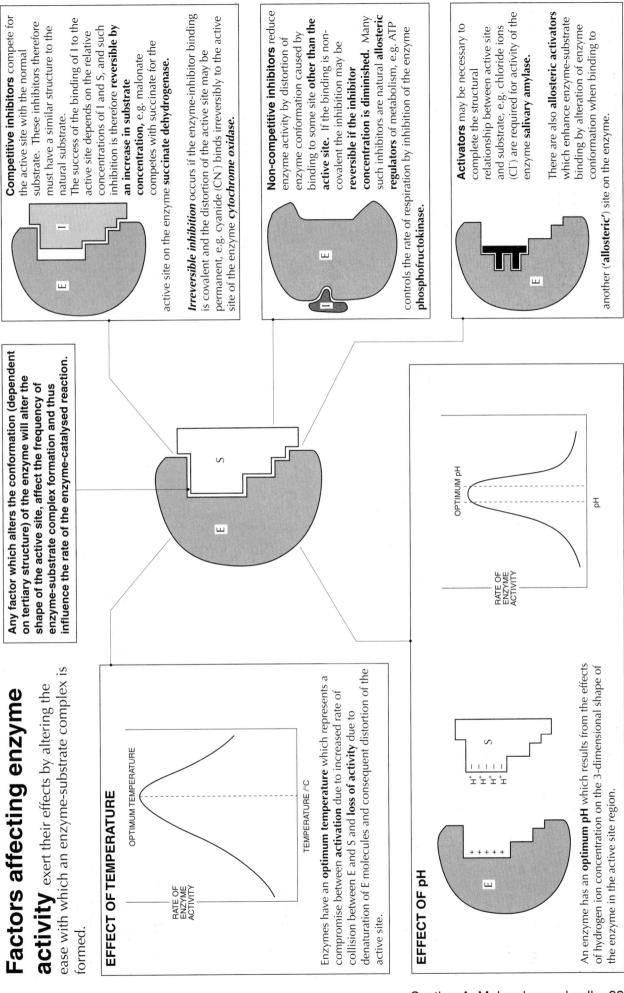

Enzymes have an **optimum temperature** which represents a compromise between **activation** due to increased rate of collision between E and S and **loss of activity** due to denaturation of E molecules and consequent distortion of the active site.

EFFECT OF pH

An enzyme has an **optimum pH** which results from the effects of hydrogen ion concentration on the 3-dimensional shape of the enzyme in the active site region.

Glucose oxidase

The reaction catalysed by glucose oxidase is

$$\beta - D - glucose + O_2 \longrightarrow gluconic\ acid + H_2O_2$$

The quick and accurate measurement of glucose is of great importance both medically (in sufferers from diabetes, for example) and industrially (in fermentation reactions, for example). A simple quantitative procedure can be devised by coupling the production of hydrogen peroxide to the activity of the enzyme **peroxidase.**

$$DH_2 + H_2O_2 \xrightarrow{\text{peroxidase}} 2H_2O + D$$

chromagen   coloured
a hydrogen donor   compound
(colourless)   (colour)

Peroxidase can oxidize an organic chromagen (DH_2) to a coloured compound (D) utilizing the hydrogen peroxide – the amount of the coloured compound D produced is a direct measure of the amount of glucose which has reacted. It can be measured quantitatively using a colorimeter or, more subjectively, by comparison with a colour reference card.

This method of glucose analysis is **highly specific** and has the enormous advantage over chemical methods in that this specificity allows glucose to be assayed **in the presence of other sugars**, e.g. in a biological fluid such as blood or urine, without the need for an initial separation.

Both of the enzymes glucose oxidase and peroxidase, and the chromagen DH_2, can be immobilized on a cellulose fibre pad. This forms the basis of the glucose dipsticks ('Clinistix') which were developed to enable diabetics to monitor their own blood or urine glucose levels.

CLINISTIX

CLINISTIX CHART
LOW NORM HIGH DANG

ANALYSIS

Commercial applications of enzymes

PHARMACEUTICALS *Papain* (protein → peptides) is used to remove stains from false teeth.

DIAGNOSIS Testing for unusual enzyme levels can indicate disease conditions, e.g. **pancreatitis** causes elevated levels of digestive enzymes in the blood.

TEXTILES *Lipase* (fats → fatty acids) is used in biological washing powders.

There are many applications of enzyme technology to industry. Enzyme technology has several advantages over 'whole-organism' technology.

a. **No loss of substrate due to increased biomass.** For example, when whole yeast is used to ferment sugar to alcohol it always 'wastes' some of the sugar by converting it into cell wall material and protoplasm for its own growth.

b. **Elimination of wasteful side reactions.** Whole organisms may convert some of the substrate into irrelevant compounds or even contain enzymes for degrading the desired product into something else.

c. **Optimum conditions for a particular enzyme may be used.** These conditions may not be optimal for the whole organism – in some organisms particular enzymes might be working at less than maximum efficiency.

d. **Purification of the product is easier.** This is especially true using immobilized enzymes.

MEDICINE

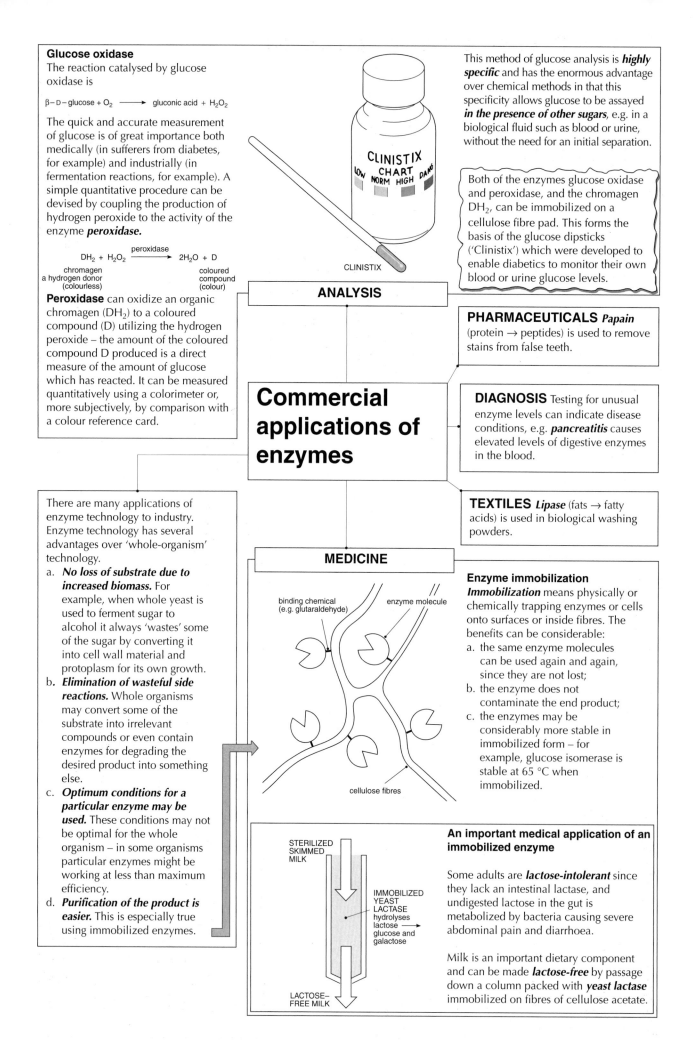

binding chemical (e.g. glutaraldehyde)   enzyme molecule

cellulose fibres

Enzyme immobilization
Immobilization means physically or chemically trapping enzymes or cells onto surfaces or inside fibres. The benefits can be considerable:
a. the same enzyme molecules can be used again and again, since they are not lost;
b. the enzyme does not contaminate the end product;
c. the enzymes may be considerably more stable in immobilized form – for example, glucose isomerase is stable at 65 °C when immobilized.

STERILIZED SKIMMED MILK

IMMOBILIZED YEAST LACTASE hydrolyses lactose → glucose and galactose

LACTOSE–FREE MILK

An important medical application of an immobilized enzyme

Some adults are **lactose-intolerant** since they lack an intestinal lactase, and undigested lactose in the gut is metabolized by bacteria causing severe abdominal pain and diarrhoea.

Milk is an important dietary component and can be made **lactose-free** by passage down a column packed with **yeast lactase** immobilized on fibres of cellulose acetate.

Section B Energy and life processes

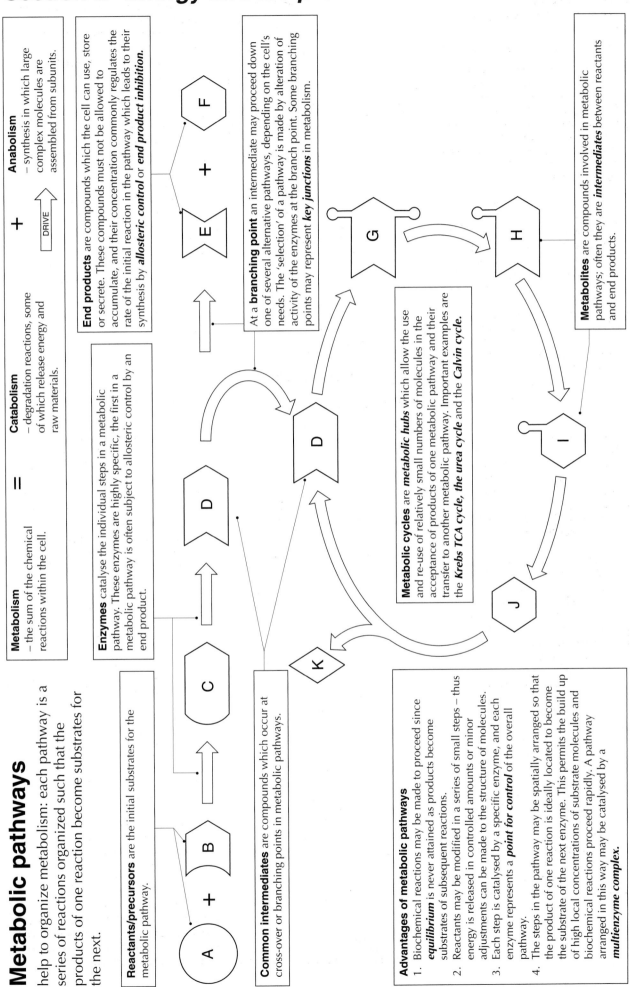

Metabolic pathways

help to organize metabolism: each pathway is a series of reactions organized such that the products of one reaction become substrates for the next.

Metabolism
– the sum of the chemical reactions within the cell.

Catabolism
– degradation reactions, some of which release energy and raw materials.

Anabolism
– synthesis in which large complex molecules are assembled from subunits.

DRIVE

Reactants/precursors are the initial substrates for the metabolic pathway.

Enzymes catalyse the individual steps in a metabolic pathway. These enzymes are highly specific, the first in a metabolic pathway is often subject to allosteric control by an end product.

End products are compounds which the cell can use, store or secrete. These compounds must not be allowed to accumulate, and their concentration commonly regulates the rate of the initial reaction in the pathway which leads to their synthesis by *allosteric control* or *end product inhibition.*

Common intermediates are compounds which occur at cross-over or branching points in metabolic pathways.

At a **branching point** an intermediate may proceed down one of several alternative pathways, depending on the cell's needs. The 'selection' of a pathway is made by alteration of activity of the enzymes at the branch point. Some branching points may represent *key junctions* in metabolism.

Metabolic cycles are *metabolic hubs* which allow the use and re-use of relatively small numbers of molecules in the acceptance of products of one metabolic pathway and their transfer to another metabolic pathway. Important examples are the *Krebs TCA cycle, the urea cycle* and the *Calvin cycle.*

Metabolites are compounds involved in metabolic pathways; often they are *intermediates* between reactants and end products.

Advantages of metabolic pathways
1. Biochemical reactions may be made to proceed since *equilibrium* is never attained as products become substrates of subsequent reactions.
2. Reactants may be modified in a series of small steps – thus energy is released in controlled amounts or minor adjustments can be made to the structure of molecules.
3. Each step is catalysed by a specific enzyme, and each enzyme represents a *point for control* of the overall pathway.
4. The steps in the pathway may be spatially arranged so that the product of one reaction is ideally located to become the substrate of the next enzyme. This permits the build up of high local concentrations of substrate molecules and biochemical reactions proceed rapidly. A pathway arranged in this way may be catalysed by a *multienzyme complex.*

Cellular respiration

(glycolysis) occurs in a series of localized stages.

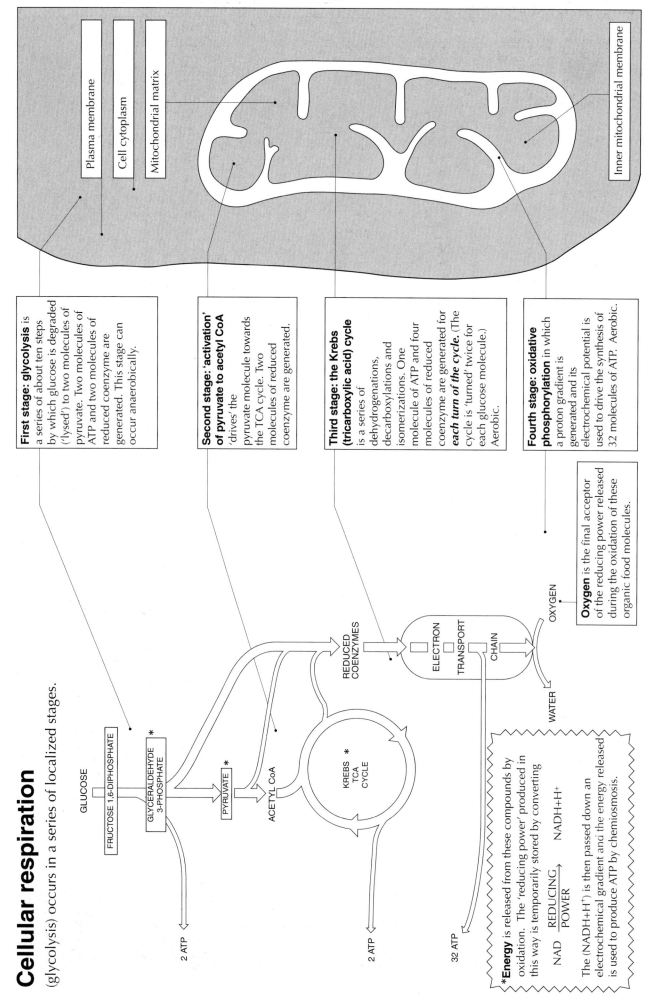

Plasma membrane

Cell cytoplasm

Mitochondrial matrix

Inner mitochondrial membrane

First stage: glycolysis is a series of about ten steps by which glucose is degraded ('lysed') to two molecules of pyruvate. Two molecules of ATP and two molecules of reduced coenzyme are generated. This stage can occur anaerobicallly.

Second stage: 'activation' of pyruvate to acetyl CoA 'drives' the pyruvate molecule towards the TCA cycle. Two molecules of reduced coenzyme are generated.

Third stage: the Krebs (tricarboxylic acid) cycle is a series of dehydrogenations, decarboxylations and isomerizations. One molecule of ATP and four molecules of reduced coenzyme are generated for *each turn of the cycle*. (The cycle is 'turned' twice for each glucose molecule.) Aerobic.

Fourth stage: oxidative phosphorylation in which a proton gradient is generated and its electrochemical potential is used to drive the synthesis of 32 molecules of ATP. Aerobic.

Oxygen is the final acceptor of the reducing power released during the oxidation of these organic food molecules.

GLUCOSE

FRUCTOSE 1,6-DIPHOSPHATE

GLYCERALDEHYDE 3-PHOSPHATE *

PYRUVATE *

ACETYL CoA

KREBS TCA CYCLE *

REDUCED COENZYMES

ELECTRON TRANSPORT CHAIN

OXYGEN

WATER

2 ATP

2 ATP

32 ATP

*Energy is released from these compounds by oxidation. The 'reducing power' produced in this way is temporarily stored by converting

NAD $\xrightarrow{\text{REDUCING POWER}}$ NADH+H⁺

The (NADH+H⁺) is then passed down an electrochemical gradient and the energy released is used to produce ATP by chemiosmosis.

ATP: the energy currency of the cell

ATP hydrolysis is favoured

$ATP^{4-} + H_2O \rightleftharpoons ADP^{3-} + P^{2-} + H^+ + 30.5\ kJ\ mol^{-1}$

i.e. ATP has a strong tendency to transfer its terminal phosphoryl group, a reaction associated with the release of 30.5 kJ mol^{-1} of ATP, because

1. the repulsion between the four negative charges in ATP^{4-} is reduced when ATP is hydrolysed because two negative charges are removed with phosphate.
2. the H$^+$ ion which is released when ATP is hydrolysed reacts with OH$^-$ ions to form water – this is a highly favoured reaction.
3. the charge distribution on ADP + P is more stable than that on ATP.

This part of the molecule acts like a 'handle' – it has a shape which can be recognized by highly specific enzymes.

This part of the molecule contains anhydride bonds (O-P) which can be hydrolysed in reactions which are **exergonic** (energy-yielding) and can be coupled to **endergonic** (energy-demanding) reactions.

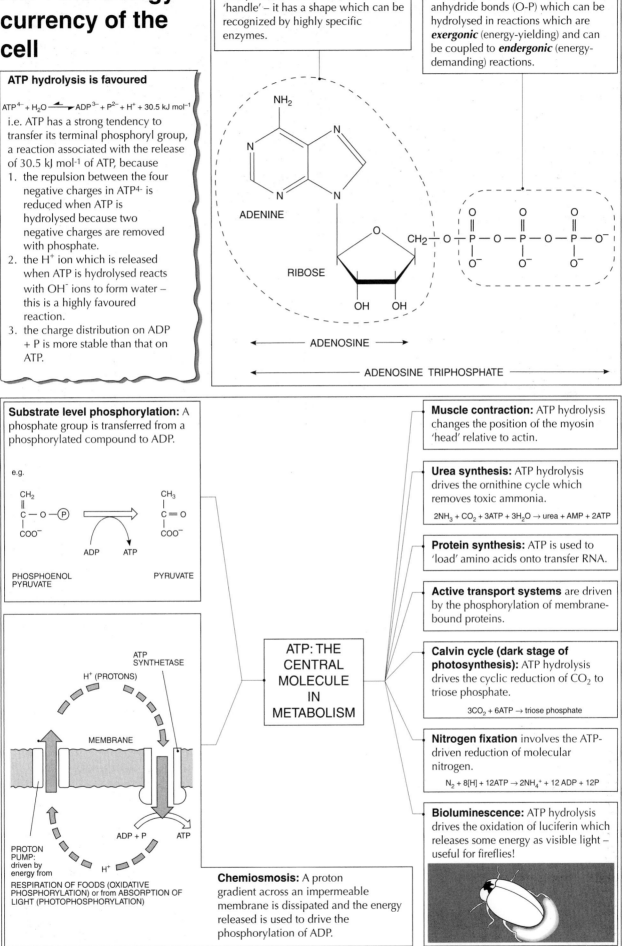

NH$_2$

ADENINE

RIBOSE

OH   OH

CH$_2$

ADENOSINE

ADENOSINE TRIPHOSPHATE

Substrate level phosphorylation: A phosphate group is transferred from a phosphorylated compound to ADP.

e.g.

PHOSPHOENOL PYRUVATE

ADP   ATP

PYRUVATE

Chemiosmosis: A proton gradient across an impermeable membrane is dissipated and the energy released is used to drive the phosphorylation of ADP.

ATP SYNTHETASE

H$^+$ (PROTONS)

MEMBRANE

ADP + P   ATP

PROTON PUMP: driven by energy from

RESPIRATION OF FOODS (OXIDATIVE PHOSPHORYLATION) or from ABSORPTION OF LIGHT (PHOTOPHOSPHORYLATION)

H$^+$

ATP: THE CENTRAL MOLECULE IN METABOLISM

Muscle contraction: ATP hydrolysis changes the position of the myosin 'head' relative to actin.

Urea synthesis: ATP hydrolysis drives the ornithine cycle which removes toxic ammonia.

$2NH_3 + CO_2 + 3ATP + 3H_2O \rightarrow urea + AMP + 2ATP$

Protein synthesis: ATP is used to 'load' amino acids onto transfer RNA.

Active transport systems are driven by the phosphorylation of membrane-bound proteins.

Calvin cycle (dark stage of photosynthesis): ATP hydrolysis drives the cyclic reduction of CO$_2$ to triose phosphate.

$3CO_2 + 6ATP \rightarrow triose\ phosphate$

Nitrogen fixation involves the ATP-driven reduction of molecular nitrogen.

$N_2 + 8[H] + 12ATP \rightarrow 2NH_4^+ + 12\ ADP + 12P$

Bioluminescence: ATP hydrolysis drives the oxidation of luciferin which releases some energy as visible light – useful for fireflies!

Glycolysis generates ATP, reduced electron carriers and pyruvate.

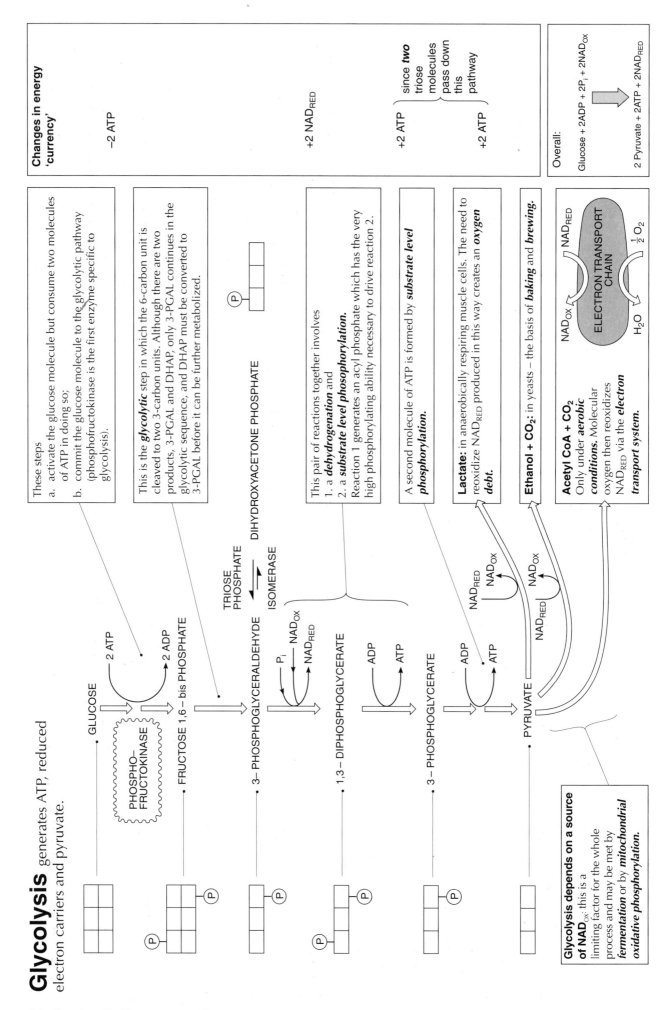

These steps
a. activate the glucose molecule but consume two molecules of ATP in doing so;
b. commit the glucose molecule to the glycolytic pathway (phosphofructokinase is the first enzyme specific to glycolysis).

This is the *glycolytic* step in which the 6-carbon unit is cleaved to two 3-carbon units. Although there are two products, 3-PGAL and DHAP, only 3-PGAL continues in the glycolytic sequence, and DHAP must be converted to 3-PGAL before it can be further metabolized.

This pair of reactions together involves
1. a *dehydrogenation* and
2. a *substrate level phosophorylation*.
Reaction 1 generates an acyl phosphate which has the very high phosphorylating ability necessary to drive reaction 2.

A second molecule of ATP is formed by *substrate level phosphorylation*.

Lactate: in anaerobically respiring muscle cells. The need to reoxidize NAD$_{RED}$ produced in this way creates an *oxygen debt*.

Ethanol + CO$_2$: in yeasts – the basis of *baking* and *brewing*.

Acetyl CoA + CO$_2$
Only under *aerobic conditions.* Molecular oxygen then reoxidizes NAD$_{RED}$ via the *electron transport system.*

GLUCOSE

2 ATP
2 ADP

PHOSPHO-FRUCTOKINASE

FRUCTOSE 1,6– bis PHOSPHATE

TRIOSE PHOSPHATE
ISOMERASE

DIHYDROXYACETONE PHOSPHATE

3– PHOSPHOGLYCERALDEHYDE

P$_i$ NAD$_{OX}$
NAD$_{RED}$

1,3 – DIPHOSPHOGLYCERATE

ADP
ATP

3 – PHOSPHOGLYCERATE

ADP
ATP

PYRUVATE

NAD$_{RED}$ NAD$_{OX}$
NAD$_{RED}$ NAD$_{OX}$
NAD$_{RED}$

ELECTRON TRANSPORT CHAIN

NAD$_{OX}$ NAD$_{RED}$
H$_2$O $\frac{1}{2}$ O$_2$

Glycolysis depends on a source of NAD$_{OX}$: this is a limiting factor for the whole process and may be met by *fermentation* or by *mitochondrial oxidative phosphorylation.*

Changes in energy 'currency'

−2 ATP

+2 NAD$_{RED}$

+2 ATP $\left.\begin{array}{l} \\ \\ \\ \\ \end{array}\right\}$ since *two* triose molecules pass down this pathway

+2 ATP

Overall:
Glucose + 2ADP + 2P$_i$ + 2NAD$_{OX}$
$\Rightarrow$
2 Pyruvate + 2ATP + 2NAD$_{RED}$

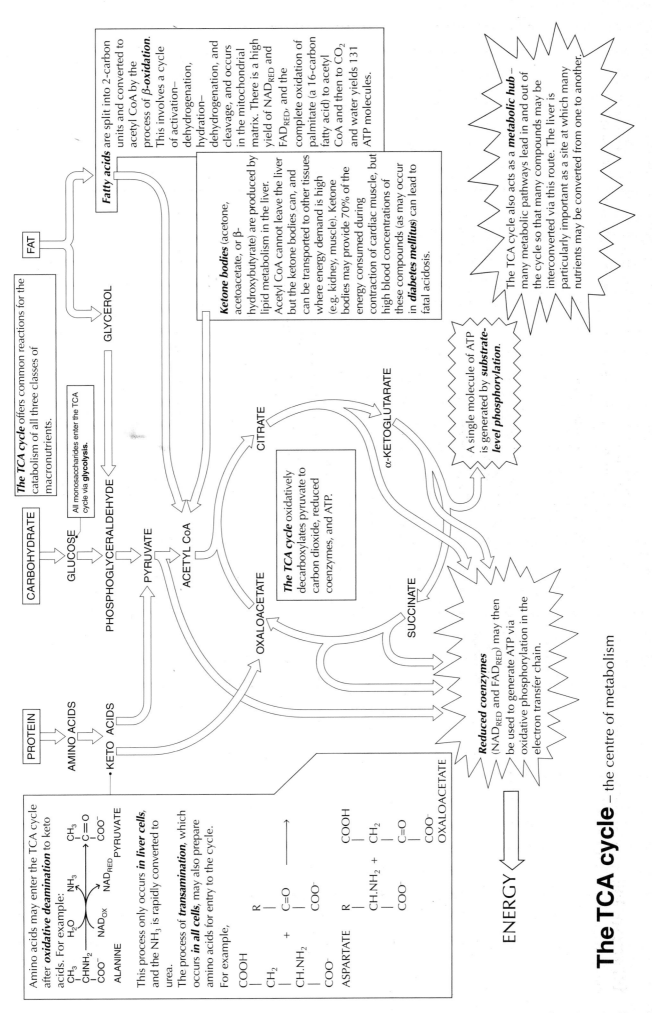

Amino acids may enter the TCA cycle after **oxidative deamination** to keto acids. For example:

$$CH_3$$
$$CHNH_2$$
$$COO^-$$
ALANINE

$$H_2O \quad NH_3$$
$$NAD_{OX} \quad NAD_{RED}$$

$$CH_3$$
$$C=O$$
$$COO^-$$
PYRUVATE

This process only occurs **in liver cells**, and the NH_3 is rapidly converted to urea.

The process of **transamination**, which occurs **in all cells**, may also prepare amino acids for entry to the cycle. For example,

$$COOH$$
$$CH_2$$
$$CH.NH_2$$
$$COO^-$$
ASPARTATE

$$+$$

$$R$$
$$C=O$$
$$COO^-$$

$$\longrightarrow$$

$$R$$
$$CH.NH_2$$
$$COO^-$$

$$+$$

$$COOH$$
$$CH_2$$
$$C=O$$
$$COO^-$$
OXALOACETATE

Fatty acids are split into 2-carbon units and converted to acetyl CoA by the process of **β-oxidation**. This involves a cycle of activation–dehydrogenation, hydration–dehydrogenation, and cleavage, and occurs in the mitochondrial matrix. There is a high yield of NAD_{RED} and FAD_{RED}, and the complete oxidation of palmitate (a 16-carbon fatty acid) to acetyl CoA and water yields 131 ATP molecules.

Ketone bodies (acetone, acetoacetate, or β-hydroxybutyrate) are produced by lipid metabolism in the liver. Acetyl CoA cannot leave the liver but the ketone bodies can, and can be transported to other tissues where energy demand is high (e.g. kidney, muscle). Ketone bodies may provide 70% of the energy consumed during contraction of cardiac muscle, but high blood concentrations of these compounds (as may occur in **diabetes mellitus**) can lead to fatal acidosis.

The TCA cycle offers common reactions for the catabolism of all three classes of macronutrients.

All monosaccharides enter the TCA cycle via **glycolysis**.

The TCA cycle also acts as a **metabolic hub** – many metabolic pathways lead in and out of the cycle so that many compounds may be interconverted via this route. The liver is particularly important as a site at which many nutrients may be converted from one to another.

A single molecule of ATP is generated by **substrate-level phosphorylation**.

The TCA cycle oxidatively decarboxylates pyruvate to carbon dioxide, reduced coenzymes, and ATP.

Reduced coenzymes (NAD_{RED} and FAD_{RED}) may then be used to generate ATP via oxidative phosphorylation in the electron transfer chain.

FAT

GLYCEROL

CARBOHYDRATE

GLUCOSE

PHOSPHOGLYCERALDEHYDE

PYRUVATE

ACETYL CoA

CITRATE

α-KETOGLUTARATE

OXALOACETATE

SUCCINATE

PROTEIN

AMINO ACIDS

KETO ACIDS

ENERGY

The TCA cycle – the centre of metabolism

Respirometers and the measurement of respiratory quotient

PRINCIPLE: Carbon dioxide evolved during respiration is absorbed by potassium hydroxide solution. If the system is closed to the atmosphere a change in volume of the gas within the chamber must be due to the consumption of oxygen. The change in volume, i.e. the oxygen consumption, is measured as the movement of a drop of coloured fluid along a capillary tube.

Spring clip. When open permits equilibration of contents of chamber with atmosphere.

Coloured fluid. Narrow bore capillary tube ensures maximum movement of fluid with any change in volume of gas in chamber.

Rubber stopper. Airtightness (which is essential) can be improved by smearing vaseline along seal between chamber and stopper.

Graduated scale against which movement of coloured fluid may be measured.

Respiratory chamber has relatively low volume to ensure that volume changes due to respiratory activity are significant enough to be measured.

Gauze basket to hold respiring material must be porous to permit free exchange of gases.

Filter paper wick to ensure maximum surface area of potassium hydroxide solution is available to contents of chamber.

Weighed amount of respiring material. The mass, m, of the respiring material should be known so that results can be made comparative.

Potassium hydroxide solution to absorb carbon dioxide evolved during respiration.

Glass rod to keep respiring material out of direct contact with potassium hydroxide solution.

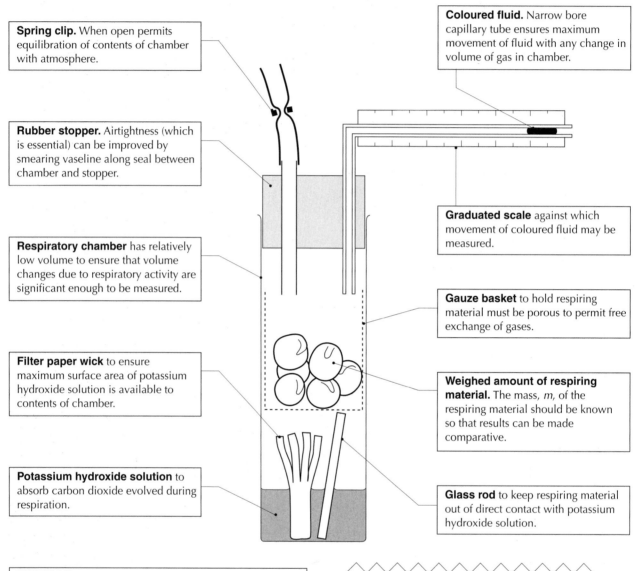

METHOD

1 The apparatus, containing a known mass of respiring material, is set up with the spring clip open to permit equilibration.

2 With the spring clip closed the time, t, in minutes, for the coloured fluid to move a distance l mm. along the capillary tube is noted.

  Oxygen Consumed = l units

3 Remove KOH solution and note fluid movement in same time, t

  Oxygen Consumed – Carbon Dioxide Released = m units

  Then R.Q. = $\dfrac{l-m}{l}$

RESPIRATORY QUOTIENT (R.Q.)

= $\dfrac{CO_2 \text{ evolved}}{O_2 \text{ consumed}}$ during respiration of a particular food

Aerobic respiration of carbohydrate

$$C_6H_{12}O_6 + 6O_2 \rightarrow 6CO_2 + 6H_2O$$

Thus R.Q. = $\dfrac{6}{6}$ = 1.0

For lipid R.Q. $\approx$ 0.7, for protein R.Q. $\approx$ 0.95

Thus the nature of the respiratory substrate can be determined by measurement of R.Q.

Mixed diet (ChO/lipid) R.Q. $\approx$ 0.85

Starvation (body protein reserves) R.Q. $\approx$ 0.9 – 1.0

Energy for exercise – the source of energy for muscle contraction

The immediate source of energy for the contraction of muscle (via the formation of actino-myosin cross bridges) is adenosine triphosphate (ATP). A problem is encountered when stimulation is continuous, since there is only a low concentration of ATP in a muscle cell – enough for about 7–8 'twitches' only. ATP is replaced by the transfer of a phosphate group from *creatine phosphate* to ADP.

Many athletes consume creatine to improve their capacity for work.

CREATINE ℗ ⟶ CREATINE

ADP ⟶ ATP

℗ and 'energy' ⟶ contraction of muscle

OXYGEN INTAKE AND EXERCISE

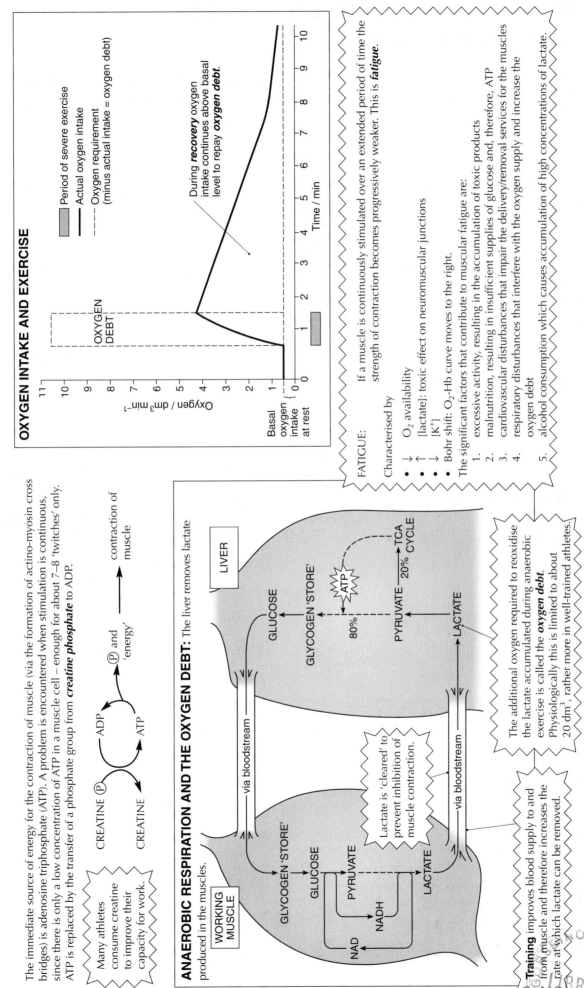

- ▨ Period of severe exercise
- —— Actual oxygen intake
- - - - Oxygen requirement (minus actual intake = oxygen debt)

Oxygen / dm³ min⁻¹ $Oxygen / dm^3\ min^{-1}$

OXYGEN DEBT

Basal oxygen intake at rest

During *recovery* oxygen intake continues above basal level to repay *oxygen debt*.

Time / min

FATIGUE: If a muscle is continuously stimulated over an extended period of time the strength of contraction becomes progressively weaker. This is *fatigue*.

Characterised by
- → O_2 availability
- ← [lactate]: toxic effect on neuromuscular junctions
- → [K$^+$]
- Bohr shift: O_2-Hb curve moves to the right.

The significant factors that contribute to muscular fatigue are:
1. excessive activity, resulting in the accumulation of toxic products
2. malnutrition, resulting in insufficient supplies of glucose and, therefore, ATP
3. cardiovascular disturbances that impair the delivery/removal services for the muscles
4. respiratory disturbances that interfere with the oxygen supply and increase the oxygen debt
5. alcohol consumption which causes accumulation of high concentrations of lactate.

ANAEROBIC RESPIRATION AND THE OXYGEN DEBT: The liver removes lactate produced in the muscles.

WORKING MUSCLE

GLYCOGEN 'STORE' → GLUCOSE → PYRUVATE

NAD ⟶ NADH

PYRUVATE ⟶ LACTATE

Lactate is 'cleared' to prevent inhibition of muscle contraction.

via bloodstream

LIVER

GLUCOSE

GLYCOGEN 'STORE'

ATP

80%

20% ⟶ TCA CYCLE

PYRUVATE ← LACTATE

via bloodstream

The additional oxygen required to reoxidise the lactate accumulated during anaerobic exercise is called the *oxygen debt*. Physiologically this is limited to about 20 dm³, rather more in well-trained athletes.

Training improves blood supply to and from muscle and therefore increases the rate at which lactate can be removed.

Nucleic acids I: DNA
(deoxyribonucleic acid)

The Watson–Crick model for DNA suggests that the molecule is a double helix of two complementary, anti-parallel polynucleotide chains

Nucleotides are the subunits of nucleic acids, including DNA. Each of these subunits is made up of:

AN ORGANIC
(NITROGENOUS) BASE
+
A PENTOSE SUGAR
+
A PHOSPHATE GROUP
(PHOSPHORIC ACID)

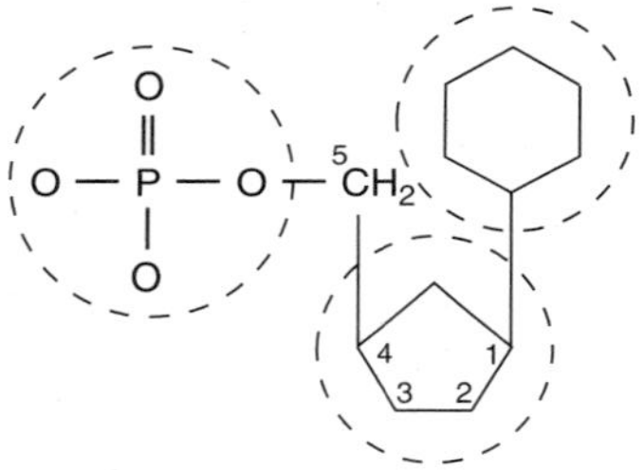

Note that the phosphate group is bonded to the C_5 atom of the pentose sugar.

There are *four different nucleotides* in a DNA molecule; they differ only in the organic (nitrogen) base present.

There are two different *pyrimidine (single ring) bases*, called *cytosine (C)* and *thymine (T)*

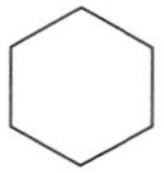

and

two different *purine (double ring) bases* called *adenine (A)* and *guanine (G)*.

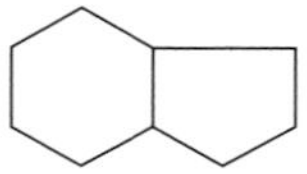

The different dimensions of the purine and pyrimidine bases is extremely important in the formation of the double-stranded DNA molecule.

Nucleotides are linked to form a *polynucleotide* by the formation of *3′ 5′ phosphodiester links* in which a phosphate group forms a bridge between the C_3 of one sugar molecule and the C_5 of the next sugar molecule.

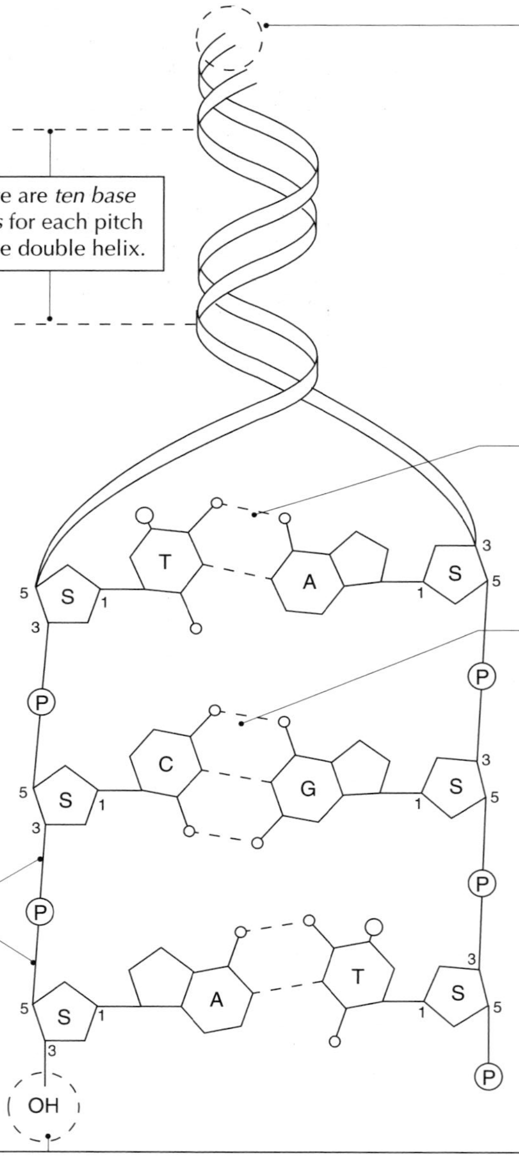

This is the 5′ *end of the chain* since the terminal phosphate group is only bonded to the C_5 of the sugar molecule.

Base pairing in DNA was proposed to explain how two polynucleotide chains could be held together by hydrogen bonds. To accommodate the measured dimensions of the molecule each base pair comprises *one purine - one pyrimidine*.

The double helix is most stable, that is the greatest number of hydrogen bonds is formed, when the base pairs

A :::::: T   (two hydrogen bonds)

and G :::::: C   (three hydrogen bonds)

are formed. These are *complementary base pairs*.

Note that in order to form and maintain this number of hydrogen bonds the nucleotides are inverted with respect to one another so that the phosphate groups (here shown as Ⓟ) *face in opposite directions*.

The chains are *anti-parallel*, that is one chain runs 5′→3′ whilst the other runs 3′→5′.

The chains are *complementary*: because of base pairing, the base sequence on one of the chains automatically dictates the base sequence on the other.

There are *ten base pairs* for each pitch of the double helix.

This is the 3′ *end of the chain* since the C_3 atom of the sugar molecule of the final nucleotide has a 'free' - OH group which is not part of a phosphodiester link.

Nucleic acids II: RNA Ribonucleic acid has a number of functions in protein synthesis.

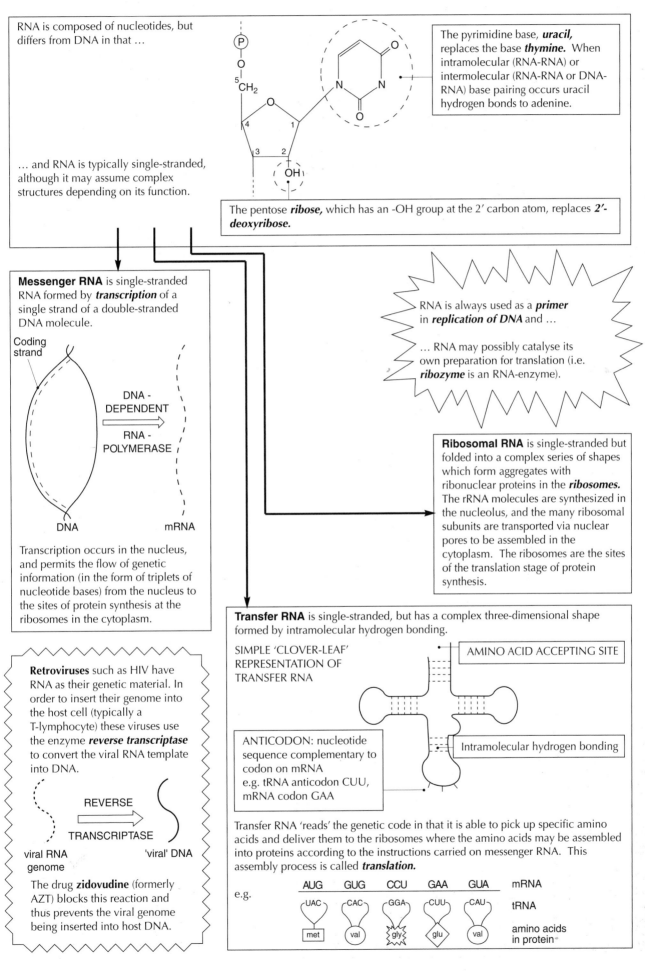

RNA is composed of nucleotides, but differs from DNA in that …

The pyrimidine base, **uracil,** replaces the base **thymine.** When intramolecular (RNA-RNA) or intermolecular (RNA-RNA or DNA-RNA) base pairing occurs uracil hydrogen bonds to adenine.

… and RNA is typically single-stranded, although it may assume complex structures depending on its function.

The pentose **ribose,** which has an -OH group at the 2' carbon atom, replaces **2'-deoxyribose.**

Messenger RNA is single-stranded RNA formed by **transcription** of a single strand of a double-stranded DNA molecule.

Coding strand

DNA-DEPENDENT
RNA-POLYMERASE

DNA   mRNA

Transcription occurs in the nucleus, and permits the flow of genetic information (in the form of triplets of nucleotide bases) from the nucleus to the sites of protein synthesis at the ribosomes in the cytoplasm.

RNA is always used as a **primer** in **replication of DNA** and …

… RNA may possibly catalyse its own preparation for translation (i.e. **ribozyme** is an RNA-enzyme).

Ribosomal RNA is single-stranded but folded into a complex series of shapes which form aggregates with ribonuclear proteins in the **ribosomes.** The rRNA molecules are synthesized in the nucleolus, and the many ribosomal subunits are transported via nuclear pores to be assembled in the cytoplasm. The ribosomes are the sites of the translation stage of protein synthesis.

Retroviruses such as HIV have RNA as their genetic material. In order to insert their genome into the host cell (typically a T-lymphocyte) these viruses use the enzyme **reverse transcriptase** to convert the viral RNA template into DNA.

REVERSE
TRANSCRIPTASE

viral RNA genome   'viral' DNA

The drug **zidovudine** (formerly AZT) blocks this reaction and thus prevents the viral genome being inserted into host DNA.

Transfer RNA is single-stranded, but has a complex three-dimensional shape formed by intramolecular hydrogen bonding.

SIMPLE 'CLOVER-LEAF' REPRESENTATION OF TRANSFER RNA

AMINO ACID ACCEPTING SITE

Intramolecular hydrogen bonding

ANTICODON: nucleotide sequence complementary to codon on mRNA e.g. tRNA anticodon CUU, mRNA codon GAA

Transfer RNA 'reads' the genetic code in that it is able to pick up specific amino acids and deliver them to the ribosomes where the amino acids may be assembled into proteins according to the instructions carried on messenger RNA. This assembly process is called **translation.**

e.g.

AUG	GUG	CCU	GAA	GUA	mRNA
UAC	CAC	GGA	CUU	CAU	tRNA
met	val	gly	glu	val	amino acids in protein

Section C Levels of organisation

The **human organism** contains ten different body systems.

Cells, tissues, and organs

The component parts of the human body are arranged in increasing levels of complexity.

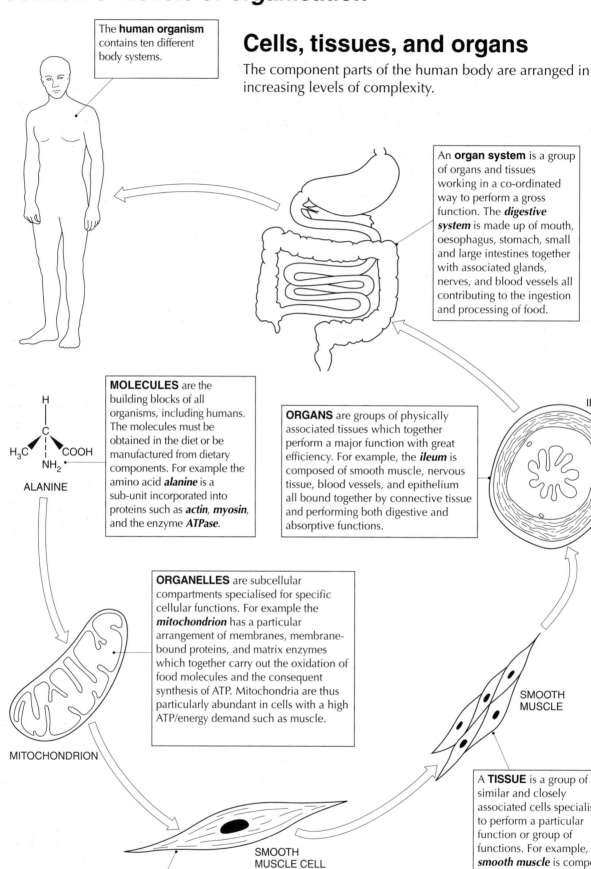

An **organ system** is a group of organs and tissues working in a co-ordinated way to perform a gross function. The *digestive system* is made up of mouth, oesophagus, stomach, small and large intestines together with associated glands, nerves, and blood vessels all contributing to the ingestion and processing of food.

ILEUM

MOLECULES are the building blocks of all organisms, including humans. The molecules must be obtained in the diet or be manufactured from dietary components. For example the amino acid *alanine* is a sub-unit incorporated into proteins such as *actin*, *myosin*, and the enzyme *ATPase*.

ALANINE

ORGANS are groups of physically associated tissues which together perform a major function with great efficiency. For example, the *ileum* is composed of smooth muscle, nervous tissue, blood vessels, and epithelium all bound together by connective tissue and performing both digestive and absorptive functions.

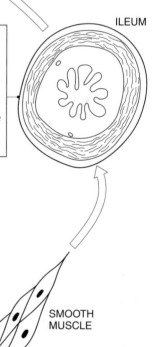

ORGANELLES are subcellular compartments specialised for specific cellular functions. For example the *mitochondrion* has a particular arrangement of membranes, membrane-bound proteins, and matrix enzymes which together carry out the oxidation of food molecules and the consequent synthesis of ATP. Mitochondria are thus particularly abundant in cells with a high ATP/energy demand such as muscle.

SMOOTH MUSCLE

MITOCHONDRION

A **TISSUE** is a group of similar and closely associated cells specialised to perform a particular function or group of functions. For example, *smooth muscle* is composed of many identical cells specialised for contraction.

SMOOTH MUSCLE CELL

CELLS are units of living matter surrounded by a plasma membrane. Cells are commonly specialised structurally and biochemically – for example *smooth muscle* cells are elongated along their axis of contraction and contain many molecules of the contractile proteins *actin* and *myosin*.

Epithelia and connective tissue

EPITHELIA are tissues which cover or line body surfaces.

Simple epithelia are only one cell thick – they are usually present in locations where the transfer of materials through the tissue is of primary importance. There are four types – named according to their shape which is a reflection of their function. A basement membrane, to which one surface of the cell is attached, is always present.

Ciliated epithelium is columnar in shape with the free surface covered in cilia. The rhythmic beating of the cilia propels the contents of the tube they line in one direction only. Located in much of the upper respiratory tract (where they propel mucus towards the throat) and in the fallopian tubes where they propel ova/zygotes towards the uterus.

Squamous pavement epithelium is formed of a single layer of flattened cells which fit together like paving stones, and provides a thin, smooth inactive layer for surfaces which need only permit rapid diffusion. Examples are alveoli of the lung, blood vessels, and lymph vessels. The heart is also lined with simple squamous epithelium.

Cuboidal epithelium is composed of cells with a greater volume needed to accommodate the organelles necessary for metabolic processes such as active uptake. Examples are found in the kidney tubules, thyroid follicles, and in many glands (e.g. Brunner's glands). At sites of high absorption there may be microvilli lining the lumen surface.

Columnar epithelium is highly elongated since it is polarized, with markedly different functions at its two surfaces. It may be absorptive, in which case microvilli are present to increase the surface area, and usually includes simple glandular cells (goblet cells) which secrete mucus as a lubricant or protective. Found in much of the alimentary canal.

Epithelia may suffer enormous cell losses both by shedding and by apoptosis (programmed cell death). For example, about 10^{11} cells from the intestinal lining are lost every day, as are a similar number from the surface of the skin. These losses are normally precisely balanced by cell division, but any malfunction in the processes which regulate cell division might lead to an imbalance between loss and replacement – it is therefore not surprising that epithelia are the sites of more than 85% of all cancers.

CONNECTIVE TISSUES join together the other tissues of the body – each is composed of cells and an intercellular matrix.

White fibrous tissue is made up mainly of densely packed bundles of white collagen fibres. There are fibroblasts lying in rows along the bundles of collagen. This tissue is almost non-extensible so that it is able to transmit mechanical forces, and may also form strong attachments between organs. White fibrous (dense connective) tissue is found in tendons. An increase in the number of elastin fibres can increase the elasticity/extensibility of fibrous tissue with little sacrifice of structural strength. Such tissue is found in ligaments.

Adipose tissue consists of ring-shaped cells filled with fat globules, in a matrix of areolar tissue. It serves as a food store, insulation against mechanical and thermal shock, and support of organs such as kidneys and eyes.

Blood has different functions to other connective tissues, but it does satisfy the criterion of having cells within an extensive extracellular matrix. In this case the matrix is the watery plasma; suspended within it are the erythrocytes (red blood cells – transport of respiratory gases), leucocytes (white blood cells – many roles in defence against disease) and platelets (clotting of blood).

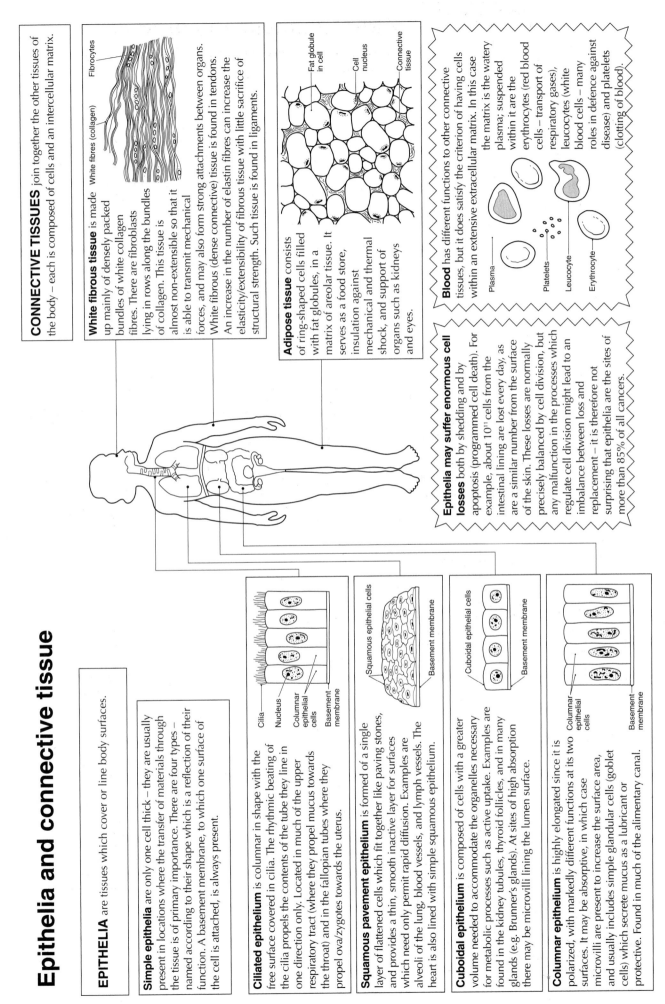

Section D Physiology of plants

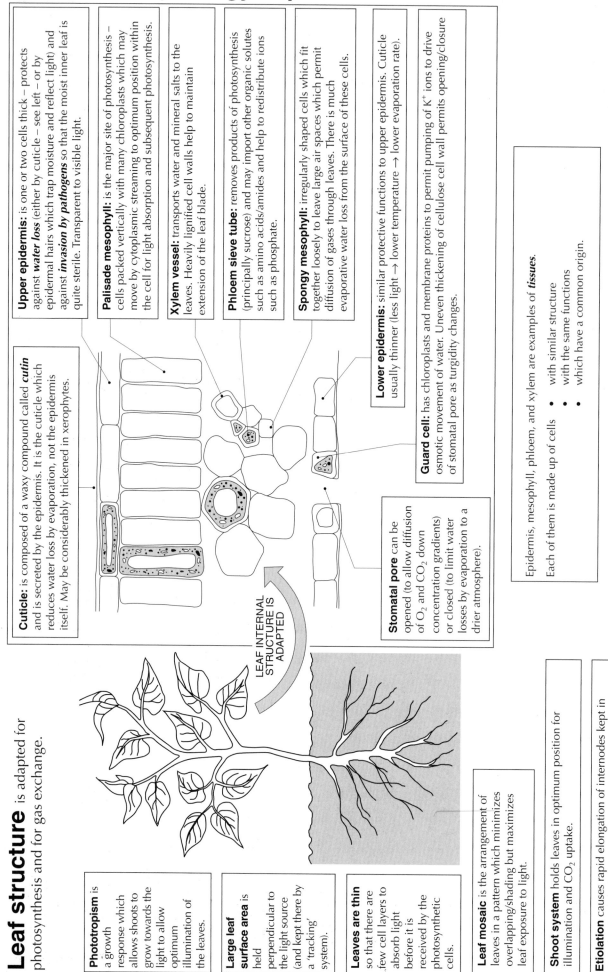

Leaf structure is adapted for photosynthesis and for gas exchange.

Upper epidermis: is one or two cells thick – protects against *water loss* (either by cuticle – see left – or by epidermal hairs which trap moisture and reflect light) and against *invasion by pathogens* so that the moist inner leaf is quite sterile. Transparent to visible light.

Palisade mesophyll: is the major site of photosynthesis – cells packed vertically with many chloroplasts which may move by cytoplasmic streaming to optimum position within the cell for light absorption and subsequent photosynthesis.

Xylem vessel: transports water and mineral salts to the leaves. Heavily lignified cell walls help to maintain extension of the leaf blade.

Phloem sieve tube: removes products of photosynthesis (principally sucrose) and may import other organic solutes such as amino acids/amides and help to redistribute ions such as phosphate.

Spongy mesophyll: irregularly shaped cells which fit together loosely to leave large air spaces which permit diffusion of gases through leaves. There is much evaporative water loss from the surface of these cells.

Lower epidermis: similar protective functions to upper epidermis. Cuticle usually thinner (less light → lower temperature → lower evaporation rate).

Guard cell: has chloroplasts and membrane proteins to permit pumping of K⁺ ions to drive osmotic movement of water. Uneven thickening of cellulose cell wall permits opening/closure of stomatal pore as turgidity changes.

Epidermis, mesophyll, phloem, and xylem are examples of *tissues*.

Each of them is made up of cells
- with similar structure
- with the same functions
- which have a common origin.

Cuticle: is composed of a waxy compound called *cutin* and is secreted by the epidermis. It is the cuticle which reduces water loss by evaporation, not the epidermis itself. May be considerably thickened in xerophytes.

LEAF INTERNAL STRUCTURE IS ADAPTED

Stomatal pore can be opened (to allow diffusion of O_2 and CO_2 down concentration gradients) or closed (to limit water losses by evaporation to a drier atmosphere).

Phototropism is a growth response which allows shoots to grow towards the light to allow optimum illumination of the leaves.

Large leaf surface area is held perpendicular to the light source (and kept there by a 'tracking' system).

Leaves are thin so that there are few cell layers to absorb light before it is received by the photosynthetic cells.

Leaf mosaic is the arrangement of leaves in a pattern which minimizes overlapping/shading but maximizes leaf exposure to light.

Shoot system holds leaves in optimum position for illumination and CO_2 uptake.

Etiolation causes rapid elongation of internodes kept in darkness to extend shoot.

Light reaction: non-cyclic photophosphorylation

Light energy excites electrons, resulting in the splitting of water and the synthesis of ATP and NADPH$_2$.

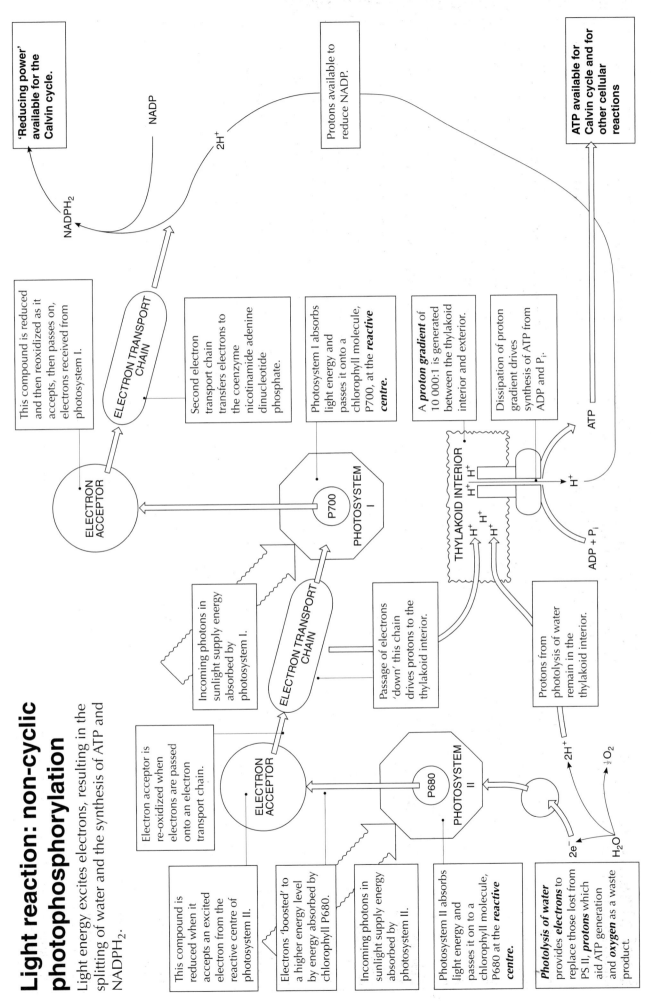

'Reducing power' available for the Calvin cycle.

NADP

NADPH$_2$

2H$^+$

Protons available to reduce NADP.

ATP available for Calvin cycle and for other cellular reactions

ELECTRON TRANSPORT CHAIN

This compound is reduced and then reoxidized as it accepts, then passes on, electrons received from photosystem I.

Second electron transport chain transfers electrons to the coenzyme nicotinamide adenine dinucleotide phosphate.

Photosystem I absorbs light energy and passes it onto a chlorophyll molecule, P700, at the *reactive centre*.

A *proton gradient* of 10 000:1 is generated between the thylakoid interior and exterior.

Dissipation of proton gradient drives synthesis of ATP from ADP and P$_i$.

ELECTRON ACCEPTOR

P700

PHOTOSYSTEM I

THYLAKOID INTERIOR

H$^+$ H$^+$

H$^+$

H$^+$ H$^+$

H$^+$ H$^+$

ATP

ADP + P$_i$

Incoming photons in sunlight supply energy absorbed by photosystem I.

Electron acceptor is re-oxidized when electrons are passed onto an electron transport chain.

ELECTRON TRANSPORT CHAIN

Passage of electrons 'down' this chain drives protons to the thylakoid interior.

Protons from photolysis of water remain in the thylakoid interior.

2H$^+$

$\frac{1}{2}$O$_2$

This compound is reduced when it accepts an excited electron from the reactive centre of photosystem II.

Electrons 'boosted' to a higher energy level by energy absorbed by chlorophyll P680.

Incoming photons in sunlight supply energy absorbed by photosystem II.

Photosystem II absorbs light energy and passes it on to a chlorophyll molecule, P680 at the *reactive centre*.

Photolysis of water provides *electrons* to replace those lost from PS II, *protons* which aid ATP generation and *oxygen* as a waste product.

ELECTRON ACCEPTOR

P680

PHOTOSYSTEM II

2e$^-$

H$_2$O

Dark reaction: the Calvin cycle

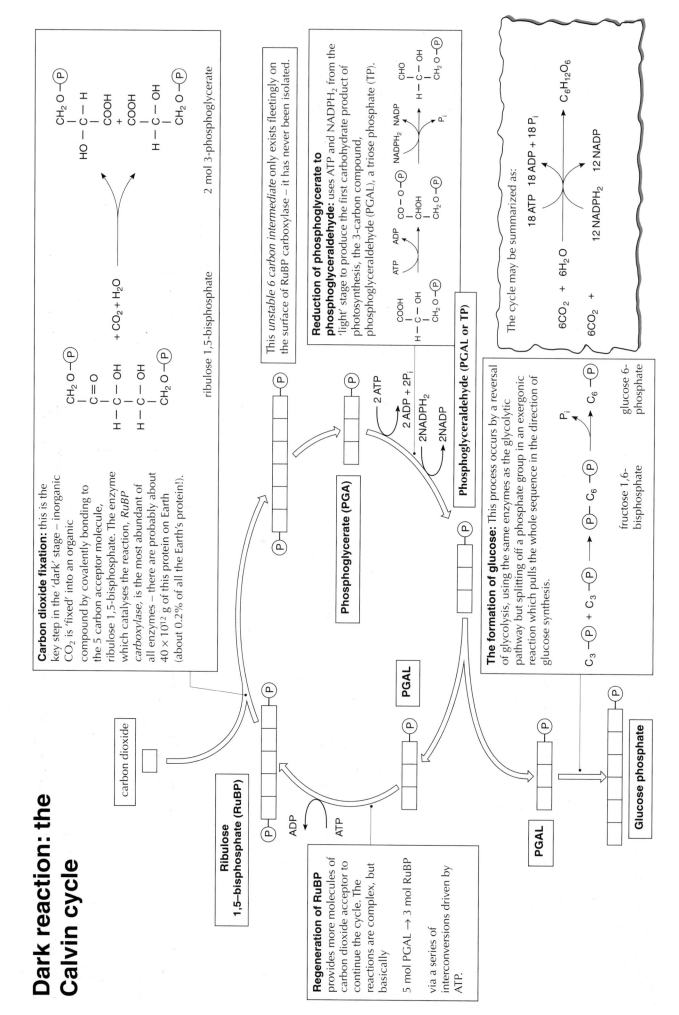

Carbon dioxide fixation: this is the key step in the 'dark' stage – inorganic CO_2 is 'fixed' into an organic compound by covalently bonding to the 5 carbon acceptor molecule, ribulose 1,5-bisphosphate. The enzyme which catalyses the reaction, *RuBP carboxylase*, is the most abundant of all enzymes – there are probably about 40×10^{12} g of this protein on Earth (about 0.2% of all the Earth's protein!).

ribulose 1,5-bisphosphate

2 mol 3-phosphoglycerate

This *unstable 6 carbon intermediate* only exists fleetingly on the surface of RuBP carboxylase – it has never been isolated.

Reduction of phosphoglycerate to phosphoglyceraldehyde: uses ATP and $NADPH_2$ from the 'light' stage to produce the first carbohydrate product of photosynthesis, the 3-carbon compound, phosphoglyceraldehyde (PGAL), a triose phosphate (TP).

Phosphoglyceraldehyde (PGAL or TP)

The cycle may be summarized as:

$$6CO_2 + 6H_2O \xrightarrow[\substack{12 NADPH_2 \quad 12 NADP}]{\substack{18 ATP \quad 18 ADP + 18 P_i}} C_6H_{12}O_6$$

$$6CO_2 +$$

carbon dioxide

Ribulose 1,5–bisphosphate (RuBP)

Phosphoglycerate (PGA)

2 ATP

2 ADP + $2P_i$

$2NADPH_2$

$2NADP$

PGAL

ADP

ATP

Regeneration of RuBP provides more molecules of carbon dioxide acceptor to continue the cycle. The reactions are complex, but basically

5 mol PGAL → 3 mol RuBP

via a series of interconversions driven by ATP.

PGAL

The formation of glucose: This process occurs by a reversal of glycolysis, using the same enzymes as the glycolytic pathway but splitting off a phosphate group in an exergonic reaction which pulls the whole sequence in the direction of glucose synthesis.

$$C_3 \!-\! \textcircled{P} + C_3 \!-\! \textcircled{P} \longrightarrow \textcircled{P}\!-\! C_6 \!-\! \textcircled{P} \xrightarrow{\quad P_i \quad} C_6 \!-\! \textcircled{P}$$

fructose 1,6-bisphosphate

glucose 6-phosphate

Glucose phosphate

Chloroplasts: sites of photosynthesis contain photosensitive pigments

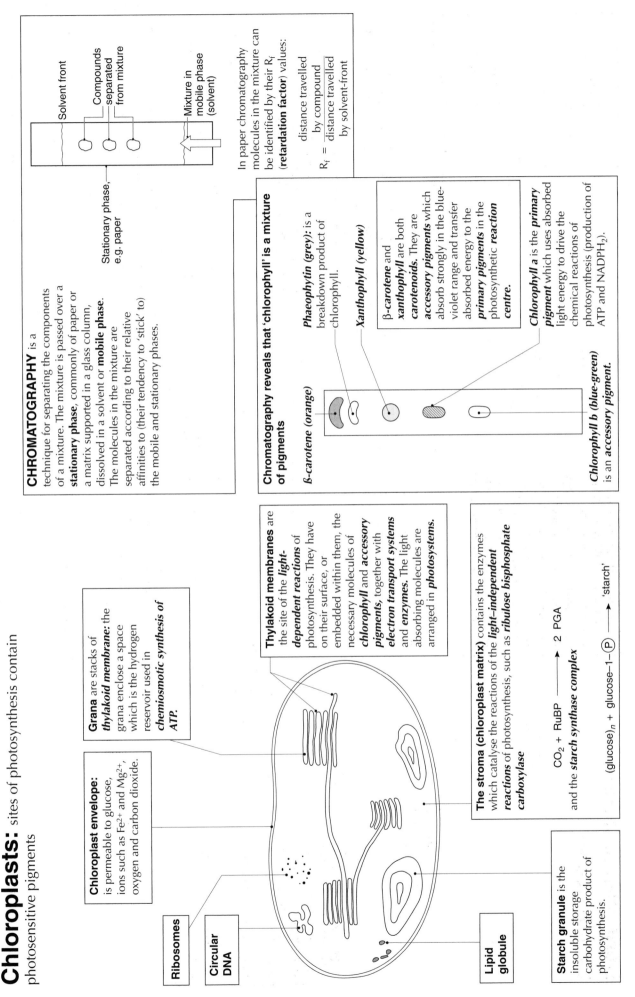

CHROMATOGRAPHY

CHROMATOGRAPHY is a technique for separating the components of a mixture. The mixture is passed over a **stationary phase**, commonly of paper or a matrix supported in a glass column, dissolved in a solvent or **mobile phase.** The molecules in the mixture are separated according to their relative affinities to (their tendency to 'stick' to) the mobile and stationary phases.

In paper chromatography molecules in the mixture can be identified by their R_f (**retardation factor**) values:

$$R_f = \frac{\text{distance travelled by compound}}{\text{distance travelled by solvent-front}}$$

Solvent front

Compounds separated from mixture

Mixture in mobile phase (solvent)

Stationary phase, e.g. paper

Chromatography reveals that 'chlorophyll' is a mixture of pigments

Phaeophytin (grey): is a breakdown product of chlorophyll.

Xanthophyll (yellow)

β-*carotene* and *xanthophyll* are both *carotenoids*. They are *accessory pigments* which absorb strongly in the blue-violet range and transfer absorbed energy to the *primary pigments* in the photosynthetic *reaction centre*.

Chlorophyll a is the *primary pigment* which uses absorbed light energy to drive the chemical reactions of photosynthesis (production of ATP and NADPH₂).

ß-*carotene (orange)*

Chlorophyll b (blue-green) is an *accessory pigment.*

Chloroplast structure

Grana are stacks of *thylakoid membrane:* the grana enclose a space which is the hydrogen reservoir used in *chemiosmotic synthesis of ATP.*

Chloroplast envelope: is permeable to glucose, ions such as Fe^{2+} and Mg^{2+}, oxygen and carbon dioxide.

Ribosomes

Circular DNA

Thylakoid membranes are the site of the *light-dependent reactions* of photosynthesis. They have on their surface, or embedded within them, the necessary molecules of *chlorophyll* and *accessory pigments,* together with *electron transport systems* and *enzymes.* The light absorbing molecules are arranged in *photosystems.*

The stroma (chloroplast matrix) contains the enzymes which catalyse the reactions of the *light-independent reactions* of photosynthesis, such as *ribulose bisphosphate carboxylase*

$$CO_2 + RuBP \longrightarrow 2\ PGA$$

and the *starch synthase complex*

$$(glucose)_n + glucose\text{-}1\text{-}\textcircled{P} \longrightarrow \text{'starch'}$$

Lipid globule

Starch granule is the insoluble storage carbohydrate product of photosynthesis.

Principle: bubbles of evolved gas are collected over a known period of time.

$$\textbf{Rate of photosynthesis} = \frac{\text{collected volume / mm}^3}{\text{time / min}}$$

= mm³ O_2 evolved/min at a known temperature, $t\,°C$.

Thermometer to check temperature of water bath. Temperature should be recorded and rate of photosynthesis defined *at a temperature t*.

Light source: for most applications the wavelength should be fixed (e.g. a white light source) but the apparatus can be used to investigate the effect of wavelength on rate of photosynthesis, in which case the wavelength can be varied by the insertion of appropriate filters between the lamp and the photosynthesising plant. If *light intensity* is to be the manipulated variable then the light may be moved to a known distance from the plant (in which case light intensity ∝ $1/d^2$) or a rheostat may be incorporated into the lamp circuit.

Problems:
1. Are *controlled variables* at their optimum?
2. Are all plant samples comparable – same size, age, activity?
3. Is apparatus reliable – clean, leakproof?
4. Are collected bubbles all oxygen, all produced by photosynthesis?

Graduated scale: allows length (*l*) of collected bubble to be measured. If diameter (*d*) of the capillary tube is known the volume (*v*) of gas evolved can be calculated from:
$$v = \frac{\pi d^2}{4} l$$

Capillary tube: the narrow diameter means that a small volume of evolved oxygen will register as a long bubble in the tube – more accurate measurement possible.

Syringe: the plunger should be pushed well in at the start of the experiment. The syringe can be used to draw the collected bubbles of evolved gas along the tube so that the length can be measured against the scale, and then towards the syringe where it will not interfere with further collections and measurements of gas.

Flared end of capillary tube: aids attachment of plastic tubing and consequent capture of bubbles.

Plastic tubing: ensures that bubbles released from the cut end of the plant are trapped so that their volume can be recorded.

Aquatic plant (often *Elodea*): when photosynthesising, releases oxygen from a cut end of the stem. The plant specimen can be induced to photosynthesise actively by prior illumination, and gentle aeration of the solution for about an hour before the experiment. Any adjustment of the manipulated variable (for example, a change in light intensity), should be followed by a short period (ca. 5 min) to allow the plant to re-adjust (it allows the system to equilibrate).

Hydrogen carbonate solution is a source of carbon dioxide of known concentration. The concentration may be varied if *concentration of carbon dioxide* is to be the manipulated variable, or should be fixed at about 5 × optimum [CO_2] if it is to be a controlled variable but not a limiting factor. The [HCO_3^-] should not be great enough to markedly alter pH.

Water bath: may serve as a heat filter if an incandescent light source is used (so that *temperature* remains a controlled variable) or as a thermostatically controlled water bath if temperature is to be the manipulated variable.

Measurement of photosynthesis:

Audus' photosynthometer

Law of limiting factors

Blackman stated: 'when a process is affected by more than one factor its rate is limited by the factor which is nearest its minimum value: it is that *limiting factor* which directly affects a process if its magnitude is changed.'

Photosynthesis is a multi-stage process – for example the Calvin cycle is dependent on the supply of ATP and reducing power from the light reactions – and the principle of limiting factors can be applied.

The rate of a multi-stage process may be subject to different limiting factors at different times. Photosynthesis may be limited by *temperature* during the early part of a summer's day, by *light intensity* during cloudy or overcast conditions or by *carbon dioxide concentration* at other times. The principal limiting factor in Britain during the summer is *carbon dioxide concentration*: the atmospheric [CO_2] is typically only 0.04%. Increased CO_2 emissions from combustion of fossil fuels may stimulate photosynthesis.

Here an increase in the availability of A does not affect the rate of photosynthesis: some other factor becomes the *limiting factor*

RATE OF PS is not ∝ [A]
RATE OF PS ∝ [B] or [C] etc.

The *mechanism of photosynthesis* is made clearer by studies of limiting factors – the fact that *light* is a limiting factor indicates a *light-dependent stage*, the effect of *temperature* suggests that there are *enzyme-catalysed* reactions, the *interaction of [CO_2] and *temperature* suggests an enzyme catalysed *fixation of carbon dioxide*. The existence of more than one limiting factor suggests that *photosynthesis is a multi-stage process.*

Here the rate of photosynthesis is limited by the availability of factor A: A is the *limiting factor* and a change in the availability of A will directly influence the rate of photosynthesis.

RATE OF PS ∝ [A]

AVAILABILITY OF FACTOR A

RATE OF PHOTOSYNTHESIS (arbitrary units)

The study of limiting factors has *commercial and horticultural applications*. Since [CO_2] is a limiting factor, crop production in greenhouses is readily stimulated by raising local carbon dioxide concentrations (from gas cylinders or by burning fossil fuels). It is also clear to horticulturalists that expensive increases in energy consumption for lighting and heating are not economically justified if neither of these is the limiting factor applying under any particular set of conditions.

The *limiting factors* which affect *photosynthesis* are:

Light intensity: light energy is necessary to generate ATP and $NADPH_2$ during the light dependent stages of photosynthesis.

Carbon dioxide concentration: CO_2 is 'fixed' by reaction with ribulose bisphosphate in the initial reaction of the Calvin cycle.

Temperature: the enzymes catalysing the reactions of the Calvin cycle and some of the light-dependent stages are affected by temperature.

Water availability and **chlorophyll concentration** are not normally limiting factors in photosynthesis.

Mineral requirements of plants

Calcium is absorbed as Ca^{2+}. It forms junctions between the molecules of pectate in the middle lamella, strengthening the binding of adjacent cells to one another. Deficiency leads to die-back of shoots due to death of apical buds.

Phosphorus is absorbed from the soil solution as $H_2PO_4^-$ (a type of phosphate). Its availability is probably **the major limiting factor** in plant growth in **uncultivated soils**. It is a component of nucleic acids, phospholipids and ATP. Lack of phosphorus usually affects processes which consume ATP, particularly the active uptake of minerals by roots.

Magnesium is absorbed from the soil solution as Mg^{2+}. It is the central atom in the porphyrin ring of the chlorophyll molecule. Magnesium deficiency is characterized by a distinct paleness due to absence of chlorophyll (**chlorosis**), usually beginning around the veins of the older leaves.

Nitrogen is absorbed as NO_3^- or as NH_4^+, and usually converted to amino acids or amides for transport through the plant. Nitrogen is essential for the synthesis of amino acids, proteins, plant hormones, nucleic acids, nucleotides and chlorophyll. Deficiency causes reduced growth of all organs, particularly the leaves, and a marked chlorosis. Nitrogen availability is probably the **major limiting factor** in plant growth in **cultivated soils**.

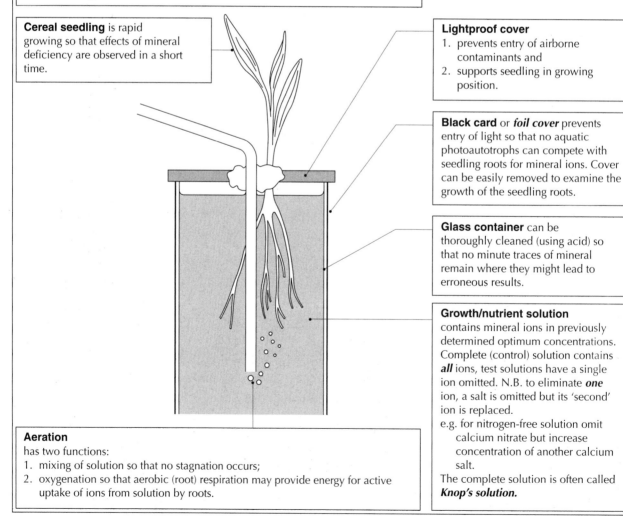

HYDROPONICS ('growing in water') permits study of plant deficiency symptoms in controlled conditions which eliminate the many variables associated with soil as a growth medium.

Cereal seedling is rapid growing so that effects of mineral deficiency are observed in a short time.

Lightproof cover
1. prevents entry of airborne contaminants and
2. supports seedling in growing position.

Black card or **foil cover** prevents entry of light so that no aquatic photoautotrophs can compete with seedling roots for mineral ions. Cover can be easily removed to examine the growth of the seedling roots.

Glass container can be thoroughly cleaned (using acid) so that no minute traces of mineral remain where they might lead to erroneous results.

Growth/nutrient solution contains mineral ions in previously determined optimum concentrations. Complete (control) solution contains **all** ions, test solutions have a single ion omitted. N.B. to eliminate **one** ion, a salt is omitted but its 'second' ion is replaced.
e.g. for nitrogen-free solution omit calcium nitrate but increase concentration of another calcium salt.
The complete solution is often called **Knop's solution.**

Aeration
has two functions:
1. mixing of solution so that no stagnation occurs;
2. oxygenation so that aerobic (root) respiration may provide energy for active uptake of ions from solution by roots.

Tissue distribution in a herbaceous stem

The tissue location provides *mechanical support* and *transport*.

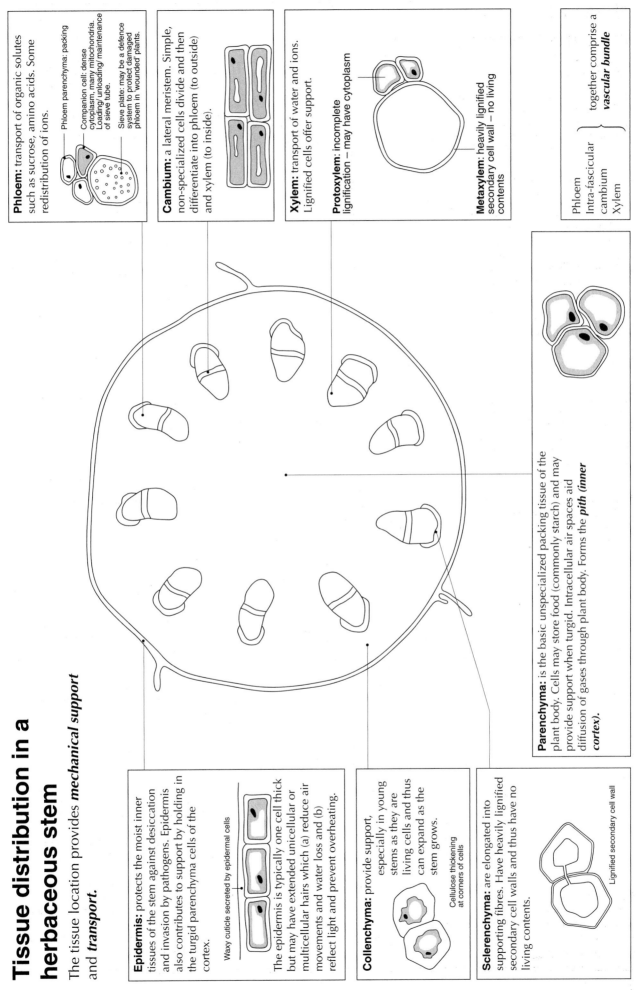

Phloem: transport of organic solutes such as sucrose, amino acids. Some redistribution of ions.

Phloem parenchyma: packing

Companion cell: dense cytoplasm, many mitochondria. Loading/unloading/maintenance of sieve tube.

Sieve plate: may be a defence system to protect damaged phloem in 'wounded' plants.

Cambium: a lateral meristem. Simple, non-specialized cells divide and then differentiate into phloem (to outside) and xylem (to inside).

Xylem: transport of water and ions. Lignified cells offer support.

Protoxylem: incomplete lignification – may have cytoplasm

Metaxylem: heavily lignified secondary cell wall – no living contents

Phloem
Intra-fascicular cambium
Xylem
} together comprise a *vascular bundle*

Epidermis: protects the moist inner tissues of the stem against desiccation and invasion by pathogens. Epidermis also contributes to support by holding in the turgid parenchyma cells of the cortex.

Waxy cuticle secreted by epidermal cells

The epidermis is typically one cell thick but may have extended unicellular or multicellular hairs which (a) reduce air movements and water loss and (b) reflect light and prevent overheating.

Collenchyma: provide support, especially in young stems as they are living cells and thus can expand as the stem grows.

Cellulose thickening at corners of cells

Sclerenchyma: are elongated into supporting fibres. Have heavily lignified secondary cell walls and thus have no living contents.

Lignified secondary cell wall

Parenchyma: is the basic unspecialized packing tissue of the plant body. Cells may store food (commonly starch) and may provide support when turgid. Intracellular air spaces aid diffusion of gases through plant body. Forms the *pith (inner cortex).*

Tissue distribution in a dicotyledonous root

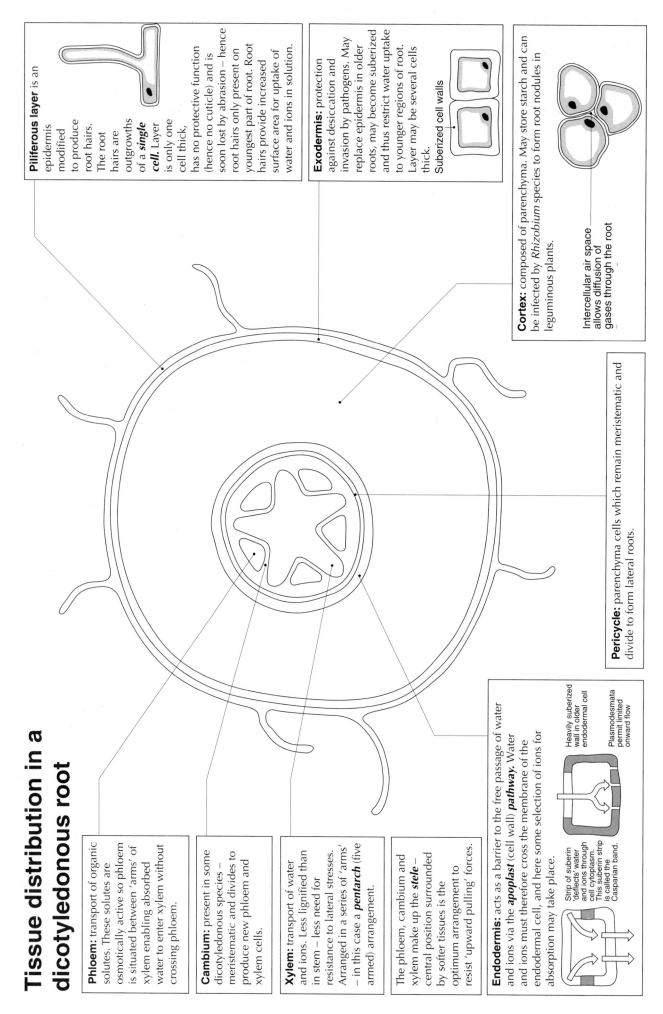

Piliferous layer is an epidermis modified to produce root hairs. The root hairs are outgrowths of a **single cell**. Layer is only one cell thick, has no protective function (hence no cuticle) and is soon lost by abrasion – hence root hairs only present on youngest part of root. Root hairs provide increased surface area for uptake of water and ions in solution.

Exodermis: protection against desiccation and invasion by pathogens. May replace epidermis in older roots, may become suberized and thus restrict water uptake to younger regions of root. Layer may be several cells thick.
Suberized cell walls

Cortex: composed of parenchyma. May store starch and can be infected by *Rhizobium* species to form root nodules in leguminous plants.

Intercellular air space allows diffusion of gases through the root

Pericycle: parenchyma cells which remain meristematic and divide to form lateral roots.

Phloem: transport of organic solutes. These solutes are osmotically active so phloem is situated between 'arms' of xylem enabling absorbed water to enter xylem without crossing phloem.

Cambium: present in some dicotyledonous species – meristematic and divides to produce new phloem and xylem cells.

Xylem: transport of water and ions. Less lignified than in stem – less need for resistance to lateral stresses. Arranged in a series of 'arms' – in this case a **pentarch** (five armed) arrangement.

The phloem, cambium and xylem make up the **stele** – central position surrounded by softer tissues is the optimum arrangement to resist 'upward pulling' forces.

Endodermis: acts as a barrier to the free passage of water and ions via the **apoplast** (cell wall) **pathway**. Water and ions must therefore cross the membrane of the endodermal cell, and here some selection of ions for absorption may take place.

Heavily suberized wall in older endodermal cell

Plasmodesmata permit limited onward flow

Strip of suberin 'deflects' water and ions through cell cytoplasm. This suberin strip is called the Casparian band.

Evidence for phloem as the tissue for translocation comes

from the use of radioactive tracers, aphids and metabolic poisons.

The pattern of movement of these solutes within the plant body has also been investigated using radioisotopes, and it has been shown that the pattern of movement may be modified as the plant ages. Up to maturity the lower leaves of an actively photosynthesizing plant may pass their products to the roots for consumption and storage, but once fruit formation begins, ever-increasing numbers of leaves pass their products up to the fruits and eventually even the lower leaves are doing so. Minerals are often remobilized – having been delivered to the photosynthetic leaves via the xylem they may be re-exported through the phloem as the leaves age prior to abscission. The direction of solute movement is under the control of plant growth substances, particularly IAA and the cytokinins.

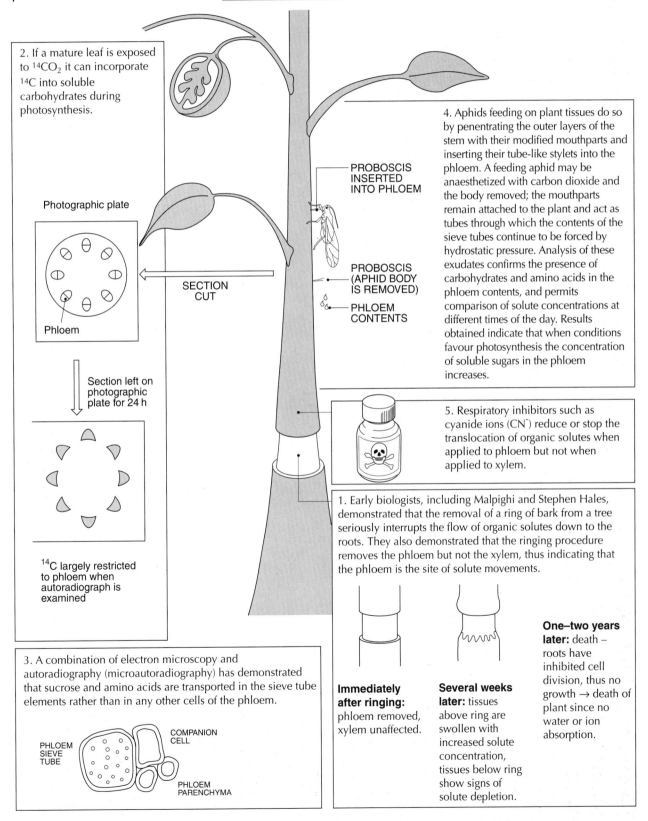

2. If a mature leaf is exposed to $^{14}CO_2$ it can incorporate ^{14}C into soluble carbohydrates during photosynthesis.

Photographic plate

Phloem

SECTION CUT

Section left on photographic plate for 24 h

^{14}C largely restricted to phloem when autoradiograph is examined

PROBOSCIS INSERTED INTO PHLOEM

PROBOSCIS (APHID BODY IS REMOVED)

PHLOEM CONTENTS

4. Aphids feeding on plant tissues do so by penentrating the outer layers of the stem with their modified mouthparts and inserting their tube-like stylets into the phloem. A feeding aphid may be anaesthetized with carbon dioxide and the body removed; the mouthparts remain attached to the plant and act as tubes through which the contents of the sieve tubes continue to be forced by hydrostatic pressure. Analysis of these exudates confirms the presence of carbohydrates and amino acids in the phloem contents, and permits comparison of solute concentrations at different times of the day. Results obtained indicate that when conditions favour photosynthesis the concentration of soluble sugars in the phloem increases.

5. Respiratory inhibitors such as cyanide ions (CN^-) reduce or stop the translocation of organic solutes when applied to phloem but not when applied to xylem.

1. Early biologists, including Malpighi and Stephen Hales, demonstrated that the removal of a ring of bark from a tree seriously interrupts the flow of organic solutes down to the roots. They also demonstrated that the ringing procedure removes the phloem but not the xylem, thus indicating that the phloem is the site of solute movements.

3. A combination of electron microscopy and autoradiography (microautoradiography) has demonstrated that sucrose and amino acids are transported in the sieve tube elements rather than in any other cells of the phloem.

PHLOEM SIEVE TUBE

COMPANION CELL

PHLOEM PARENCHYMA

Immediately after ringing: phloem removed, xylem unaffected.

Several weeks later: tissues above ring are swollen with increased solute concentration, tissues below ring show signs of solute depletion.

One–two years later: death – roots have inhibited cell division, thus no growth → death of plant since no water or ion absorption.

Mass flow theory – transport in the phloem: movements of water and sucrose generate a gradient of hydrostatic pressure which drives the translocation of organic solutes from leaf to root.

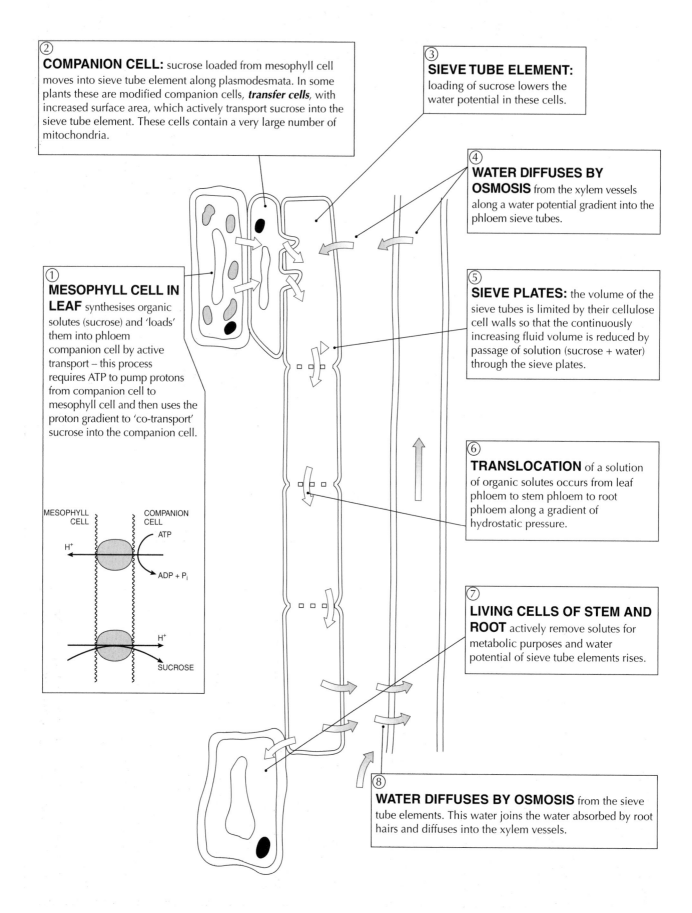

② COMPANION CELL: sucrose loaded from mesophyll cell moves into sieve tube element along plasmodesmata. In some plants these are modified companion cells, *transfer cells*, with increased surface area, which actively transport sucrose into the sieve tube element. These cells contain a very large number of mitochondria.

③ SIEVE TUBE ELEMENT: loading of sucrose lowers the water potential in these cells.

④ WATER DIFFUSES BY OSMOSIS from the xylem vessels along a water potential gradient into the phloem sieve tubes.

① MESOPHYLL CELL IN LEAF synthesises organic solutes (sucrose) and 'loads' them into phloem companion cell by active transport – this process requires ATP to pump protons from companion cell to mesophyll cell and then uses the proton gradient to 'co-transport' sucrose into the companion cell.

⑤ SIEVE PLATES: the volume of the sieve tubes is limited by their cellulose cell walls so that the continuously increasing fluid volume is reduced by passage of solution (sucrose + water) through the sieve plates.

⑥ TRANSLOCATION of a solution of organic solutes occurs from leaf phloem to stem phloem to root phloem along a gradient of hydrostatic pressure.

MESOPHYLL CELL   COMPANION CELL
ATP
H^+
ADP + P_i
H^+
SUCROSE

⑦ LIVING CELLS OF STEM AND ROOT actively remove solutes for metabolic purposes and water potential of sieve tube elements rises.

⑧ WATER DIFFUSES BY OSMOSIS from the sieve tube elements. This water joins the water absorbed by root hairs and diffuses into the xylem vessels.

Water potential of plant cells

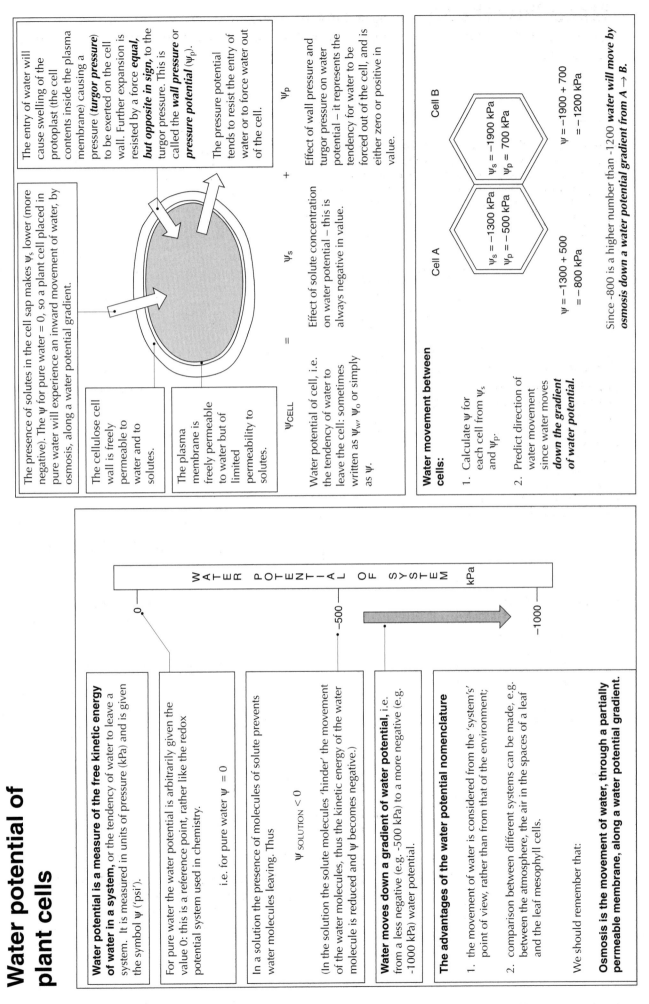

Water potential is a measure of the free kinetic energy of water in a system, or the tendency of water to leave a system. It is measured in units of pressure (kPa) and is given the symbol ψ ('psi').

For pure water the water potential is arbitrarily given the value 0: this is a reference point, rather like the redox potential system used in chemistry.

i.e. for pure water $\psi = 0$

$$\psi_{SOLUTION} < 0$$

In a solution the presence of molecules of solute prevents water molecules leaving. Thus

(In the solution the solute molecules 'hinder' the movement of the water molecules, thus the kinetic energy of the water molecule is reduced and ψ becomes negative.)

Water moves down a gradient of water potential, i.e. from a less negative (e.g. -500 kPa) to a more negative (e.g. -1000 kPa) water potential.

The advantages of the water potential nomenclature

1. the movement of water is considered from the 'system's' point of view, rather than from that of the environment;

2. comparison between different systems can be made, e.g. between the atmosphere, the air in the spaces of a leaf and the leaf mesophyll cells.

We should remember that:

Osmosis is the movement of water, through a partially permeable membrane, along a water potential gradient.

0
WATER POTENTIAL OF SYSTEM
-500
-1000
kPa

The presence of solutes in the cell sap makes ψ_s lower (more negative). The ψ for pure water = 0, so a plant cell placed in pure water will experience an inward movement of water, by osmosis, along a water potential gradient.

The cellulose cell wall is freely permeable to water and to solutes.

The plasma membrane is freely permeable to water but of limited permeability to solutes.

The entry of water will cause swelling of the protoplast (the cell contents inside the plasma membrane) causing a pressure (***turgor pressure***) to be exerted on the cell wall. Further expansion is resisted by a force **equal, but opposite in sign,** to the turgor pressure. This is called the ***wall pressure*** or ***pressure potential*** (ψ_p).

The pressure potential tends to resist the entry of water or to force water out of the cell.

$$\psi_{CELL} = \psi_s + \psi_p$$

Water potential of cell, i.e. the tendency of water to leave the cell: sometimes written as ψ_w, ψ_o or simply as ψ.

Effect of solute concentration on water potential – this is always negative in value.

Effect of wall pressure and turgor pressure on water potential – it represents the tendency for water to be forced out of the cell, and is either zero or positive in value.

Water movement between cells:

1. Calculate ψ for each cell from ψ_s and ψ_p.

2. Predict direction of water movement since water moves **down the gradient of water potential.**

Cell A

$\psi_s = -1300$ kPa
$\psi_p = -500$ kPa

$\psi = -1300 + 500$
$= -800$ kPa

Cell B

$\psi_s = -1900$ kPa
$\psi_p = 700$ kPa

$\psi = -1900 + 700$
$= -1200$ kPa

Since -800 is a higher number than -1200 **water will move by osmosis down a water potential gradient from A → B.**

Water relationships of plant cells

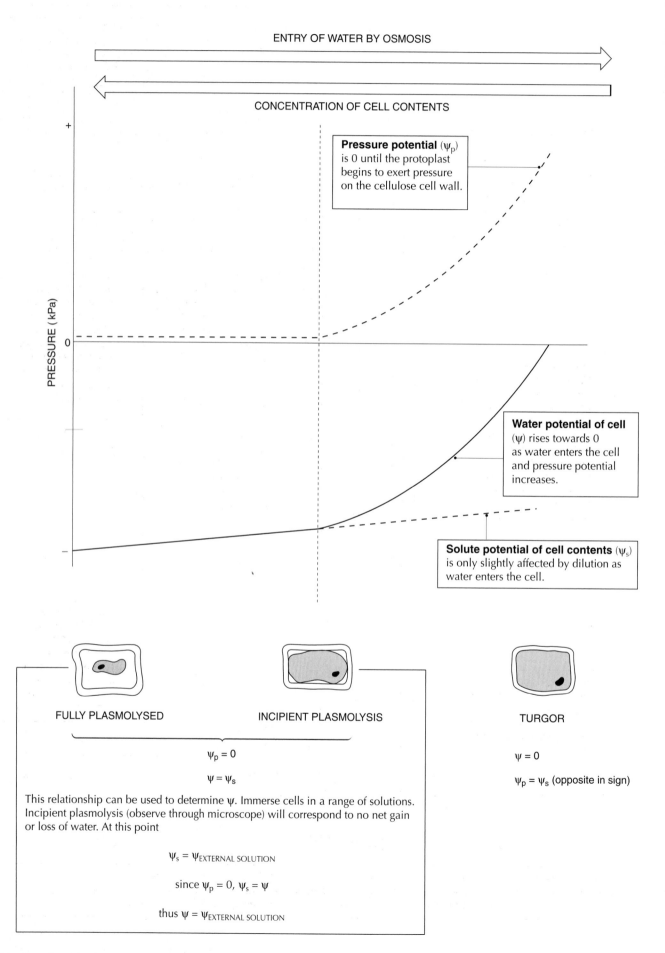

ENTRY OF WATER BY OSMOSIS

CONCENTRATION OF CELL CONTENTS

Pressure potential (ψ_p) is 0 until the protoplast begins to exert pressure on the cellulose cell wall.

Water potential of cell (ψ) rises towards 0 as water enters the cell and pressure potential increases.

Solute potential of cell contents (ψ_s) is only slightly affected by dilution as water enters the cell.

PRESSURE (kPa)

FULLY PLASMOLYSED

INCIPIENT PLASMOLYSIS

TURGOR

$\psi_p = 0$

$\psi = \psi_s$

This relationship can be used to determine ψ. Immerse cells in a range of solutions. Incipient plasmolysis (observe through microscope) will correspond to no net gain or loss of water. At this point

$\psi_s = \psi_{\text{EXTERNAL SOLUTION}}$

since $\psi_p = 0$, $\psi_s = \psi$

thus $\psi = \psi_{\text{EXTERNAL SOLUTION}}$

$\psi = 0$

$\psi_p = \psi_s$ (opposite in sign)

Cohesion–tension theory of transpiration

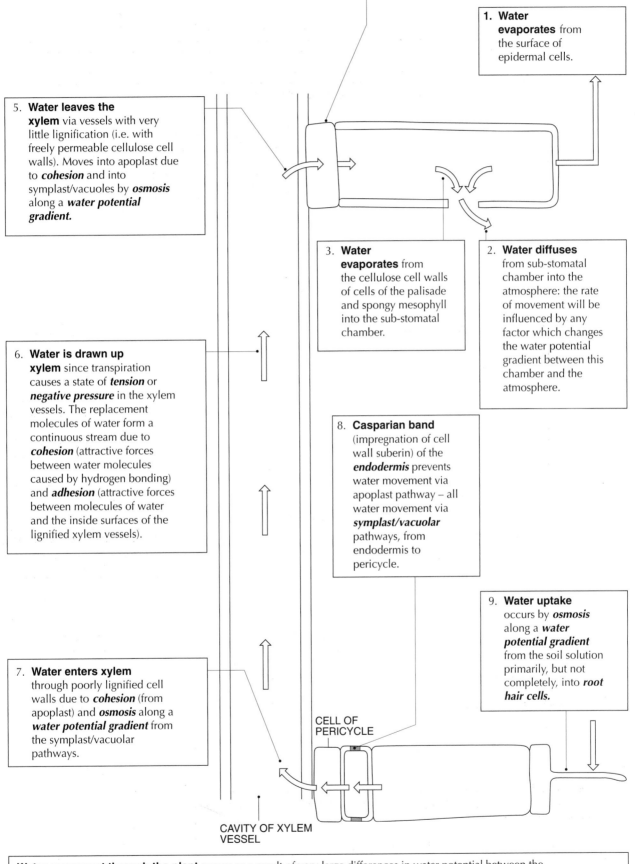

4. Water moves through leaf tissues via *apoplast* (cell wall), *symplast* (cytoplasmic) and *vacuolar* pathways.

1. Water evaporates from the surface of epidermal cells.

5. Water leaves the xylem via vessels with very little lignification (i.e. with freely permeable cellulose cell walls). Moves into apoplast due to *cohesion* and into symplast/vacuoles by *osmosis* along a *water potential gradient*.

3. Water evaporates from the cellulose cell walls of cells of the palisade and spongy mesophyll into the sub-stomatal chamber.

2. Water diffuses from sub-stomatal chamber into the atmosphere: the rate of movement will be influenced by any factor which changes the water potential gradient between this chamber and the atmosphere.

6. Water is drawn up xylem since transpiration causes a state of *tension* or *negative pressure* in the xylem vessels. The replacement molecules of water form a continuous stream due to *cohesion* (attractive forces between water molecules caused by hydrogen bonding) and *adhesion* (attractive forces between molecules of water and the inside surfaces of the lignified xylem vessels).

8. Casparian band (impregnation of cell wall suberin) of the *endodermis* prevents water movement via apoplast pathway – all water movement via *symplast/vacuolar* pathways, from endodermis to pericycle.

9. Water uptake occurs by *osmosis* along a *water potential gradient* from the soil solution primarily, but not completely, into *root hair cells*.

7. Water enters xylem through poorly lignified cell walls due to *cohesion* (from apoplast) and *osmosis* along a *water potential gradient* from the symplast/vacuolar pathways.

CELL OF PERICYCLE

CAVITY OF XYLEM VESSEL

Water movement through the plant occurs as a result of very large differences in water potential between the atmosphere and the soil solution. The process begins with *evaporation* from the leaf surfaces, is continued due to *cohesion* between water molecules and *tension* in the xylem vessels and is completed by *osmosis* from the soil solution.

The bubble potometer

measures **water uptake** (= water loss by transpiration + water consumption for cell expansion and photosynthesis).

Water uptake

- Water 'lost' by transpiration (98%)
- Water used in photosynthesis (1%)
- Water used in building of protoplasm (1%)

Reservoir of water: may be connected to capillary tubing if the tap is opened. This is used to prevent the air bubble entering the plant, and to move the bubble back along the capillary tube.

Rubber tubing – to connect cut shoot to the potometer. The tube should be **greased** and **wired** to prevent any leakage of air into the apparatus.

Atmometer control: The atmometer is an instrument which can measure evaporation from a non-living surface. When subjected to the same conditions as a potometer the changes in the rate of evaporation from a plant and from a purely physical system can be compared – for example, a reduction in light intensity will show a decrease in water loss **only from a potometer** (due to stomatal closure). The atmometer control indicates when the potometer is acting as a free evaporator and when it is affected by physiological factors such as photosynthesis and stomatal closure.

Capillary tube: must be kept horizontal to prevent the bubble moving due to its density compared with water.

Air bubble: inserted by removal of tube end from beaker of water. Movement corresponds to water uptake by the cut shoot.

Graduated scale: permits direct reading of bubble movement/water uptake.

This porous pot replaces the cut shoot

PROCEDURE

1. The leafy shoot must be cut **under water,** the apparatus must be filled **under water** and the shoot fixed to the potometer **under water** to prevent air locks in the system.

2. Allow plant to equilibrate (5 min) before introduction of air bubble. Take at least three readings of rate of bubble movement, and use reservoir to return bubble to zero on each occasion. Calculate mean of readings. Record air temperature.

3. Scale can be calibrated by introducing a known mass of mercury into the capillary tubing and using $p=m/v$ (p for mercury is known, m can be measured, thus v corresponding to a measured distance of bubble movement can be determined).

4. Rate of water uptake per unit area of leaves can be calculated by measurement of leaf area.

EXTERNAL FACTORS AFFECTING TRANSPIRATION

Light intensity: use bench lamp (with water bath to act as heat filter) to increase light intensity. To simulate 'darkness' enclose shoot in black polythene bag.

Humidity: enclose shoot in clear plastic bag to **increase** relative humidity of atmosphere – include water absorbant such as calcium chloride to **decrease** relative humidity.

Wind: use small electric fan with 'cool' control to mimic air movements whilst avoiding effects of temperature changes.

May also determine relative importance of upper surface/lower surface/stem/petiole in water loss by smearing with vaseline (acts like a waxy cuticle) as appropriate.

N.B. It is sometimes difficult to change only one condition at a time, e.g. enclosure in a black bag to eliminate light will also increase the relative humidity of the atmosphere.

Xerophytes are plants adapted for water conservation: they show many *xeromorphic adaptations*.

MARRAM GRASS is adapted to drying conditions of the seashore.

Thick waxy cuticle reduces cuticular transpiration.

Absence of stomata on upper epidermis reduces stomatal transpiration.

Leaf rolling traps humid layer close to lower epidermis. Caused by *hinge cells* which become flaccid when water is in short supply.

Leaf hairs limit air movement over leaf surface.

Sunken stomata keep humid air close to stomatal pore and so reduce water potential gradient.

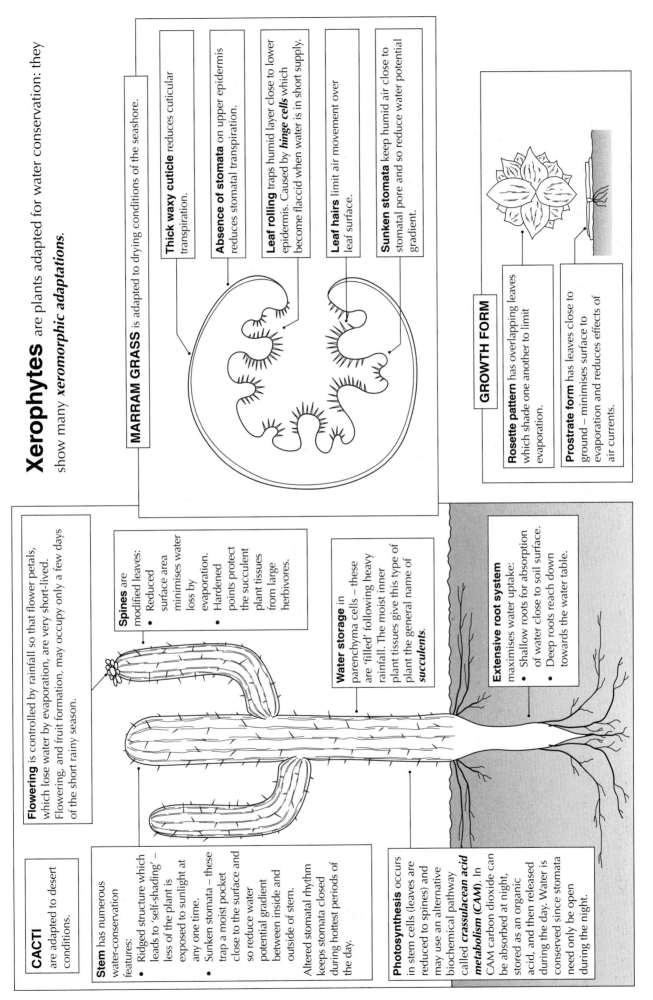

GROWTH FORM

Rosette pattern has overlapping leaves which shade one another to limit evaporation.

Prostrate form has leaves close to ground – minimises surface to evaporation and reduces effects of air currents.

CACTI are adapted to desert conditions.

Stem has numerous water-conservation features:
- Ridged structure which leads to 'self-shading' – less of the plant is exposed to sunlight at any one time.
- Sunken stomata – these trap a moist pocket close to the surface and so reduce water potential gradient between inside and outside of stem.

Altered stomatal rhythm keeps stomata closed during hottest periods of the day.

Photosynthesis occurs in stem cells (leaves are reduced to spines) and may use an alternative biochemical pathway called *crassulacean acid metabolism (CAM)*. In CAM carbon dioxide can be absorbed at night, stored as an organic acid, and then released during the day. Water is conserved since stomata need only be open during the night.

Flowering is controlled by rainfall so that flower petals, which lose water by evaporation, are very short-lived. Flowering, and fruit formation, may occupy only a few days of the short rainy season.

Spines are modified leaves:
- Reduced surface area minimises water loss by evaporation.
- Hardened points protect the succulent plant tissues from large herbivores.

Water storage in parenchyma cells – these are 'filled' following heavy rainfall. The moist inner plant tissues give this type of plant the general name of *succulents*.

Extensive root system maximises water uptake:
- Shallow roots for absorption of water close to soil surface.
- Deep roots reach down towards the water table.

Stomata represent an important adaptation to life in a terrestrial environment.

Adjacent epidermal cells: note that there is no **symplastic** connection (i.e. no **plasmodesmata**) between these cells and the guard cells. The adjacent epidermal cells do not have the chloroplasts or the dense cytoplasm typical of guard cells.

Control of stomatal aperture: must permit
a. entry of enough CO_2 to permit photosynthesis;
b. control water loss to prevent desiccation of plant tissues

and has both **internal** (plant growth regulators) and **environmental** (e.g. air humidity) **signals.**

Stomatal opening is promoted by:
1. **low intercellular [CO_2]:** sensed by guard cells and corresponds to the need for more CO_2 to maintain rates of photosynthesis.
2. **high light intensity:** light absorbed by chlorophyll (the PAR photosystem) provides ATP by photophosphorylation – this increases proton pumping and 'opens' stomata when light is available – 'anticipating' the need for more CO_2 for photosynthesis. This is a **feedforward response.** There is a second (blue-light photosystem) response for opening stomata in shady conditions, or at dawn.

Stomatal closure is triggered by:
1. **low environmental humidity;**
2. **increasing leaf temperature;**

both of which are signals that the need to conserve water must override the need to allow CO_2 uptake. This response is triggered by the water content of leaf epidermal and mesophyll cells and it is a **feedback response.** (N.B. at very high temperatures stomata may open very wide to allow maximum leaf cooling by evapotranspiration from mesophyll cells.)
3. **Abscisic acid secretion:** severe drought stress is detected in epidermal cells which secrete ABA into the apoplast causing rapid and immediate stomatal closure (possibly by inhibiting the proton pump).

Guard cells

1. Change **shape** as their **degree of turgor** is altered. The reason for this is twofold:
   a. the inner guard cell wall has microfibrils orientated so that longitudinal expansion is not easy;
   b. the outer wall is less thickened with cellulose so that it elongates much more readily than the inner wall.

2. May alter their **solute potential,** and thus their **water potential,** by the movement of ions, principally potassium ions, K^+.

An ATP-dependent proton pump moves H^+ **out of** the guard cells, creating an electrochemical gradient (inside **negative** with respect to **outside**) so that K^+ can flow passively through K^+ channels down this electrochemical gradient.

K^+ movement **in** → reduced water potential → water movement **in.**

CHLOROPLAST – generator of ATP

THIN OUTER WALL

THICKENED CELLULOSE CELL WALL ON INNER SIDE

STOMATAL PORE

DENSE CYTOPLASM WITH STARCH GRAINS

Closed: guard cells are **flaccid** because water has been lost to the **apoplast system** of the epidermis.

Reason: [K^+] in guard cells has fallen – high water potential thus allows water to leave the guard cells.

Open: Guard cells are **turgid** because water has been gained from the **apoplast system** of the epidermis.

Reason: [K^+] in guard cells has risen – low water potential thus allows water to enter the guard cells.

Plant growth substances

NATURAL PLANT GROWTH HORMONES

Lateral bud development is inhibited by *auxin* but promoted by *cytokinin (antagonism)*.

Stomatal closure under stress may be promoted by *abscisic acid*.

Flowering may be triggered by *florigen*.

Root growth of adventitious roots is promoted by *auxin*.

Growth of stem: cell enlargement is promoted by *auxin* and *gibberellin*. Redistribution of *auxin* causes phototropism.

Seed dormancy is maintained by *abscisic acid* but is broken by *gibberellic acid*.

Leaf fall is promoted by *abscisic acid*.

SHOOTS ARE POSITIVELY PHOTOTROPIC

They show growth *movements towards light*.

Auxin produced at shoot tip diffuses and accumulates on shaded side

LIGHT

Auxin is inactivated on 'light' side of shoot

- **Mechanism:** the plant growth substance called *auxin* is inhibited by light. It therefore accumulates on the shaded side of shoots *causing faster growth on the shaded side* and thus *'bending' towards light*.

- **Benefit:** leafy shoots tilt towards light to improve rate of photosynthesis and therefore increase rate of growth.

ROOTS ARE POSITIVELY GEOTROPIC

– they show growth *movements towards gravity*.

- **Mechanism:** gravity causes large starch grains called *statoliths* to sink towards cells on lower side of root. This causes *auxin* to accumulate on lower side of root. Unlike shoot growth, root growth is *inhibited* by this concentration of auxin so that *lower side grows more slowly* and root *'bends' towards gravity*.

- **Benefit:** root grows into soil to anchor plant firmly and improve uptake of minerals and water.

Auxin produced at root tip is affected by gravity and accumulates on lower side of root.

N.B. Many commercial applications of these growth phenomena rely on *plant growth regulators*, which are synthetic derivatives of the natural compounds, but are usually more effective in lower concentrations because they are degraded less rapidly by the plant.

Ethene is used to accelerate ripening – ideal for grapes which can be picked earlier and thus have a longer drying period for forming raisins. Ripening can be delayed by keeping fruits in an oxygen-free atmosphere: ethene can then induce ripening as required.

Auxins are used as defoliants, e.g. during the Vietnam War to clear areas of vegetation and make bombing of bridges, roads, and troops easier. Also used to remove vegetation from overhead power lines – manual removal would be costly and dangerous.

Gibberellic acid may mimic red light: control of flowering time (promote long-day species, inhibit short-day species) means flowers can be available 'out of season'.

A mixture of *auxin*, *cytokinin* and *gibberellin* will inhibit apical growth and allow limited development of lateral buds. This mixture applied to hedges promotes dense, bushy growth and limits the need for mechanical trimming to one or two occasions per year.

Ethene sprayed onto day-neutral species such as pineapple can synchronize flowering/fruiting so that crop picking can be more efficient.

Auxin can prevent premature fruit drop (windfall losses) since it is antagonistic to *abscisic acid*.

Auxin can act as a selective lawn weed killer since broad leaved 'weed' species are killed by auxin concentrations which do not affect monocotyledons.

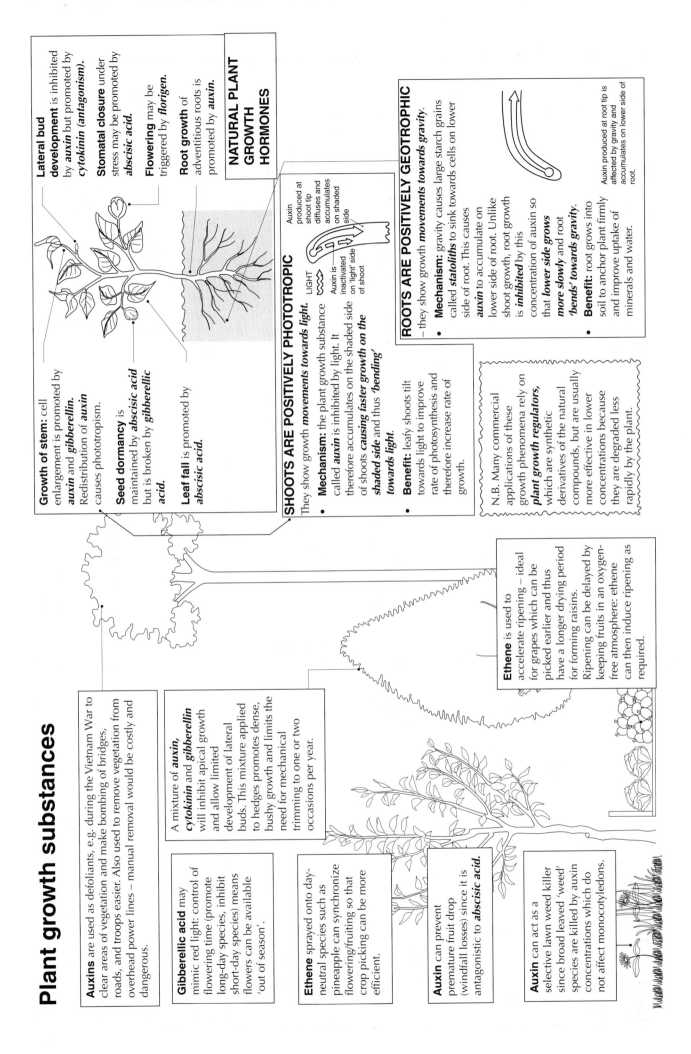

Structure of a typical flower

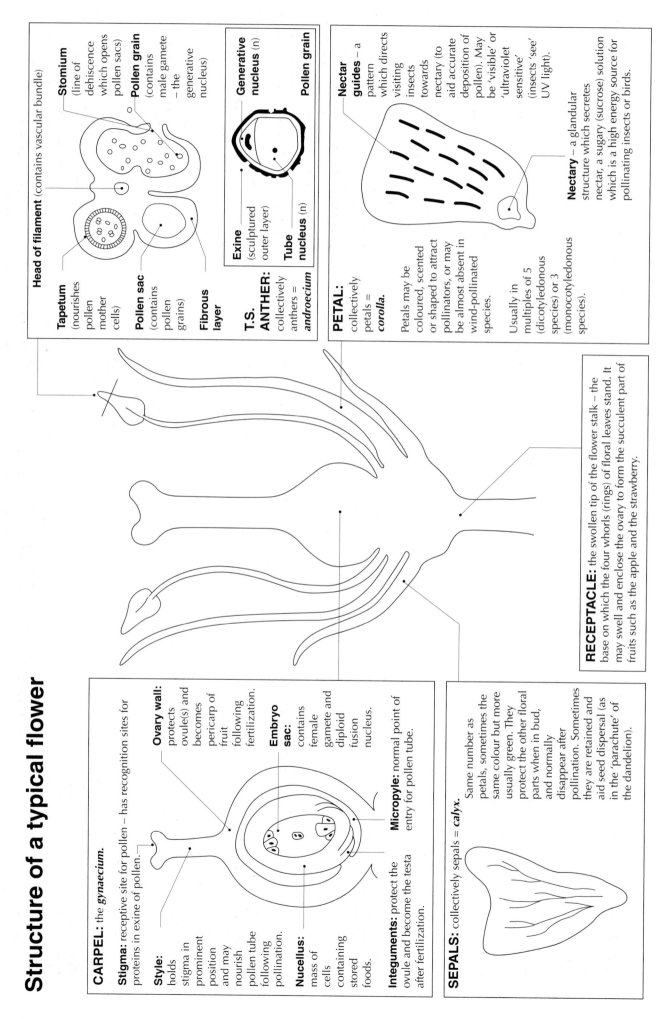

Head of filament (contains vascular bundle)

Stomium (line of dehiscence which opens pollen sacs)

Pollen grain (contains male gamete – the generative nucleus)

Tapetum (nourishes pollen mother cells)

Pollen sac (contains pollen grains)

Fibrous layer

T.S. ANTHER: collectively anthers = *androecium*

Generative nucleus (n)

Pollen grain

Exine (sculptured outer layer)

Tube nucleus (n)

Nectar guides – a pattern which directs visiting insects towards nectary (to aid accurate deposition of pollen). May be 'visible' or 'ultraviolet sensitive' (insects 'see' UV light).

Nectary – a glandular structure which secretes nectar, a sugary (sucrose) solution which is a high energy source for pollinating insects or birds.

PETAL: collectively petals = *corolla*.

Petals may be coloured, scented or shaped to attract pollinators, or may be almost absent in wind-pollinated species.

Usually in multiples of 5 (dicotyledonous species) or 3 (monocotyledonous species).

CARPEL: the *gynaecium.*

Stigma: receptive site for pollen – has recognition sites for proteins in exine of pollen.

Style: holds stigma in prominent position and may nourish pollen tube following pollination.

Ovary wall: protects ovule(s) and becomes pericarp of fruit following fertilization.

Embryo sac: contains female gamete and diploid fusion nucleus.

Micropyle: normal point of entry for pollen tube.

Integuments: protect the ovule and become the testa after fertilization.

Nucellus: mass of cells containing stored foods.

SEPALS: collectively sepals = *calyx.*

Same number as petals, sometimes the same colour but more usually green. They protect the other floral parts when in bud, and normally disappear after pollination. Sometimes they are retained and aid seed dispersal (as in the 'parachute' of the dandelion).

RECEPTACLE: the swollen tip of the flower stalk – the base on which the four whorls (rings) of floral leaves stand. It may swell and enclose the ovary to form the succulent part of fruits such as the apple and the strawberry.

Wind-pollinated (anemophilous) flowers

are typically grasses or forest tree species which occur in dense groups covering very large areas.

Bract: leaf-like structure which encloses and protects floral structures.

Filaments are long and flexible so that anthers may be held out in an exposed position.

Anthers are versatile – hinged at the mid-point – so that pollen is readily shaken out of flower by the wind.

Pollen is produced in enormous quantities since transfer by wind is very wasteful. The pollen is smooth, with wing-like extensions, light and small to promote transfer by wind.

Stigmas are long and feathery giving a large surface area to receive pollen, and often protrude outside the flower into the pollen-bearing atmosphere.

Petals are dull in colour and much reduced in size, since they need not attract insects and must not obstruct pollen access to stigma. Together with the very small **sepals** the petals form the **lodicule** which swells to open the flower by forcing apart two **bracts.**

Neither nectar nor scent is produced: the production of these compounds would be biochemically 'expensive' and would offer no advantage since there is no need to attract pollinating insects.

Receptacle

Ovary contains a single ovule.

Insect-pollinated (entomophilous) flowers

typically belong to species which are solitary or exist in small groups.

Filaments are stiff to resist buffeting by pollinating insects, short enough to keep anther within the corolla and may be hinged to aid deposition of pollen on body of visiting insect.

Anthers are held within corolla and may have a sticky surface to hold pollen ready for visiting insect.

Pollen is large with sticky projections to adhere to body of insect. Less pollen is produced as transfer by insect is very efficient.

Stigma is held within the corolla in a position which ensures contact with body of visiting insect. Stigma surface is relatively small, since insect transfer is accurate, and often the style is lignified/stiffened to avoid damage by pollinating insect.

Petals are brightly coloured and large. They must attract pollinating insects and often have markings or hairs to act as *nectar guides* (these may be visible in the ultraviolet light to which insects are sensitive). Petals may be shaped to attract pollinators (by mimicry of female insects in some orchids) or may be reinforced to act as *landing platforms* for pollinators or as *tunnels* to direct insects towards the reproductive parts.

Scent and/or nectar may be produced to attract or reward visiting insects. Scents are often derivatives of fatty acids and nectar is a dilute sugar (sucrose/fructose) solution with a high energy value.

Receptacle

Ovary contains ovules.

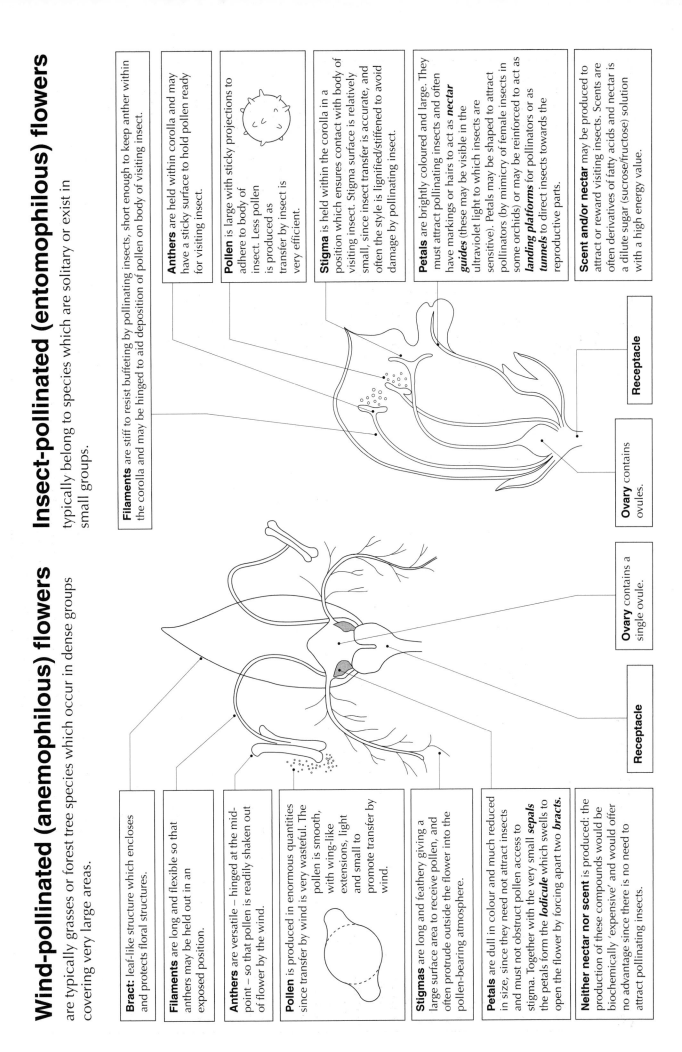

The seed is a fertilised ovule

1. It has the advantages of genetic variation.
2. It can remain dormant and survive adverse conditions.
3. It contains a food store for the developing embryo.

Flower: the most obvious of all the adaptations of the flowering plants. Four whorls (rings) of leaves adapted to produce spores and allow fusion of their products. Cells are *diploid* (2n).

MEIOSIS IN OVULE OF OVARY

Megaspores (unripe embryo sacs) which are *haploid* (n).

Megagametophyte is the *ripe embryosac* which is *haploid* (n).

MEIOSIS IN POLLEN SAC OF ANTHER

Microspores (unripe pollen grains) which are *haploid* (n).

Mature sporophyte: typical flowering plant. All cells are *diploid* (2n).

GERMINATION, GROWTH AND DEVELOPMENT

Zygote (2n) undergoes *mitosis* to produce *embryo*.

LIFE CYCLE OF A FLOWERING PLANT
involves *meiosis* to produce *spores* and a unique *double fertilisation.*

Male gamete is the *generative nucleus* (n) within the pollen grain.

Microgametophyte (lit. 'small gamete producing plant') is the *ripe pollen grain* which is *haploid* (n).

Diploid fusion nucleus will become *endosperm* following fertilization.

Female gamete is the *oosphere* (n) contained within the embryo sac awaiting fertilisation.

Embryo (2n) contained within *testa* is the *seed.*

Unique *double fertilisation* produces *diploid zygote* (2n) and *triploid endosperm nucleus* (3n).

Endosperm contains *insoluble food stores* (particularly starch in cereals) which are hydrolysed by enzymes:

STARCH $\xrightarrow{\text{Amylase}}$ MALTOSE $\xrightarrow{\text{Maltase}}$ GLUCOSE

PROTEINS $\xrightarrow{\text{Protease}}$ AMINO ACIDS

LIPIDS $\xrightarrow{\text{Lipase}}$ FATTY ACIDS + GLYCEROL

Aleurone layer is three cells thick in cereals: contains protein – may be inactive enzymes or may be source of amino acids for synthesis of enzymes under influence of GA.

Hydrolytic enzymes are released from *aleurone layer* (in cereals) or from *lysosomes* (in other seeds).

Embryo: manufactures and secretes *gibberellic acid*. Other hormones, notably *cytokinin* and *indole acetic acid* promote cell division at the growing apices of the embryo.

Scutellum: modified cotyledon which functions as an *absorptive organ:* transfer of the soluble products of hydrolysis of food store in endosperm to developing embryo.

Gibberellic acid is secreted from the embryo to the aleurone layer where it promotes the mobilisation of amino acids and the condensation of these amino acids into *hydrolytic enzymes* such as α-amylase and a *protease.*

SUGARS
AMINO ACIDS
FATTY ACIDS
IONS

Oxygen is required for *aerobic respiration.* The transport of nutrients and processes of cell division are very energy demanding - aerobic respiration represents the most efficient energy release from oxidizable food stores. Measurement of *respiratory quotient* (=CO_2 released/O_2 consumed) may indicate which food is being used as a respiratory substrate.

Aerobic carbohydrate respiration   RQ = 1.0

Aerobic fat respiration   RQ = 0.7

Fat carbohydrate conversion (in lipid-rich seeds)   RQ = 0.4

SEED GERMINATION
is regulated by both *internal* and *environmental factors.*

Water is required as a reagent in *hydrolysis of food stores,* in the *mobilisation of enzymes* and in the *transport of the products of hydrolysis.* It also causes the *rupture of the testa* by the swelling of colloidal substances such as *proteins* and *cell wall materials* as water is absorbed by imbibition through the micropyle and the testa.

Temperature affects germination because of its influence on the rate of enzyme-catalysed reactions, so that the optimum temperature for germination is commonly in the range 20-40 °C. Many seeds require a period of low temperature - *stratification* - before germination will begin. The low temperature may *increase permeability of the testa* or may reduce the concentration of *inhibitory compounds* such as *abscisic acid.*

Light may be necessary to *break seed dormancy.* Light of the appropriate wavelength is absorbed by *phytochrome* and raises the *gibberellic acid: abscisic acid* ratio.

Plant life cycles

show alternation of generations.

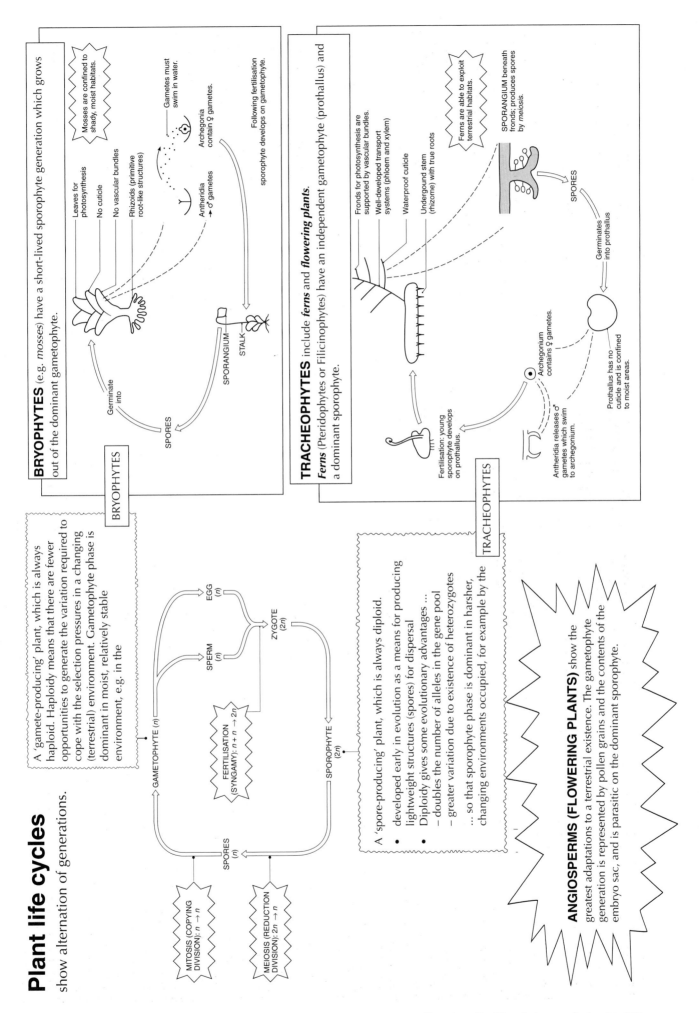

BRYOPHYTES (e.g. *mosses*) have a short-lived sporophyte generation which grows out of the dominant gametophyte.

Mosses are confined to shady, moist habitats.

Gametes must swim in water.

Archegonia contain ♀ gametes.

Antheridia → ♂ gametes

Leaves for photosynthesis

No cuticle

No vascular bundles

Rhizoids (primitive root-like structures)

Following fertilisation sporophyte develops on gametophyte.

SPORANGIUM

STALK

Germinate into

SPORES

BRYOPHYTES

TRACHEOPHYTES include *ferns* and *flowering plants*.

Ferns (Pteridophytes or Filicinophytes) have an independent gametophyte (prothallus) and a dominant sporophyte.

Fronds for photosynthesis are supported by vascular bundles.

Well-developed transport systems (phloem and xylem)

Waterproof cuticle

Underground stem (rhizome) with true roots

Ferns are able to exploit terrestrial habitats.

SPORANGIUM beneath fronds; produces spores by *meiosis*.

SPORES

Germinates into prothallus

Archegonium contains ♀ gametes.

Prothallus has no cuticle and is confined to moist areas.

Antheridia releases ♂ gametes which swim to archegonium.

Fertilisation: young sporophyte develops on prothallus.

TRACHEOPHYTES

A 'gamete-producing' plant, which is always haploid. Haploidy means that there are fewer opportunities to generate the variation required to cope with the selection pressures in a changing (terrestrial) environment. Gametophyte phase is dominant in moist, relatively stable environment, e.g. in the

GAMETOPHYTE (n)

EGG (n)

SPERM (n)

FERTILISATION (SYNGAMY): $n + n \rightarrow 2n$

ZYGOTE ($2n$)

SPOROPHYTE ($2n$)

SPORES (n)

MITOSIS (COPYING DIVISION): $n \rightarrow n$

MEIOSIS (REDUCTION DIVISION): $2n \rightarrow n$

A 'spore-producing' plant, which is always diploid.

• developed early in evolution as a means for producing lightweight structures (spores) for dispersal
• Diploidy gives some evolutionary advantages ...
  – doubles the number of alleles in the gene pool
  – greater variation due to existence of heterozygotes
  ... so that sporophyte phase is dominant in harsher, changing environments occupied, for example by the

ANGIOSPERMS (FLOWERING PLANTS) show the greatest adaptations to a terrestrial existence. The gametophyte generation is represented by pollen grains and the contents of the embryo sac, and is parasitic on the dominant sporophyte.

Angiosperm adaptations
- more than 80% of all plant species are angiosperms (i.e. plants with enclosed seeds).

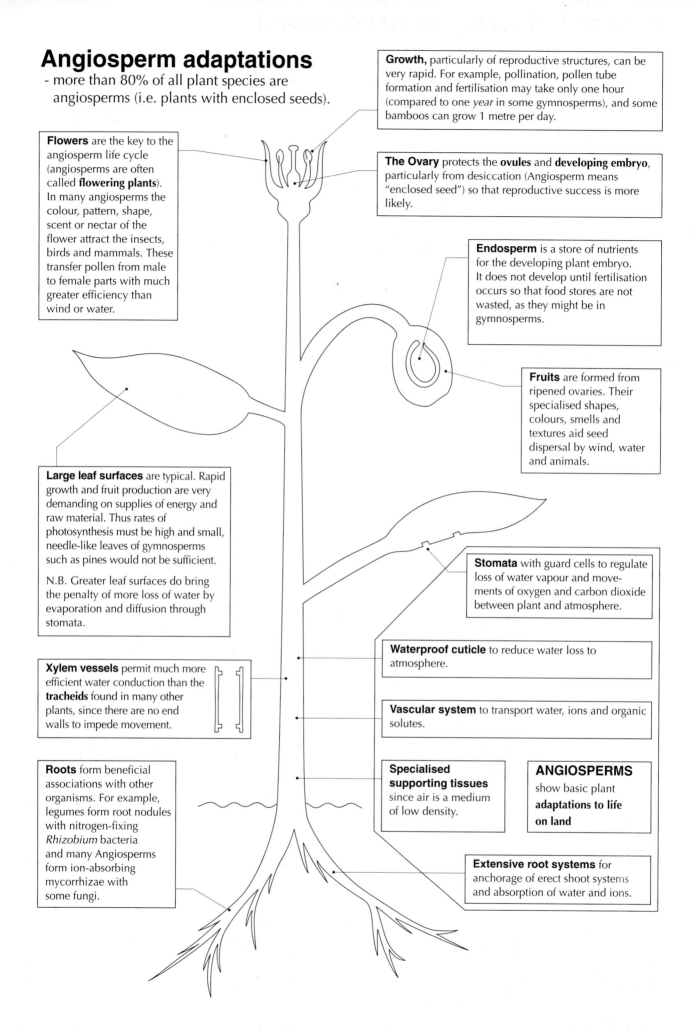

Growth, particularly of reproductive structures, can be very rapid. For example, pollination, pollen tube formation and fertilisation may take only one hour (compared to one *year* in some gymnosperms), and some bamboos can grow 1 metre per day.

Flowers are the key to the angiosperm life cycle (angiosperms are often called **flowering plants**). In many angiosperms the colour, pattern, shape, scent or nectar of the flower attract the insects, birds and mammals. These transfer pollen from male to female parts with much greater efficiency than wind or water.

The Ovary protects the **ovules** and **developing embryo,** particularly from desiccation (Angiosperm means "enclosed seed") so that reproductive success is more likely.

Endosperm is a store of nutrients for the developing plant embryo. It does not develop until fertilisation occurs so that food stores are not wasted, as they might be in gymnosperms.

Fruits are formed from ripened ovaries. Their specialised shapes, colours, smells and textures aid seed dispersal by wind, water and animals.

Large leaf surfaces are typical. Rapid growth and fruit production are very demanding on supplies of energy and raw material. Thus rates of photosynthesis must be high and small, needle-like leaves of gymnosperms such as pines would not be sufficient.

N.B. Greater leaf surfaces do bring the penalty of more loss of water by evaporation and diffusion through stomata.

Stomata with guard cells to regulate loss of water vapour and move-ments of oxygen and carbon dioxide between plant and atmosphere.

Waterproof cuticle to reduce water loss to atmosphere.

Xylem vessels permit much more efficient water conduction than the **tracheids** found in many other plants, since there are no end walls to impede movement.

Vascular system to transport water, ions and organic solutes.

Roots form beneficial associations with other organisms. For example, legumes form root nodules with nitrogen-fixing *Rhizobium* bacteria and many Angiosperms form ion-absorbing mycorrhizae with some fungi.

Specialised supporting tissues since air is a medium of low density.

ANGIOSPERMS show basic plant **adaptations to life on land**

Extensive root systems for anchorage of erect shoot systems and absorption of water and ions.

Section E Ecology and environment

Ecology is the 'study of living organisms in relation to their environment'.

A more recent definition is 'the scientific study of the interactions that determine the distribution and abundance of organisms'.

Synecology is the study of groups of organisms associated to form a functional unit of the environment.

Two useful terms are

Community (biotic community): all of the populations occupying a given, defined physical area, e.g. all the organisms within a rock pool.

Ecosystem: the biotic community together with the physical (non-living or abiotic) environment, e.g. a rock pool.

Autecology is the study of *single organisms* or *populations of single species* and their relationship to their environment, e.g. Common limpet (*Patella vulgaris*) – an animal of the rocky shore.

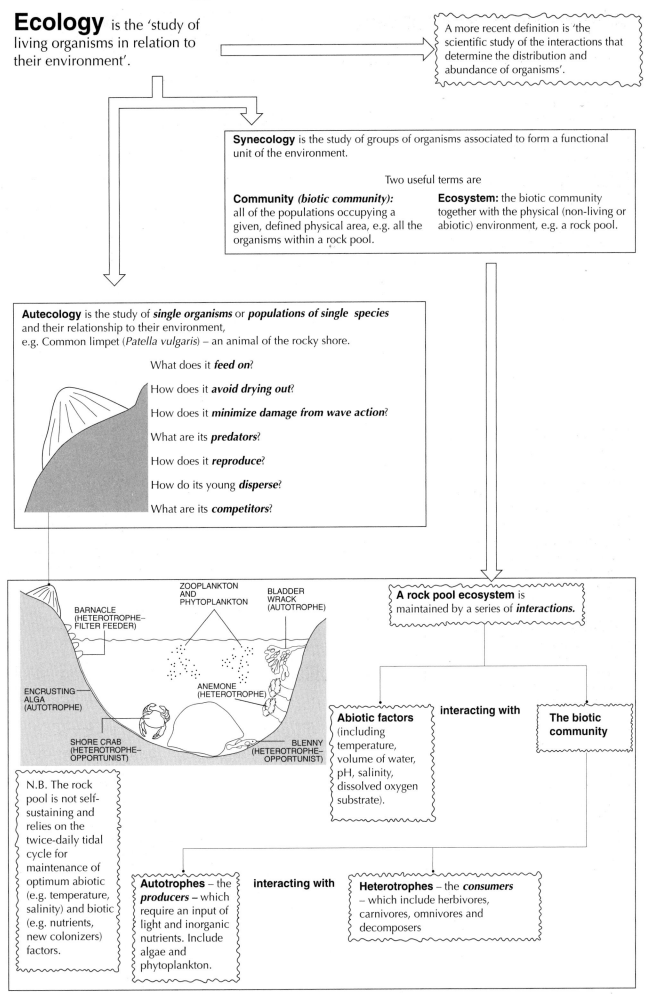

What does it *feed on*?

How does it *avoid drying out*?

How does it *minimize damage from wave action*?

What are its *predators*?

How does it *reproduce*?

How do its young *disperse*?

What are its *competitors*?

ZOOPLANKTON AND PHYTOPLANKTON

BLADDER WRACK (AUTOTROPHE)

BARNACLE (HETEROTROPHE– FILTER FEEDER)

ENCRUSTING ALGA (AUTOTROPHE)

ANEMONE (HETEROTROPHE)

SHORE CRAB (HETEROTROPHE– OPPORTUNIST)

BLENNY (HETEROTROPHE– OPPORTUNIST)

A rock pool ecosystem is maintained by a series of *interactions.*

N.B. The rock pool is not self-sustaining and relies on the twice-daily tidal cycle for maintenance of optimum abiotic (e.g. temperature, salinity) and biotic (e.g. nutrients, new colonizers) factors.

Abiotic factors (including temperature, volume of water, pH, salinity, dissolved oxygen substrate).

interacting with

The biotic community

Autotrophes – the *producers* – which require an input of light and inorganic nutrients. Include algae and phytoplankton.

interacting with

Heterotrophes – the *consumers* – which include herbivores, carnivores, omnivores and decomposers

Estimating populations

To study the dynamics of a population, or how the distribution of the members of a population is influenced by a biotic or an abiotic factor, it is necessary to estimate the population size. In other words, it will be necessary to count the number of individuals in a population. Such counting is usually carried out by taking **samples** (in which the organisms are in the same proportion as in the whole population) because:

1. counting the whole population would be extremely laborious and time-consuming
2. counting the whole population might cause unacceptable levels of damage to the habitat, or to the population being studied.

The samples must be representative. They should be:

1. of the same size (e.g. a 0.25 m² area of grassland)
2. randomly selected – for example, samples may be taken at predetermined points on an imaginary grid laid over the sampling area. The coordinates of the points may be selected using random numbers generated by a calculator
3. non-overlapping.

QUADRAT SAMPLING

Quadrats are sampling units of a known area. They are most often square, and are usually constructed of wood or metal. The quadrat can be used in simple form, or it may have wire subdivisions to produce a number of sampling points.

Reliable sampling with quadrats requires answers to three questions:

1. What size of quadrat should be used?
2. How many quadrats should be used?
3. Where should the quadrats be positioned?

WHAT SIZE QUADRAT?

If individuals within a population are truly randomly dispersed, then any quadrat size should be equally efficient in the estimation of that population. However, environmental factors are rarely evenly distributed so that the living organisms dependent on them tend to occur in an aggregated distribution. Small quadrats are more efficient in estimating populations (more can be taken, and they can cover a wider range of habitat than larger ones) but there are practical considerations to be taken into account (a small quadrat might not include a dominant tree in a woodland). Optimum quadrat size is determined by counting the number of different species present in quadrats of increasing size.

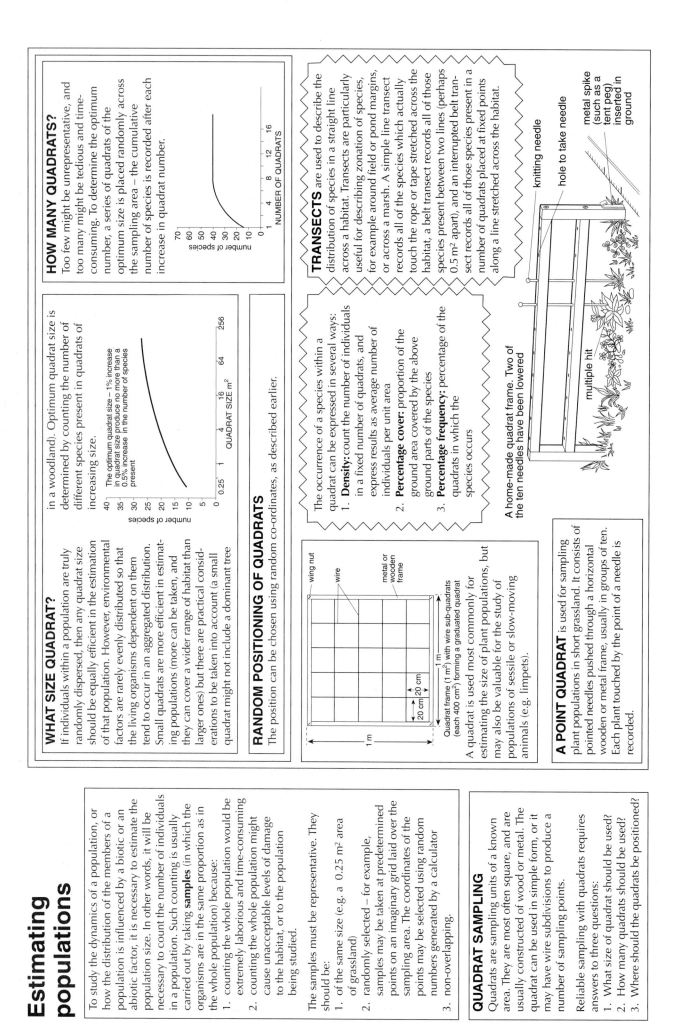

The optimum quadrat size – 1% increase in quadrat size produce no more than a 0.5% increase in the number of species present

HOW MANY QUADRATS?

Too few might be unrepresentative, and too many might be tedious and time-consuming. To determine the optimum number, a series of quadrats of the optimum size is placed randomly across the sampling area – the cumulative number of species is recorded after each increase in quadrat number.

RANDOM POSITIONING OF QUADRATS

The position can be chosen using random co-ordinates, as described earlier.

Quadrat frame (1 m²) with wire sub-quadrats (each 400 cm²) forming a graduated quadrat

A quadrat is used most commonly for estimating the size of plant populations, but may also be valuable for the study of populations of sessile or slow-moving animals (e.g. limpets).

The occurrence of a species within a quadrat can be expressed in several ways:

1. **Density:** count the number of individuals in a fixed number of quadrats, and express results as average number of individuals per unit area
2. **Percentage cover:** proportion of the ground area covered by the above ground parts of the species
3. **Percentage frequency:** percentage of the quadrats in which the species occurs

TRANSECTS

are used to describe the distribution of species in a straight line across a habitat. Transects are particularly useful for describing zonation of species, for example around field or pond margins, or across a marsh. A simple line transect records all of the species which actually touch the rope or tape stretched across the habitat, a belt transect records all of those species present between two lines (perhaps 0.5 m² apart), and an interrupted belt transect records all of those species present in a number of quadrats placed at fixed points along a line stretched across the habitat.

A home-made quadrat frame. Two of the ten needles have been lowered

A POINT QUADRAT

is used for sampling plant populations in short grassland. It consists of pointed needles pushed through a horizontal wooden or metal frame, usually in groups of ten. Each plant touched by the point of a needle is recorded.

Sampling motile species

Quadrats and line transects are ideal methods for estimating populations of plants or sedentary animals. Motile animals, however, must be captured before their populations can be estimated. Once more, a representative sample of the population will be counted and the total population estimated from the sample. One important technique is the **mark-recapture** (also known as mark-release-recapture) **method.** This method involves:

1. capturing the organism

2. marking in some way which causes no harm (e.g. beetles can be marked with a drop of waterproof paint on their wing cases, and mice may have a small mark clipped into their fur)

3. releasing the organism to rejoin its population

4. a second sample group from this population is captured and counted at a later date

5. the population size is estimated using the Lincoln Index:

Population size = $\dfrac{n_1 \times n_2}{n_m}$

where n_1 is the number of individuals marked and released ('1' because it was the first sample), n_2 is the number of individuals caught the second time round ('2' because it was the second sample) and n_m is the number of marked individuals in the second sample ('m' standing for marked).

This method depends on a number of assumptions – failure of any of these to hold up can lead to poor estimates being made. These assumptions are

a. the marked organisms mix randomly back into the normal population (allow sufficient time for this to occur, bearing in mind the mobility of the species)

b. the marked animals are no different to the unmarked ones – they are no more prone to predation, for example

c. changes in population size due to births, deaths, immigration and emigration are negligible

d. the mark does not wear off or grow out during the sampling period.

TULLGREN FUNNEL:
used to collect small organisms from the air spaces of the soil or from leaf litter. The lamp is a source of heat and dehydration – organisms move to escape from it and fall through the sieve (the mesh is fine enough to retain the soil or litter). The animals slip down the smooth-sided funnel and are immobilised in the alcohol. They may then be removed for identification.

- 25 watt bulb
- 16 mesh flour sieve
- polythene funnel
- soil sample
- 80% alcohol

With both Tullgren and Baermann funnels it is essential that samples are treated in identical fashion if results are to be comparative – for example, use fixed sample size, length of exposure to heat source and wattage of lamp.

BAERMANN FUNNEL:
works on a similar principle to the Tullgren funnel, but extracts organisms living in the soil water. The heat source drives the animals out of the muslin bag and into the surrounding water. Examples of water can be released at intervals, and the organisms in the sample collected and identified.

- 60 watt bulb
- soil sample in muslin bag
- rubber tubing
- clip
- beaker
- glass rod for supporting bag
- water
- glass funnel

POOTER:
used to collect specimens of insects and other arthropods which have been extracted from trees or bushes by beating the vegetation over a sheet or tray. Collection in the pooter does not harm the organism, and it can then be returned to its natural habitat.

- clear plastic tube
- cork or rubber bung
- gauze covering tube opening
- glass mouthpiece
- specimen tube
- glass collecting tube

PITFALL TRAPS:
used to sample arthropods moving over the soil surface.

The roof prevents rainfall from flooding the trap, and also limits access to certain predators. The activities of trapped predators can be prevented by adding a small quantity of methanol to the trap. Bait of meat or ripe fruit can be placed in the trap.

Pitfall traps are often set up on a grid system to investigate the movements of ground animals more systematically.

- flat stone
- stick support
- ground slopes away from trap for drainage
- bait
- jam jar sunk into soil

Other methods of collection are numerous. Many are based on some form of netting – for example large mist nets may be used to collect migrating birds for identification and ringing, and sweep nets may be used to capture aerial or aquatic arthropods.

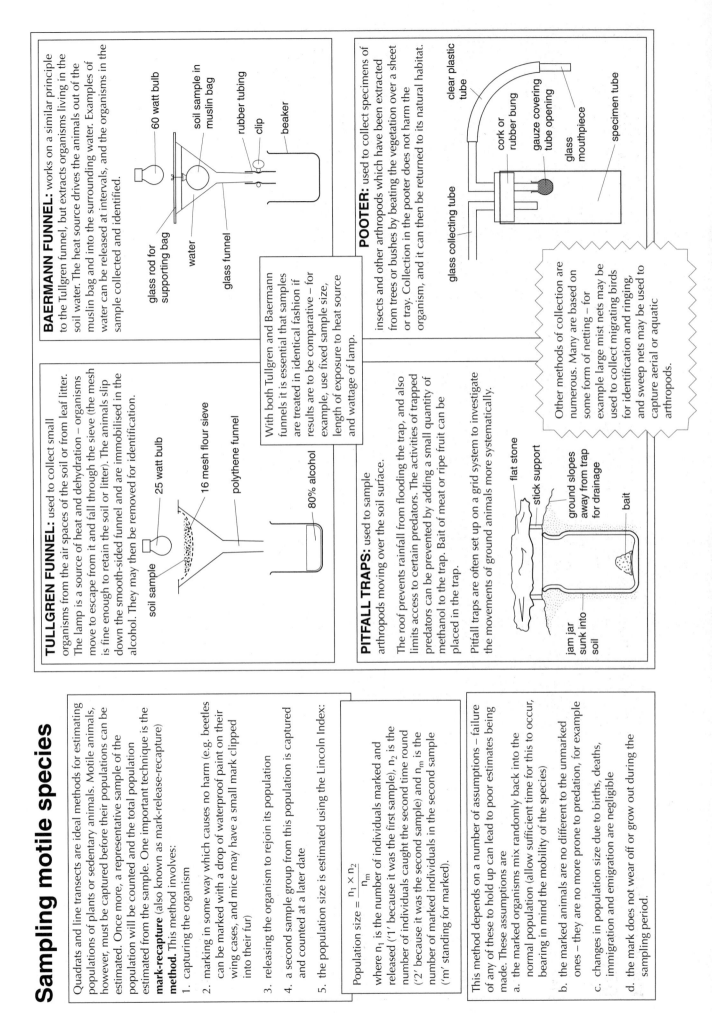

Factors affecting population growth

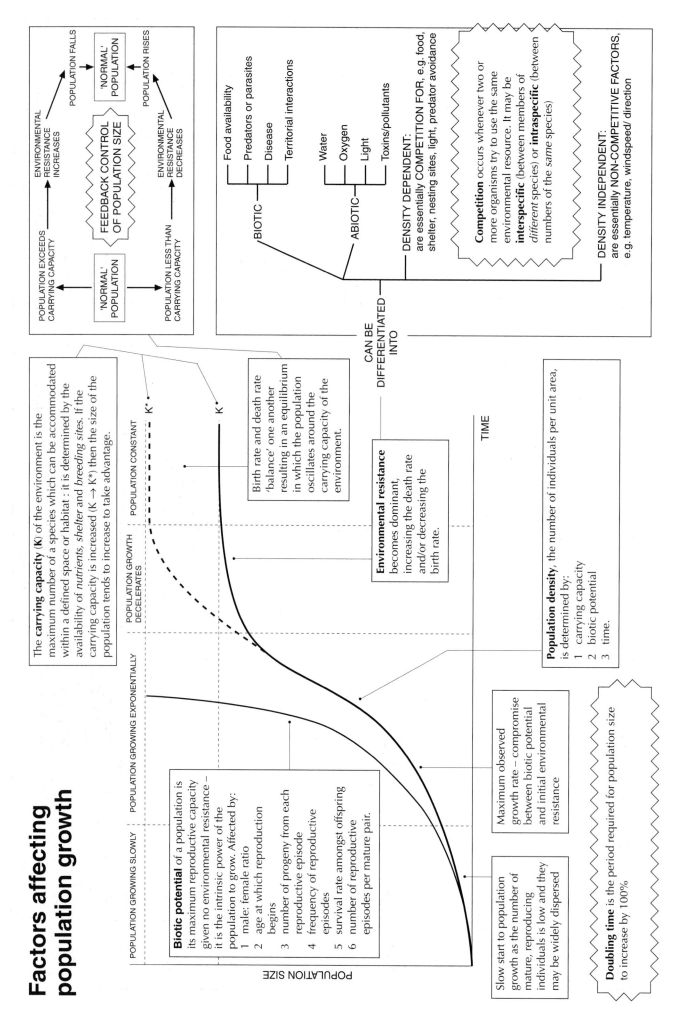

Energy flow through an ecosystem I

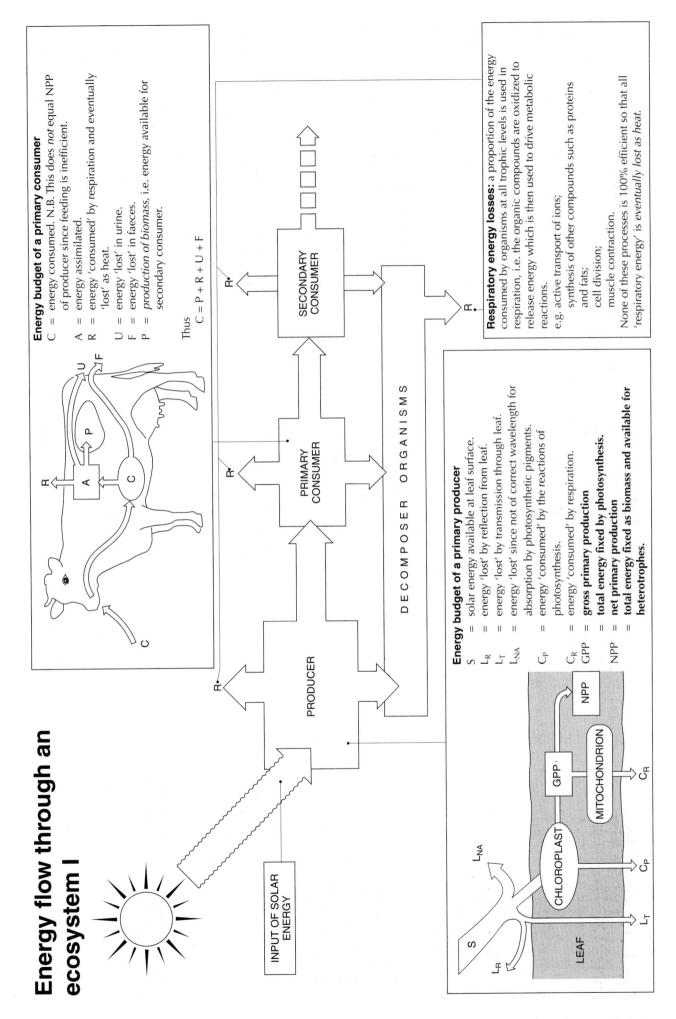

Energy budget of a primary consumer

C = energy consumed. N.B. This does *not* equal NPP of producer since feeding is inefficient.

A = energy assimilated.

R = energy 'consumed' by respiration and eventually 'lost' as heat.

U = energy 'lost' in urine.

F = energy 'lost' in faeces.

P = *production of biomass*, i.e. energy available for secondary consumer.

Thus

$$C = P + R + U + F$$

Respiratory energy losses: a proportion of the energy consumed by organisms at all trophic levels is used in respiration, i.e. the organic compounds are oxidized to release energy which is then used to drive metabolic reactions.

e.g. active transport of ions;
synthesis of other compounds such as proteins and fats;
cell division;
muscle contraction.

None of these processes is 100% efficient so that all 'respiratory energy' is *eventually lost as heat*.

Energy budget of a primary producer

S = solar energy available at leaf surface.

L_R = energy 'lost' by reflection from leaf.

L_T = energy 'lost' by transmission through leaf.

L_{NA} = energy 'lost' since not of correct wavelength for absorption by photosynthetic pigments.

C_P = energy 'consumed' by the reactions of photosynthesis.

C_R = energy 'consumed' by respiration.

GPP = **gross primary production**
= **total energy fixed by photosynthesis.**

NPP = **net primary production**
= **total energy fixed as biomass and available for heterotrophes.**

INPUT OF SOLAR ENERGY

PRODUCER

PRIMARY CONSUMER

SECONDARY CONSUMER

D E C O M P O S E R   O R G A N I S M S

LEAF

S

L_{NA}

L_R

L_T

CHLOROPLAST

GPP

NPP

C_P

MITOCHONDRION

C_R

Energy flow through an ecosystem II

A closed ecosystem is rare: migratory animals may deposit faeces, fruits and seeds may enter or leave during dispersal, and leaves may blow in from surrounding trees.

The **gross primary production (GPP)** – the total energy fixed by photosynthesis – represents only about 0.5–1% of the light energy available to the leaf.

At the equator, the **solar flux** (sunlight which reaches the Earth's upper atmosphere) is almost constant at 1.4 kJ m^{-2} s^{-1}. Most of this incoming sunlight energy is reflected by the atmosphere, heats the atmosphere and Earth's surface or causes the evaporation of water. Less than 0.1% actually falls on leaves and is thus available for photosynthesis.

The **net primary production (NPP)** is the energy available for consumption by the heterotrophes. NPP can therefore be used to compare the productivity of different ecosystems. For example:

ECOSYSTEM	NPP (arbitrary units)
Coral reef	1000
Rainforest	880
Estuaries	600
Deciduous forest	500
Grassland	260
Open ocean	50
Desert	2

Productivity may be expressed as **units of energy** (e.g. kJ m^{-2} yr^{-1}) or **units of mass** (e.g. kg m^{-2} yr^{-1}).

Energy transfer from producer to primary consumer is typically in the order of 5–10% of NPP. This is because

1. Much of plant biomass (NPP) is indigestible to herbivores – there are no animal enzymes to digest lignin and cellulose.
2. Much of the plant biomass may not be consumed by any individual herbivore species – roots may be inaccessible or trampled grass may be considered uneatable

Energy transfer from primary consumer (herbivore) to secondary consumer (carnivore) is typically 10–20% of herbivore biomass. This is more efficient than producer → consumer biomass because

1. animal tissue is more digestible than plant tissue;
2. animal tissue has a higher energy value;
3. carnivores may be extremely specialized for prey consumption;

but is still considerably less than 100% because
a. some animal tissue – bone, hooves and hide for example – is not readily digestible;
b. feeding is not 100% efficient – much digestible material (e.g. food fragments and blood) may be lost to the environment.

PRODUCER

PRIMARY CONSUMER

SECONDARY CONSUMER

D E C O M P O S E R   O R G A N I S M S

The **decomposers** are fungi and bacteria which obtain energy and raw materials from animal and plant remains. In some situations 80% or more of the productivity at any trophic level may go through a decomposer pathway (e.g. forest floors of tropical forests). In some ecosystems – peat bogs, for example – the cold, wet, acidic conditions inhibit decomposition to such an extent that only about 10% of the material entering the decomposer food chain is broken down. The remainder accumulates as peat.

The limit to the number of trophic levels is determined by:
1. the total producer biomass;
2. the efficiency of energy transfer between trophic levels (only 10%).

In practice, the energy losses limit the number of levels to 3 or 4, very rarely 5 or 6. The longest food chains can only be supported by an enormous producer biomass, e.g. a 6 level chain will only have about 10% x 10% x 10% x 10% of NPP available to the top carnivores. The enormous volume of the oceans can provide sufficient biomass to support the longest food chains.

Pyramid of energy, which represents the flow of energy through each trophic level of an ecosystem *during a fixed time period* (usually one year, to account for seasonal effects). The energy values are expressed in kJ m^{-2} yr^{-1}.

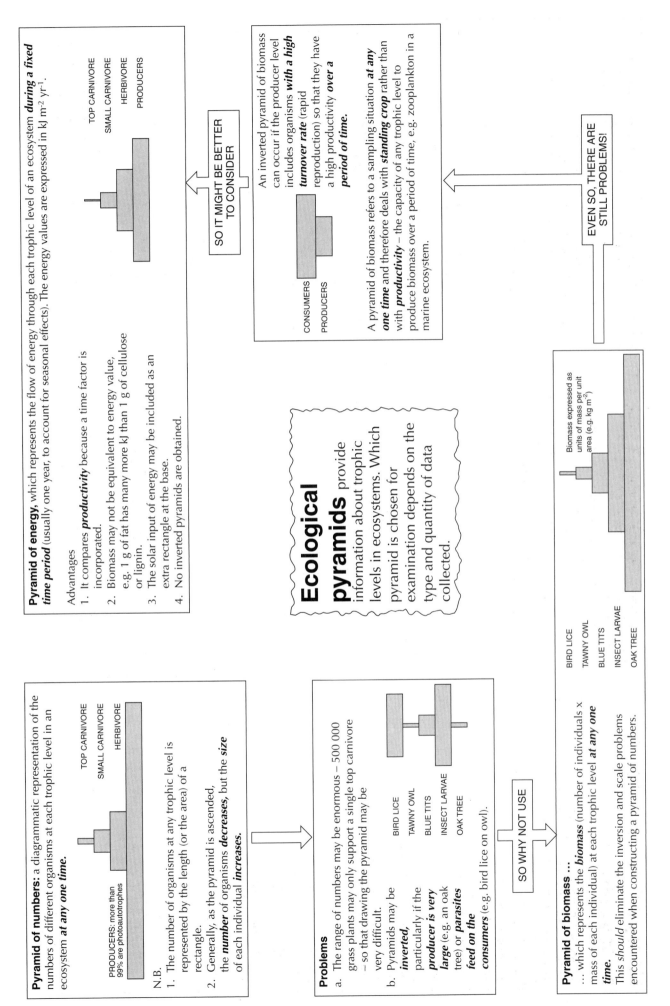

TOP CARNIVORE
SMALL CARNIVORE
HERBIVORE
PRODUCERS

Advantages
1. It compares *productivity* because a time factor is incorporated.
2. Biomass may not be equivalent to energy value, e.g. 1 g of fat has many more kJ than 1 g of cellulose or lignin.
3. The solar input of energy may be included as an extra rectangle at the base.
4. No inverted pyramids are obtained.

SO IT MIGHT BE BETTER TO CONSIDER

An inverted pyramid of biomass can occur if the producer level includes organisms *with a high turnover rate* (rapid reproduction) so that they have a high productivity *over a period of time.*

CONSUMERS
PRODUCERS

A pyramid of biomass refers to a sampling situation *at any one time* and therefore deals with **standing crop** rather than with *productivity* – the capacity of any trophic level to produce biomass over a period of time, e.g. zooplankton in a marine ecosystem.

EVEN SO, THERE ARE STILL PROBLEMS!

Biomass expressed as units of mass per unit area (e.g. kg m^{-2})

BIRD LICE
TAWNY OWL
BLUE TITS
INSECT LARVAE
OAK TREE

Ecological pyramids

provide information about trophic levels in ecosystems. Which pyramid is chosen for examination depends on the type and quantity of data collected.

Pyramid of numbers: a diagrammatic representation of the numbers of different organisms at each trophic level in an ecosystem *at any one time.*

TOP CARNIVORE
SMALL CARNIVORE
HERBIVORE
PRODUCERS: more than 99% are photoautotrophs

N.B.
1. The number of organisms at any trophic level is represented by the length (or the area) of a rectangle.
2. Generally, as the pyramid is ascended, the *number* of organisms *decreases,* but the *size* of each individual *increases.*

Problems
a. The range of numbers may be enormous – 500 000 grass plants may only support a single top carnivore – so that drawing the pyramid may be very difficult.
b. Pyramids may be *inverted,* particularly if the *producer is very large* (e.g. an oak tree) or *parasites feed on the consumers* (e.g. bird lice on owl).

BIRD LICE
TAWNY OWL
BLUE TITS
INSECT LARVAE
OAK TREE

SO WHY NOT USE

Pyramid of biomass ...
... which represents the *biomass* (number of individuals × mass of each individual) at each trophic level *at any one time.*
This *should* eliminate the inversion and scale problems encountered when constructing a pyramid of numbers.

Ecological succession proceeds via *several* *stages* to a *climax community* and is characterized by:

1. an increase in *species diversity* and in *complexity of feeding relationships*;
2. a progressive increase in *biomass*;
3. completion when *energy input (community photosynthesis) = energy loss (community respiration)*.

A COMMUNITY (all the species present in a given locality at any given time), is the group of interacting populations (all the members of a species in a place at a given time) which represents the biotic component of an ecosystem, and is seldom static. The relative abundance of different species may change, new species may enter the community and others may leave. There are reasons for these changes.

Catastrophes: may be natural (e.g. flooding, volcanic eruption) or caused by people (e.g. oil spill, deforestation).

Seasons: changes in temperature, rainfall, light intensity and windspeed, for example, may alter the suitability of a habitat for particular species.

Succession: long-term directional change in the composition of a community *brought about by the actions of the organisms themselves.*

Primary succession occurs when the community develops on bare, uncolonized ground *which has never had any vegetation growing on it ,* e.g. mud in river deltas, lava flows, sand dunes, artificial ponds and newly erupted volcanic islands.

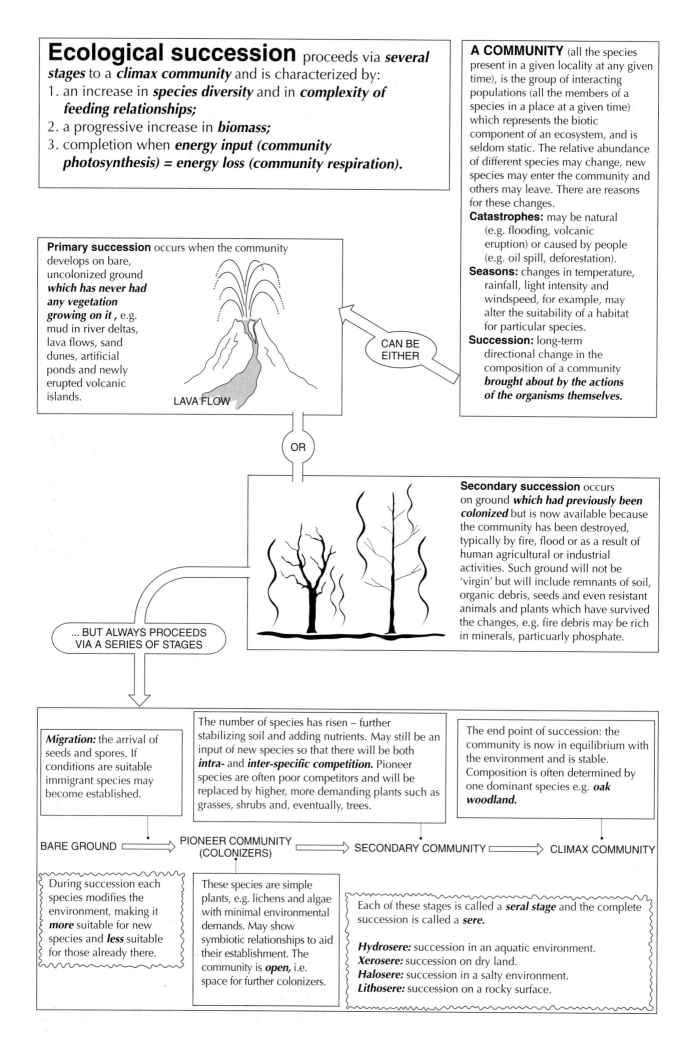

LAVA FLOW

CAN BE EITHER

OR

Secondary succession occurs on ground *which had previously been colonized* but is now available because the community has been destroyed, typically by fire, flood or as a result of human agricultural or industrial activities. Such ground will not be 'virgin' but will include remnants of soil, organic debris, seeds and even resistant animals and plants which have survived the changes, e.g. fire debris may be rich in minerals, particulary phosphate.

... BUT ALWAYS PROCEEDS VIA A SERIES OF STAGES

Migration: the arrival of seeds and spores. If conditions are suitable immigrant species may become established.

The number of species has risen – further stabilizing soil and adding nutrients. May still be an input of new species so that there will be both *intra-* and *inter-specific competition.* Pioneer species are often poor competitors and will be replaced by higher, more demanding plants such as grasses, shrubs and, eventually, trees.

The end point of succession: the community is now in equilibrium with the environment and is stable. Composition is often determined by one dominant species e.g. *oak woodland.*

BARE GROUND ⇒ PIONEER COMMUNITY (COLONIZERS) ⇒ SECONDARY COMMUNITY ⇒ CLIMAX COMMUNITY

During succession each species modifies the environment, making it *more* suitable for new species and *less* suitable for those already there.

These species are simple plants, e.g. lichens and algae with minimal environmental demands. May show symbiotic relationships to aid their establishment. The community is *open,* i.e. space for further colonizers.

Each of these stages is called a *seral stage* and the complete succession is called a *sere.*

Hydrosere: succession in an aquatic environment.
Xerosere: succession on dry land.
Halosere: succession in a salty environment.
Lithosere: succession on a rocky surface.

The carbon cycle depends upon both biochemical and physical processes.

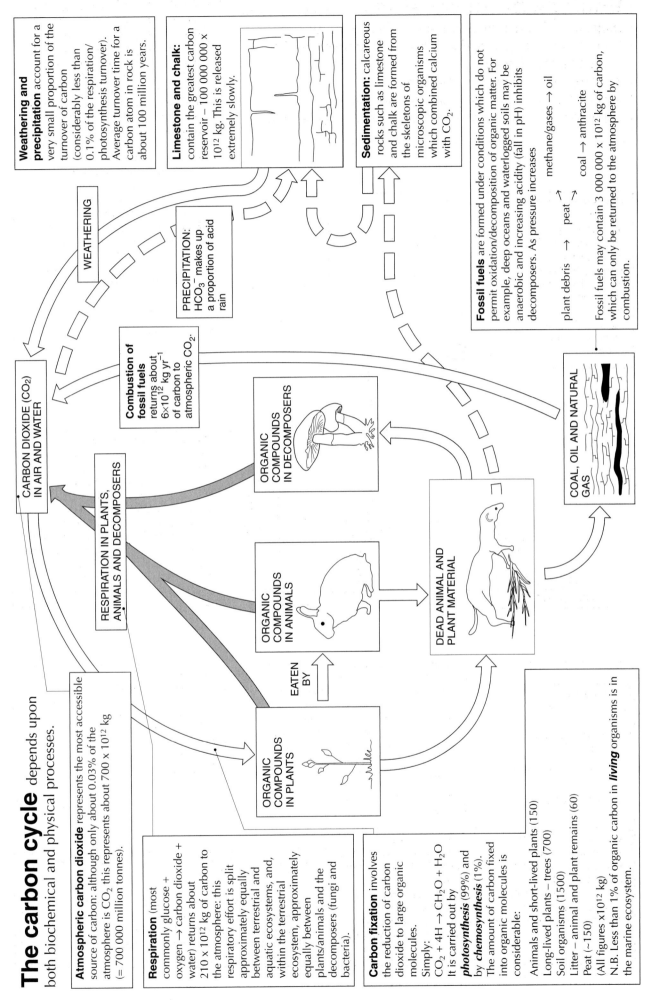

Weathering and precipitation account for a very small proportion of the turnover of carbon (considerably less than 0.1% of the respiration/photosynthesis turnover). Average turnover time for a carbon atom in rock is about 100 million years.

Limestone and chalk: contain the greatest carbon reservoir – 100 000 000 x 10^{12} kg. This is released extremely slowly.

Sedimentation: calcareous rocks such as limestone and chalk are formed from the skeletons of microscopic organisms which combined calcium with CO_2.

Fossil fuels are formed under conditions which do not permit oxidation/decomposition of organic matter. For example, deep oceans and waterlogged soils may be anaerobic and increasing acidity (fall in pH) inhibits decomposers. As pressure increases

plant debris → peat
methane/gases → oil
coal → anthracite

Fossil fuels may contain 3 000 000 x 10^{12} kg of carbon, which can only be returned to the atmosphere by combustion.

WEATHERING

PRECIPITATION: HCO_3^- makes up a proportion of acid rain

Combustion of fossil fuels returns about 6×10^{12} kg yr^{-1} of carbon to atmospheric CO_2.

CARBON DIOXIDE (CO_2) IN AIR AND WATER

RESPIRATION IN PLANTS, ANIMALS AND DECOMPOSERS

ORGANIC COMPOUNDS IN DECOMPOSERS

ORGANIC COMPOUNDS IN ANIMALS

ORGANIC COMPOUNDS IN PLANTS

EATEN BY

DEAD ANIMAL AND PLANT MATERIAL

COAL, OIL AND NATURAL GAS

Atmospheric carbon dioxide represents the most accessible source of carbon: although only about 0.03% of the atmosphere is CO_2 this represents about 700 x 10^{12} kg (= 700 000 million tonnes).

Respiration (most commonly glucose + oxygen → carbon dioxide + water) returns about 210 x 10^{12} kg of carbon to the atmosphere: this respiratory effort is split approximately equally between terrestrial and aquatic ecosystems, and, within the terrestrial ecosystem, approximately equally between plants/animals and the decomposers (fungi and bacteria).

Carbon fixation involves the reduction of carbon dioxide to large organic molecules. The amount of carbon fixed into organic molecules is considerable.
Simply:
$$CO_2 + 4H \rightarrow CH_2O + H_2O$$
It is carried out by **photosynthesis** (99%) and by **chemosynthesis** (1%).

Animals and short-lived plants (150)
Long-lived plants – trees (700)
Soil organisms (1500)
Litter – animal and plant remains (60)
Peat (~150)
(All figures x10^{12} kg)
N.B. Less than 1% of organic carbon in **living** organisms is in the marine ecosystem.

The nitrogen cycle depends on micro-organisms.

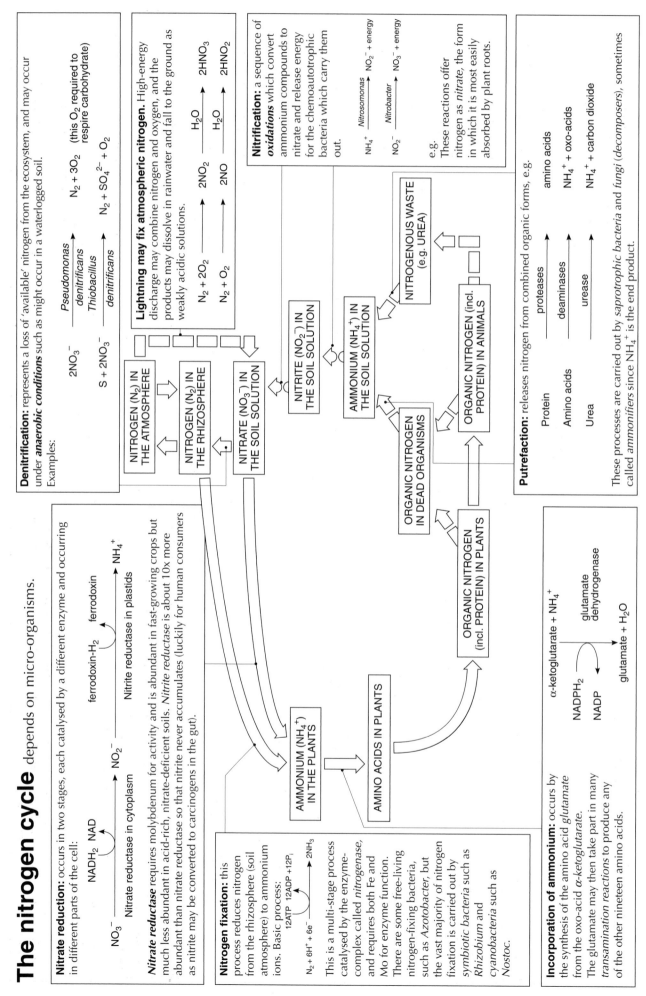

Nitrate reduction: occurs in two stages, each catalysed by a different enzyme and occurring in different parts of the cell:

$$NO_3^- \xrightarrow{\text{Nitrate reductase in cytoplasm}} NO_2^- \xrightarrow{\text{Nitrite reductase in plastids}} NH_4^+$$

$NADH_2 \to NAD$ $\qquad$ ferrodoxin-H_2 → ferrodoxin

Nitrate reductase requires molybdenum for activity and is abundant in fast-growing crops but much less abundant in acid-rich, nitrate-deficient soils. *Nitrite reductase* is about 10x more abundant than nitrate reductase so that nitrite never accumulates (luckily for human consumers as nitrite may be converted to carcinogens in the gut).

Nitrogen fixation: this process reduces nitrogen from the rhizosphere (soil atmosphere) to ammonium ions. Basic process:

$$N_2 + 6H^+ + 6e^- \xrightarrow{\text{12ATP} \quad \text{12ADP} + 12P_i} 2NH_3$$

This is a multi-stage process catalysed by the enzyme-complex called *nitrogenase*, and requires both Fe and Mo for enzyme function. There are some free-living nitrogen-fixing bacteria, such as *Azotobacter*, but the vast majority of nitrogen fixation is carried out by *symbiotic bacteria* such as *Rhizobium* and cyanobacteria such as *Nostoc*.

Incorporation of ammonium: occurs by the synthesis of the amino acid *glutamate* from the oxo-acid *α-ketoglutarate*. The glutamate may then take part in many *transamination reactions* to produce any of the other nineteen amino acids.

$$\alpha\text{-ketoglutarate} + NH_4^+ \xrightarrow[\substack{\text{glutamate} \\ \text{dehydrogenase}}]{} \text{glutamate} + H_2O$$

$NADPH_2 \to NADP$

Denitrification: represents a loss of 'available' nitrogen from the ecosystem, and may occur under *anaerobic conditions* such as might occur in a waterlogged soil. Examples:

$$2NO_3^- \xrightarrow{\substack{\textit{Pseudomonas} \\ \textit{denitrificans}}} N_2 + 3O_2 \quad (\text{this } O_2 \text{ required to respire carbohydrate})$$

$$S + 2NO_3^- \xrightarrow{\substack{\textit{Thiobacillus} \\ \textit{denitrificans}}} N_2 + SO_4^{2-} + O_2$$

Lightning may fix atmospheric nitrogen. High-energy discharge may combine nitrogen and oxygen, and the products may dissolve in rainwater and fall to the ground as weakly acidic solutions.

$$N_2 + 2O_2 \to 2NO_2 \xrightarrow{H_2O} 2HNO_3$$
$$N_2 + O_2 \to 2NO \xrightarrow{H_2O} 2HNO_2$$

Nitrification: a sequence of *oxidations* which convert ammonium compounds to nitrate and release energy for the chemoautotrophic bacteria which carry them out.

$$NH_4^+ \xrightarrow{\textit{Nitrosomonas}} NO_2^- + \text{energy}$$
$$NO_2^- \xrightarrow{\textit{Nitrobacter}} NO_3^- + \text{energy}$$

e.g.
These reactions offer nitrogen as *nitrate*, the form in which it is most easily absorbed by plant roots.

Putrefaction: releases nitrogen from combined organic forms, e.g.

$$\text{Protein} \xrightarrow{\text{proteases}} \text{amino acids}$$
$$\text{Amino acids} \xrightarrow{\text{deaminases}} NH_4^+ + \text{oxo-acids}$$
$$\text{Urea} \xrightarrow{\text{urease}} NH_4^+ + \text{carbon dioxide}$$

These processes are carried out by *saprotrophic bacteria* and *fungi (decomposers)*, sometimes called *ammonifiers* since NH_4^+ is the end product.

Boxes in diagram:
- NITROGEN (N_2) IN THE ATMOSPHERE
- NITROGEN (N_2) IN THE RHIZOSPHERE
- NITRATE (NO_3^-) IN THE SOIL SOLUTION
- NITRITE (NO_2^-) IN THE SOIL SOLUTION
- AMMONIUM (NH_4^+) IN THE SOIL SOLUTION
- NITROGENOUS WASTE (e.g. UREA)
- ORGANIC NITROGEN (incl. PROTEIN) IN ANIMALS
- ORGANIC NITROGEN IN DEAD ORGANISMS
- ORGANIC NITROGEN (incl. PROTEIN) IN PLANTS
- AMMONIUM (NH_4^+) IN THE PLANTS
- AMINO ACIDS IN PLANTS

The greenhouse effect is a
natural feature of the Earth, but when upset may lead to *global warming.*

ORIGINS OF GREENHOUSE GASES

Photosynthesis in forests and grasslands removes carbon dioxide (CO_2) from the atmosphere.

Car exhaust emissions contain much CO_2 – released to the atmosphere.

Combustion of fossil fuels by industrial plants releases large amounts of CO_2.

Ruminant fermentation produces *methane* (CH_4) which cattle release into the atmosphere. Intensive cattle ranching increases CH_4 release at the expense of CO_2 uptake by photosynthesis.

Aerosol propellants contain *chlorofluorocarbons* (CFCs) which are 10^5 × worse than carbon dioxide as greenhouse gases.

Anaerobic fermentation in swamps and paddy fields produces CH_4. Inorganic fertilizers cause release of nitric oxide (NO).

Phew!

All living organisms release carbon dioxide by respiration – the additional **greenhouse gases** contributed by humans (**anthropogenic contributions**) include methane and CFCs in addition to greater quantities of carbon dioxide.

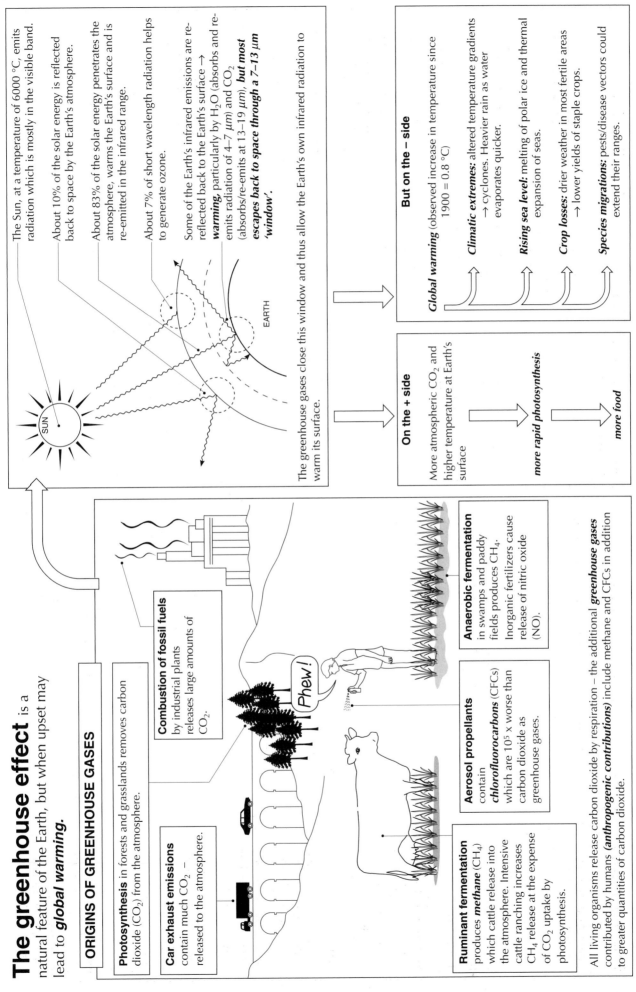

The Sun, at a temperature of 6000 °C, emits radiation which is mostly in the visible band.

About 10% of the solar energy is reflected back to space by the Earth's atmosphere.

About 83% of the solar energy penetrates the atmosphere, warms the Earth's surface and is re-emitted in the infrared range.

About 7% of short wavelength radiation helps to generate ozone.

Some of the Earth's infrared emissions are re-reflected back to the Earth's surface → *warming*, particularly by H_2O (absorbs and re-emits radiation of 4–7 μm) and CO_2 (absorbs/re-emits at 13–19 μm), **but most escapes back to space through a 7–13 μm 'window'.**

SUN

EARTH

The greenhouse gases close this window and thus allow the Earth's own infrared radiation to warm its surface.

On the + side

More atmospheric CO_2 and higher temperature at Earth's surface

more rapid photosynthesis

more food

But on the – side

Global warming (observed increase in temperature since 1900 = 0.8 °C)

Climatic extremes: altered temperature gradients → cyclones. Heavier rain as water evaporates quicker.

Rising sea level: melting of polar ice and thermal expansion of seas.

Crop losses: drier weather in most fertile areas → lower yields of staple crops.

Species migrations: pests/disease vectors could extend their ranges.

Acid rain

Sulphur dioxide is an atmospheric pollutant that contributes to acid rain.

Catalytic converters

... clean up exhaust emissions by encouraging several pollutants to react with one another to give less harmful products.

to EXHAUST

from ENGINE

CERAMIC HONEYCOMB COATED WITH Pt, Pd and Rh

CARBON MONOXIDE + NITROGEN OXIDES → CARBON DIOXIDE + NITROGEN

HYDROCARBONS + NITROGEN OXIDES → CARBON DIOXIDE + WATER + NITROGEN

+ FEWER POISONOUS EMISSIONS

USE UNLEADED FUEL (Pb DAMAGES CATALYST)

− MORE CO_2, A GREENHOUSE GAS
 EXPENSIVE
 LESS FUEL EFFICIENT

ACTIONS

Prevention ...

Involves the reduction of NO, NO_2 and SO_2 emissions from internal combustion engines

and from industrial plants.

'CLEAN GASES'

REMOVAL OF SO_2 BY A WET SCRUBBER

SPRAY TRAPS POLLUTANTS

WATER

POLLUTED GASES

WATER AND DISSOLVED POLLUTANTS (COMMERCIAL SOURCE OF H_2SO_4)

... and **cure.**

Is much more difficult and very expensive ...

1. local applications of alkali, e.g. limestone ($CaCO_3$) has been tipped into 'acid' lakes

2. addition of supplementary mineral nutrients to forest soils

... and is usually only a short-term solution.

NO_2 AND SO_2 IN ATMOSPHERE

Ca^{2+}
Mg^{2+}

Al^{3+}

Al^{3+}

PROBLEMS

Soils: H^+ ions in soils are normally stable (bound to soil particles or to HCO_3^-) but the SO_4^{2-} ion in H_2SO_4 is very mobile and so transfers H^+ to the run-off waters. (NO_3^- would have the same effect but it is rapidly absorbed by plant roots.)

Water in lakes and rivers: SO_4^{2-} displaces Al^{3+} from soil. The aluminium ions washed into the water interfere with gill function in fish - 'sticky' mucus accumulates, limits oxygen uptake → **death of fish.** Changes in pH also cause soft exoskeletons → **death of invertebrates.**

Forests: the leaching effect of the acid rain removes Ca^{2+} and Mg^{2+} from the soil (poor middle lamella and chlorophyll formation). Uptake of Al^{3+} occurs to toxic levels. Atmospheric SO_2 and NO_2 damage spongy and palisade mesophyll → **tree starvation.** All made worse by ozone, ammonia and frost – trees are stressed and sensitive to fungal infections.

SOURCES

Sulphur and nitrogen in fossil fuels are oxidized during combustion.

$S + O_2 \longrightarrow SO_2$
$N + O_2 \longrightarrow NO_2$

Further oxidation occurs in the clouds.

$SO_2 + O_2 \longrightarrow SO_4^{2-}$
$N_2 + \tfrac{1}{2}O_2 \longrightarrow NO_3^-$

These reactions are catalysed by ozone, ammonia, and by unburnt hydrocarbons

... and are followed by solution in water, to make up **acid rain.**

$SO_4^{2-} + 2H^+ \longrightarrow H_2SO_4$
$NO_3^- + H^+ \longrightarrow HNO_3$

SMOG

'PURE' RAIN

ACID RAIN

2
4
6
8
10
12
14

pH scale is logarithmic: rain of pH 4 is 100 times more acidic than rain of pH6.

Ozone in the atmosphere is essential for life

(but too much in the wrong place can be harmful!).

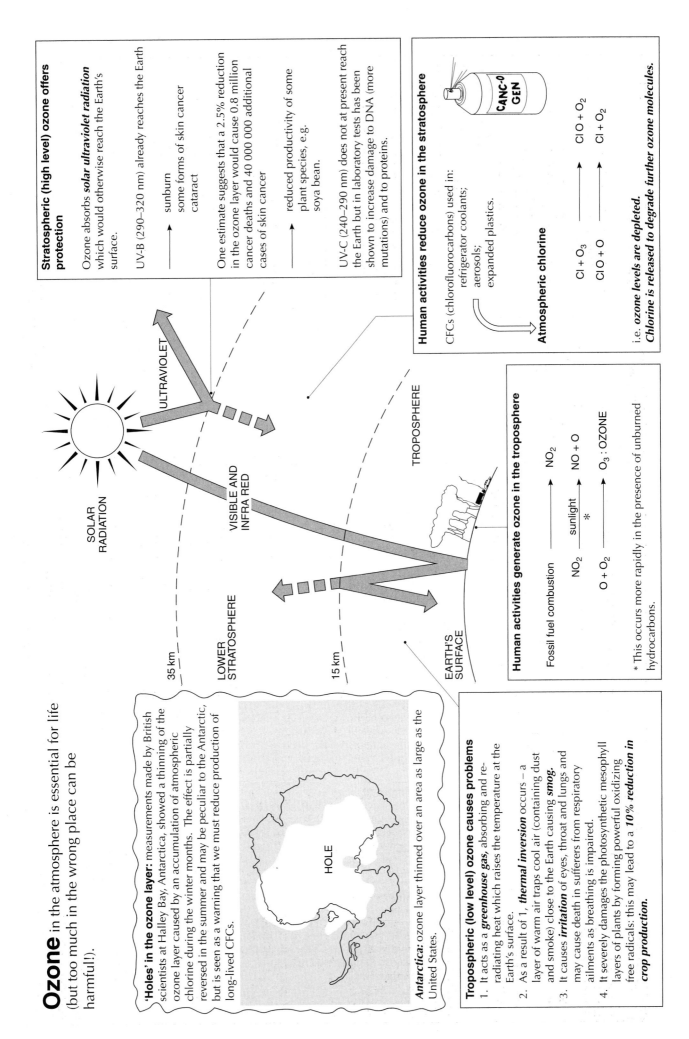

'Holes' in the ozone layer: measurements made by British scientists at Halley Bay, Antarctica, showed a thinning of the ozone layer caused by an accumulation of atmospheric chlorine during the winter months. The effect is partially reversed in the summer and may be peculiar to the Antarctic, but is seen as a warning that we must reduce production of long-lived CFCs.

HOLE

Antarctica: ozone layer thinned over an area as large as the United States.

Tropospheric (low level) ozone causes problems

1. It acts as a *greenhouse gas,* absorbing and re-radiating heat which raises the temperature at the Earth's surface.
2. As a result of 1, *thermal inversion* occurs – a layer of warm air traps cool air (containing dust and smoke) close to the Earth causing *smog.*
3. It causes *irritation* of eyes, throat and lungs and may cause death in sufferers from respiratory ailments as breathing is impaired.
4. It severely damages the photosynthetic mesophyll layers of plants by forming powerful oxidizing free radicals: this may lead to a *10% reduction in crop production.*

SOLAR RADIATION

ULTRAVIOLET

VISIBLE AND INFRA RED

35 km

LOWER STRATOSPHERE

15 km

TROPOSPHERE

EARTH'S SURFACE

Stratospheric (high level) ozone offers protection

Ozone absorbs *solar ultraviolet radiation* which would otherwise reach the Earth's surface.

UV-B (290–320 nm) already reaches the Earth

→ sunburn
  some forms of skin cancer
  cataract

One estimate suggests that a 2.5% reduction in the ozone layer would cause 0.8 million cancer deaths and 40 000 000 additional cases of skin cancer

→ reduced productivity of some plant species, e.g. soya bean.

UV-C (240–290 nm) does not at present reach the Earth but in laboratory tests has been shown to increase damage to DNA (more mutations) and to proteins.

Human activities reduce ozone in the stratosphere

CAN-C-0 GEN

CFCs (chlorofluorocarbons) used in: refrigerator coolants; aerosols; expanded plastics.

→ Atmospheric chlorine

$Cl + O_3 \longrightarrow ClO + O_2$

$ClO + O \longrightarrow Cl + O_2$

i.e. *ozone levels are depleted.*
Chlorine is released to degrade further ozone molecules.

Human activities generate ozone in the troposphere

Fossil fuel combustion → NO_2

$NO_2 \xrightarrow{\text{sunlight}} NO + O$

$O + O_2 \xrightarrow{\quad * \quad} O_3$: OZONE

* This occurs more rapidly in the presence of unburned hydrocarbons.

Deforestation: the rapid destruction of woodland.

BULLDOZE, SLASH AND BURN

Has been occurring on a major scale throughout the world.

- Between 1880 and 1980 about 40% of all tropical rainforest was destroyed.

- Britain has fallen from 85% forest cover to about 8% (probably the lowest in Europe).

- Major reasons
  - removal of hardwood for high-quality furnishings;
  - removal of softwoods for chipboards, paper and other wood products;
  - clearance for cattle ranching and for cash crop agriculture;
  - clearance for urban development (roads and towns being built).

Current losses: about 11 hectares *per minute* (that's about 40 soccer or hockey pitches!).

Reduction in soil fertility

1. Deciduous trees may contain 90% of the nutrients in a forest ecosystem: these nutrients are removed, and are thus not available to the soil, if the trees are cut down and taken away.
2. Soil erosion may be rapid since in the absence of trees
   a. wind and direct rain may remove the soil;
   b. soil structure is no longer stabilized by tree root systems.

N.B. The soil below coniferous forests is often of poor quality for agriculture because the shed pine needles contain toxic compounds which act as germination and growth inhibitors.

Flooding and landslips

Heavy rainfall on deforested land is not 'held up': normally 25% of rainfall is absorbed by foliage or evaporates and 50% is absorbed by root systems. As a result water may accumulate rapidly in river valleys, often causing landslips from steep hillsides.

Changes in recycling of materials: Fewer trees mean

1. atmospheric CO_2 concentration may rise as less CO_2 is removed for photosynthesis;
2. atmospheric O_2 – vital for aerobic respiration – is diminished as less is produced by photosynthesis;
3. the atmosphere may become drier and the soil wetter as evaporation (from soil) is slower than transpiration (from trees).

Climatic changes

1. Reduced transpiration rates and drier atmosphere affect the water cycle and reduce rainfall.
2. Rapid heat absorption by bare soil raises the temperature of the lower atmosphere in some areas, causing thermal gradients which result in more frequent and intense winds.

Species extinction: Many species are dependent on forest conditions.

e.g. mountain gorilla depends on cloud forest of Central Africa; golden lion tamarin depends on coastal rainforest of Brazil; osprey depends on mature pine forests in Northern Europe.

It is estimated that one plant and one animal species become extinct every 30 minutes due to deforestation.

Many plant species may have medicinal properties,
e.g. as tranquilizers, reproductive hormones, anticoagulants, painkillers and antibiotics.
The Madagascan periwinkle, for example, yields one of the most potent known anti-leukaemia drugs.

Crops

Crops are plants grown on a large scale for human use. They may benefit from genetic modification, and show both structural and physiological adaptations.

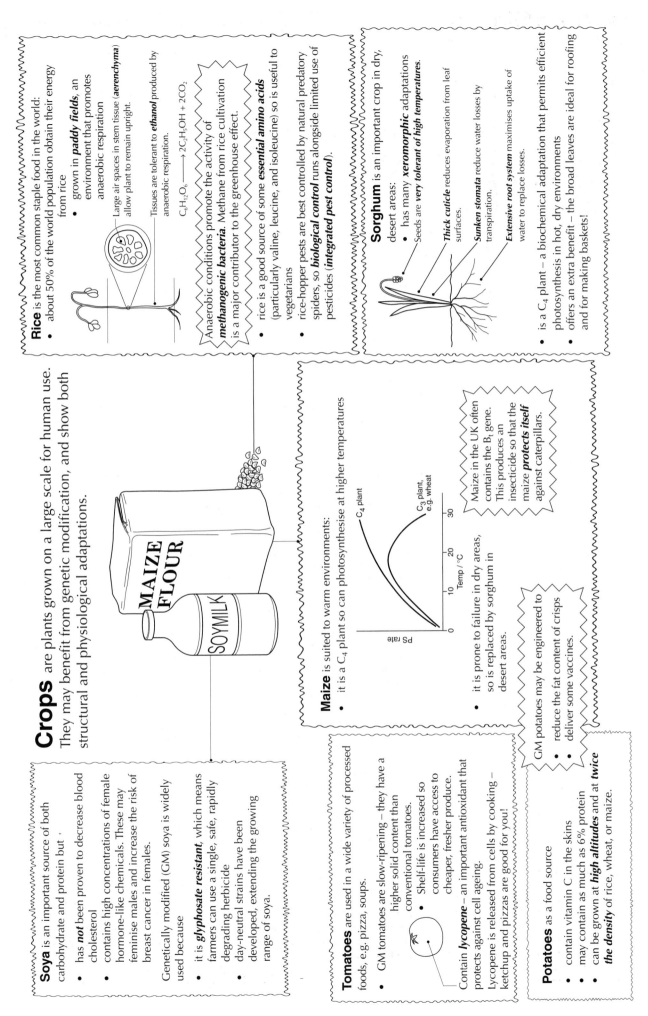

Rice

Rice is the most common staple food in the world:

- about 50% of the world population obtain their energy from rice
  - grown in **paddy fields**, an environment that promotes anaerobic respiration
  - Large air spaces in stem tissue (**aerenchyma**) allow plant to remain upright.
  - Tissues are tolerant to **ethanol** produced by anaerobic respiration.

$$C_6H_{12}O_6 \longrightarrow 2C_2H_5OH + 2CO_2$$

Anaerobic conditions promote the activity of **methanogenic bacteria**. Methane from rice cultivation is a major contributor to the greenhouse effect.

- rice is a good source of some **essential amino acids** (particularly valine, leucine, and isoleucine) so is useful to vegetarians
- rice-hopper pests are best controlled by natural predatory spiders, so **biological control** runs alongside limited use of pesticides (**integrated pest control**).

Sorghum

Sorghum is an important crop in dry, desert areas:

- has many **xeromorphic** adaptations and is **very tolerant of high temperatures.**
  - Seeds are **very tolerant of high temperatures.**
  - **Thick cuticle** reduces evaporation from leaf surfaces.
  - **Sunken stomata** reduce water losses by transpiration.
  - **Extensive root system** maximises uptake of water to replace losses.

- is a C_4 plant – a biochemical adaptation that permits efficient photosynthesis in hot, dry environments
- offers an extra benefit – the broad leaves are ideal for roofing and for making baskets!

Maize

Maize is suited to warm environments:

- it is a C_4 plant so can photosynthesise at higher temperatures

Maize in the UK often contains the B_t gene. This produces an insecticide so that the maize **protects itself** against caterpillars.

- it is prone to failure in dry areas, so is replaced by sorghum in desert areas.

GM potatoes may be engineered to
- reduce the fat content of crisps
- deliver some vaccines.

Soya

Soya is an important source of both carbohydrate and protein but

- has **not** been proven to decrease blood cholesterol
- contains high concentrations of female hormone-like chemicals. These may feminise males and increase the risk of breast cancer in females.

Genetically modified (GM) soya is widely used because

- it is **glyphosate resistant**, which means farmers can use a single, safe, rapidly degrading herbicide
- day-neutral strains have been developed, extending the growing range of soya.

Tomatoes

Tomatoes are used in a wide variety of processed foods, e.g. pizza, soups.

- GM tomatoes are slow-ripening – they have a higher solid content than conventional tomatoes.
  - Shelf-life is increased so consumers have access to cheaper, fresher produce.

Contain **lycopene** – an important antioxidant that protects against cell ageing. Lycopene is released from cells by cooking – ketchup and pizzas are good for you!

Potatoes

Potatoes as a food source
- contain vitamin C in the skins
- may contain as much as 6% protein
- can be grown at **high altitudes** and at **twice the density** of rice, wheat, or maize.

River pollution affects animal and plant populations.

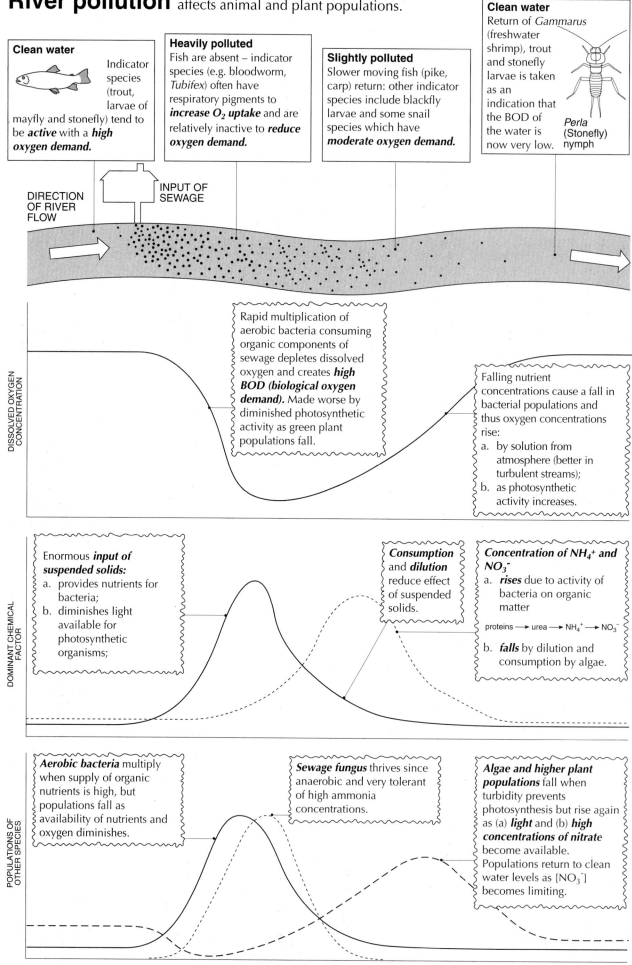

Clean water
Indicator species (trout, larvae of mayfly and stonefly) tend to be **active** with a **high oxygen demand.**

Heavily polluted
Fish are absent – indicator species (e.g. bloodworm, *Tubifex*) often have respiratory pigments to **increase O₂ uptake** and are relatively inactive to **reduce oxygen demand.**

Slightly polluted
Slower moving fish (pike, carp) return: other indicator species include blackfly larvae and some snail species which have **moderate oxygen demand.**

Clean water
Return of *Gammarus* (freshwater shrimp), trout and stonefly larvae is taken as an indication that the BOD of the water is now very low.

Perla (Stonefly) nymph

DIRECTION OF RIVER FLOW

INPUT OF SEWAGE

DISSOLVED OXYGEN CONCENTRATION

Rapid multiplication of aerobic bacteria consuming organic components of sewage depletes dissolved oxygen and creates **high BOD (biological oxygen demand).** Made worse by diminished photosynthetic activity as green plant populations fall.

Falling nutrient concentrations cause a fall in bacterial populations and thus oxygen concentrations rise:
a. by solution from atmosphere (better in turbulent streams);
b. as photosynthetic activity increases.

DOMINANT CHEMICAL FACTOR

Enormous **input of suspended solids:**
a. provides nutrients for bacteria;
b. diminishes light available for photosynthetic organisms;

Consumption and **dilution** reduce effect of suspended solids.

Concentration of NH₄⁺ and NO₃⁻
a. **rises** due to activity of bacteria on organic matter

proteins ⟶ urea ⟶ NH₄⁺ ⟶ NO₃⁻

b. **falls** by dilution and consumption by algae.

POPULATIONS OF OTHER SPECIES

Aerobic bacteria multiply when supply of organic nutrients is high, but populations fall as availability of nutrients and oxygen diminishes.

Sewage fungus thrives since anaerobic and very tolerant of high ammonia concentrations.

Algae and higher plant populations fall when turbidity prevents photosynthesis but rise again as (a) **light** and (b) **high concentrations of nitrate** become available. Populations return to clean water levels as [NO₃⁻] becomes limiting.

Nitrates are significant pollutants of water.

Effects on human health

1. In the stomach

$NO_3^- \longrightarrow$ NITROSAMINES

Nitrosamines are highly carcinogenic, and some studies have linked high $[NO_3^-]$ in water supplies with increased incidence of stomach and oesophageal cancer.

2. *Blue baby syndrome* (in children younger than 3 months)

$NO_3^- \xrightarrow[\text{or water supply}]{\text{bacteria in gut}}$ NITRITE (NO_2^-)

Haemoglobin in baby's red blood cells

methaemoglobin (has Fe^{II} oxidized to Fe^{III}) which reduces oxygen-carrying capacity of baby's blood → *'blue' appearance.*

EUTRO–MAX
ACA

N
P FERTILISER
K

The EU has set a limit of 11.3 p.p.m. total nitrogen in drinking water – this is exceeded in some parts of East Anglia and Cleveland.

THERE'S ALSO A PROBLEM WITH PESTICIDES!

Bioaccumulation of toxins: pesticides or their products may be toxic and thus alter

a. they may seriously affect micro-organisms and thus alter decomposition in soils;

b. they may pass along food chains, becoming more concentrated in organisms further up the chain.

e.g. DDT used as an insecticide accumulates in the fatty tissues of carnivorous animals, inhibiting cytochrome oxidase and limiting reproductive success (especially thin eggshells in birds of prey).

| PHYTOPLANKTON 1 | → | MAYFLY LARVAE 4 | → | TROUT 50 | → | OSPREY 800 |

Relative DDT concentration along an aquatic food chain.

Eutrophication – nutrient enrichment of ponds, lakes and rivers – is responsible for **biological oxygen demand (BOD)**

BOD is the mass of oxygen consumed by micro-organisms in a sample of water determined by measuring oxygen concentration with an oxygen electrode *before and after* a period of microbial respiration: indicates the oxygen *not available* to more advanced organisms.

Input of *raw sewage*

Leaching of *inorganic fertilizers* from farmland.

*N.B. The leaching of *phosphates* into ponds and rivers is at least as important as nitrates in causing eutrophication.

Increased concentration of *nitrate* and *phosphate* in bodies of water.

Algae and green protists use nutrients to multiply rapidly = *algal bloom.*

Reduction of light for bottom-growing plants.

DIE

Large quantities of *organic material.*

DIE

POSITIVE FEEDBACK

Aerobic decomposers (mainly bacteria) multiply and *consume oxygen.*

Aerobic organisms (fish and invertebrates) die of *lack of oxygen.*

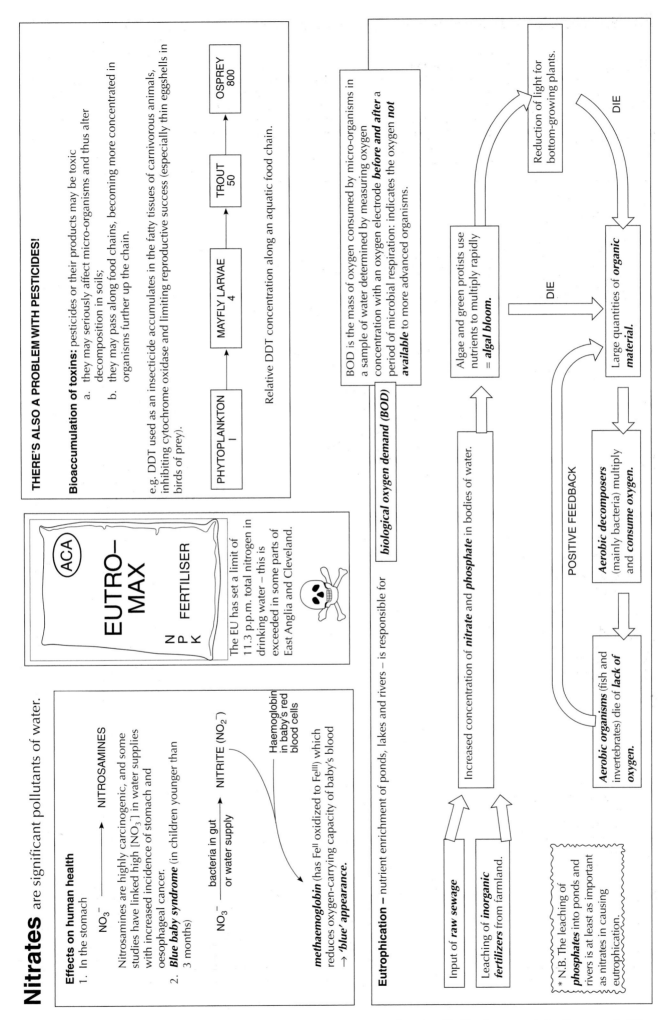

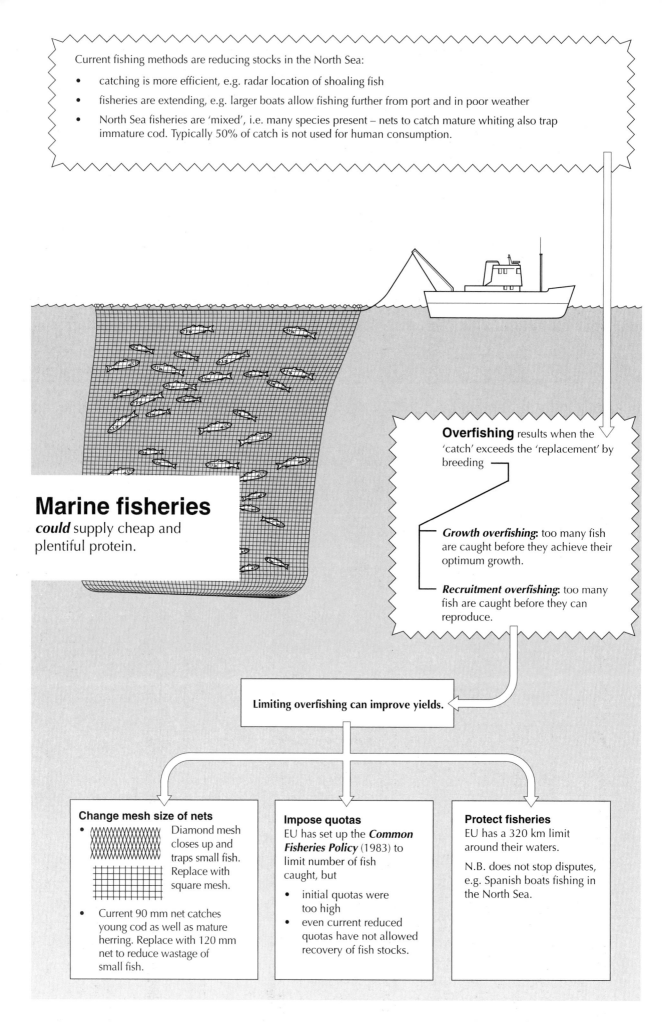

Current fishing methods are reducing stocks in the North Sea:

- catching is more efficient, e.g. radar location of shoaling fish
- fisheries are extending, e.g. larger boats allow fishing further from port and in poor weather
- North Sea fisheries are 'mixed', i.e. many species present – nets to catch mature whiting also trap immature cod. Typically 50% of catch is not used for human consumption.

Marine fisheries

could supply cheap and plentiful protein.

Overfishing results when the 'catch' exceeds the 'replacement' by breeding

Growth overfishing: too many fish are caught before they achieve their optimum growth.

Recruitment overfishing: too many fish are caught before they can reproduce.

Limiting overfishing can improve yields.

Change mesh size of nets
- Diamond mesh closes up and traps small fish. Replace with square mesh.
- Current 90 mm net catches young cod as well as mature herring. Replace with 120 mm net to reduce wastage of small fish.

Impose quotas
EU has set up the **Common Fisheries Policy** (1983) to limit number of fish caught, but
- initial quotas were too high
- even current reduced quotas have not allowed recovery of fish stocks.

Protect fisheries
EU has a 320 km limit around their waters.

N.B. does not stop disputes, e.g. Spanish boats fishing in the North Sea.

Managing ecosystems: horticulture

can be profitable if environmental resistance is reduced

ILLUMINATION: it is important to control
- **intensity** - this will influence the rate of food production by photosynthesis. Greater intensity → more photosynthesis until some other limiting factor intervenes.
- **quality** - photosynthesis is most efficient at red and blue wavelengths so that 'white' light contains some wavelengths ('green') which are not useful.
- **duration** - if fruit is the desired product the plant must flower. Flowering is controlled by daylength (the duration of light in a 24 hour period).

Sunlight provides some illumination but artificial lighting systems are more controllable (but more expensive).

PEST CONTROL: is particularly important since a pest species could spread very rapidly through a greenhouse containing only a single crop.
- **Fungi** can be a problem since high humidity/high temperature encourages spore germination and growth of fungus. Control with **fungicide spray** or by **reduction of humidity.**
- **Weed plants** could compete with the crop species for light, water and mineral nutrients. They are removed **by hand** or with a **selective herbicide.**
- **Animal pests** are herbivorous (e.g. caterpillars) or sap-sucking (e.g. aphids) insects. These may be controlled using **insecticides** but
  - these can be inefficient, especially against waxy-coated pupae
  - they may need to be reapplied, so can be expensive
  - they may leave residues so that crop will need washing

or **biological control** which
  - uses a natural predator of the pest
  - usually only requires a single application
  - does not leave residues

Important examples of biological control are **ladybirds** which eat aphids **wasp larvae** which live as parasites on **whitefly larvae**

AAGH!

YUMMY!

HUMIDITY: this affects the rate of transpiration, and therefore the rate at which the plants must be supplied with water. Thus a high humidity reduces the need for additional water and generally favours growth (NB **too** high humidity favours the growth of fungal pests). Humidity is reduced by opening ventilators and raised using an automatic mist spray.

POLLINATING INSECTS: bees may be introduced to ensure high rates of fertilisation and therefore fruit formation.

TEMPERATURE: affects plant growth because of its affect on the enzymes of photosynthesis. NB high temperatures may also speed up the life cycle of pests. Sunlight provides some heat (shading may be necessary in summer) but more control is available using thermostatically regulated heaters.

CO_2

CARBON DIOXIDE CONCENTRATION: is a major limiting factor in photosynthesis. In a greenhouse the plants may photosynthesise very quickly and CO_2 is rapidly used up. The CO_2 concentration is usually raised to about 0.1% of the atmosphere (about three times that in air). This gives a significant (about 50%) increase in crop yield. The extra CO_2 can be provided by burning paraffin (which also raises the temperature) or, more accurately, by releasing it from a cylinder.

COMPUTER CONTROL: this is widely used in large commercial greenhouses. Sensors provide information about air temperature, CO_2 concentration, water available to roots, humidity and mineral concentration, and the control centre ensures that any changes are corrected.

e.g.

temperature → sensor → heater

HIGH-YIELDING STRAINS OF CROP: selective breeding and/or genetic engineering can develop crop strains which
- have a high yield
- produce fruit of a desirable colour/texture
- produce fruit at the same time
- may have genetic pest resistance.

HERBICIDES may be

1. **Pre-emergent**, i.e. applied **before** emergence of crop.
   a. Contact herbicides, e.g. **Paraquat**, which kill all above-ground parts of all plants.
   b. Residual herbicides, e.g. **Linuron**, which bind to soil particles and kill weed seedlings as they emerge.

   Pre-emergent herbicides can be **non-selective** and are ideal for clearing ground prior to cultivation.

2. **Post-emergent** is applied to both crop and weed, and therefore must be **selective.** Many, such as 2,4-D, are growth regulators.

Systemic herbicides, such as **glyphosate,** are absorbed by weeds and individuals, and translocated to the meristems where they typically act by inhibition of cell division.

Chemical pest control may

involve the use of:

herbicides – for control of weeds;
insecticides – for control of insects;
fungicides – for control of fungi;
molluscicides – for control of slugs and snails.

IDEAL PESTICIDE

- Should be safe to store and transport.
- Should be easy to apply at the correct dosage.
- Should cost less to purchase and apply than the financial gain in the protected crop.
- Should not be dangerous to the people applying it.
- Should effectively control the pest under field growing conditions.
- Should be biodegradable so that toxic products are not left in or on crop plants.
- Should be specific so that only pest species is killed.
- Should not accumulate in food chains.

PROBLEMS WITH INSECTICIDES:

these arise since the principal idea behind chemical control is to **kill as many of the pests as possible** – the effects on harmless or beneficial organisms were not studied or were ignored.

1. **Direct killing:** accidental misuse of toxic chemicals may cause death in humans or in domestic animals.

2. **Non-specificity:** non-target species, particularly natural predators of the pest species, may be killed by some wide-spectrum insecticides, e.g. large doses of **dieldrin** killed many birds as well as the Japanese beetle pest which was the intended target organism.

3. **Pest resistance:** genetic variation means that each pest population contains a *few* resistant individuals. The pesticide eliminates the non-resistant forms and thus a resistant population is selected for and may quickly develop (since many pests reproduce rapidly).

4. **Pest replacement:** most crops are susceptible to attack by more than one species – a **pest complex** and the use of a pesticide to eliminate one species may simply allow another species to assume major pest proportions (since a pesticide may be more deadly to one species than another).

5. **Pest resurgence:** non-specific pesticides may kill natural predators as well as pests – a small residual pest population may now multiply without check, creating a worse problem than initially was present.

6. **Bioaccumulation of toxins:** pesticides or their products may be toxic
   a. they may seriously affect micro-organisms and thus alter decomposition in soils;
   b. they may pass along food chains, becoming more concentrated in organisms further up the chain.

e.g. DDT used as an insecticide accumulates in the fatty tissues of carnivorous animals, inhibiting cytochrome oxidase and limiting reproductive success (especially thin eggshells in birds of prey).

| PHYTOPLANKTON 1 | → | MAYFLY LARVAE 4 | → | TROUT 50 | → | OSPREY 800 |

Relative DDT concentration along an aquatic food chain.

Biological pest control

reduces the population of one species to levels at which it is no longer a pest by the use of one of the pest species' natural predators.

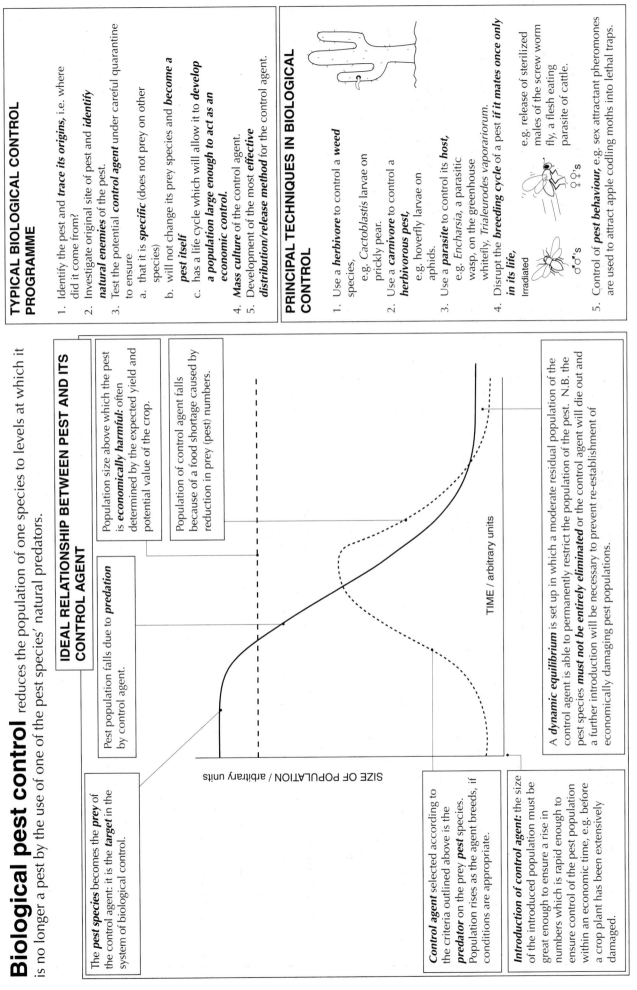

TYPICAL BIOLOGICAL CONTROL PROGRAMME

1. Identify the pest and **trace its origins**, i.e. where did it come from?
2. Investigate original site of pest and **identify natural enemies** of the pest.
3. Test the potential **control agent** under careful quarantine to ensure
   a. that it is **specific** (does not prey on other species)
   b. will not change its prey species and **become a pest itself**
   c. has a life cycle which will allow it to **develop a population large enough to act as an economic control.**
4. **Mass culture** of the control agent.
5. Development of the most **effective distribution/release method** for the control agent.

PRINCIPAL TECHNIQUES IN BIOLOGICAL CONTROL

1. Use a **herbivore** to control a **weed** species,
   e.g. *Cactoblastis* larvae on prickly pear.
2. Use a **carnivore** to control a **herbivorous pest,**
   e.g. hoverfly larvae on aphids.
3. Use a **parasite** to control its **host,**
   e.g. *Encharsia*, a parasitic wasp, on the greenhouse whitefly, *Trialeurodes vaporariorum.*
4. Disrupt the **breeding cycle** of a pest **if it mates once only in its life,**

   Irradiated

   e.g. release of sterilized males of the screw worm fly, a flesh eating parasite of cattle.
5. Control of **pest behaviour,** e.g. sex attractant pheromones are used to attract apple codling moths into lethal traps.

IDEAL RELATIONSHIP BETWEEN PEST AND ITS CONTROL AGENT

The **pest species** becomes the **prey** of the control agent: it is the **target** in the system of biological control.

Pest population falls due to **predation** by control agent.

Population size above which the pest is **economically harmful:** often determined by the expected yield and potential value of the crop.

Population of control agent falls because of a food shortage caused by reduction in prey (pest) numbers.

Control agent selected according to the criteria outlined above is the **predator** on the prey **pest** species. Population rises as the agent breeds, if conditions are appropriate.

Introduction of control agent: the size of the introduced population must be great enough to ensure a rise in numbers which is rapid enough to ensure control of the pest population within an economic time, e.g. before a crop plant has been extensively damaged.

A **dynamic equilibrium** is set up in which a moderate residual population of the control agent is able to permanently restrict the population of the pest. N.B. the pest species **must not be entirely eliminated** or the control agent will die out and a further introduction will be necessary to prevent re-establishment of economically damaging pest populations.

SIZE OF POPULATION / arbitrary units

TIME / arbitrary units

Pesticides and pest control

Techniques of integrated pest management (IPM)

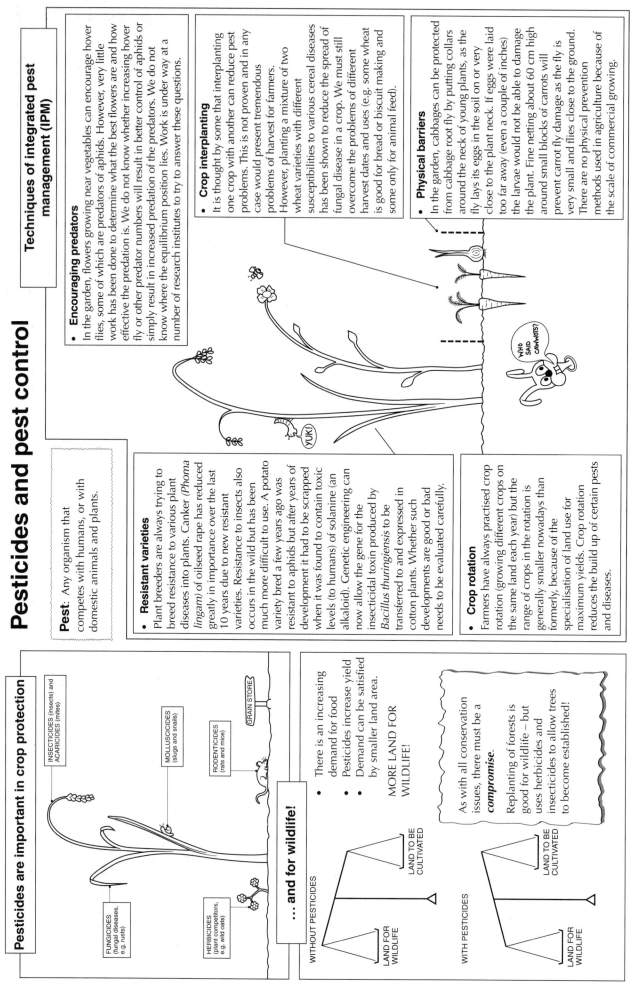

Pest: Any organism that competes with humans, or with domestic animals and plants.

- **Encouraging predators**

In the garden, flowers growing near vegetables can encourage hover flies, some of which are predators of aphids. However, very little work has been done to determine what the best flowers are and how effective the predation is. We do not know whether increasing hover fly or other predator numbers will result in better control of aphids or simply result in increased predation of the predators. We do not know where the equilibrium position lies. Work is under way at a number of research institutes to try to answer these questions.

- **Crop interplanting**

It is thought by some that interplanting one crop with another can reduce pest problems. This is not proven and in any case would present tremendous problems of harvest for farmers. However, planting a mixture of two wheat varieties with different susceptibilities to various cereal diseases has been shown to reduce the spread of fungal disease in a crop. We must still overcome the problems of different harvest dates and uses (e.g. some wheat is good for bread or biscuit making and some only for animal feed).

- **Physical barriers**

In the garden, cabbages can be protected from cabbage root fly by putting collars around the neck of young plants, as the fly lays its eggs in the soil on or very close to the plant neck. If eggs were laid too far away (even a couple of inches) the larvae would not be able to damage the plant. Fine netting about 60 cm high around small blocks of carrots will prevent carrot fly damage as the fly is very small and flies close to the ground. There are no physical prevention methods used in agriculture because of the scale of commercial growing.

- **Resistant varieties**

Plant breeders are always trying to breed resistance to various plant diseases into plants. Canker (*Phoma lingam*) of oilseed rape has reduced greatly in importance over the last 10 years due to new resistant varieties. Resistance to insects also occurs in the wild but has been much more difficult to use. A potato variety bred a few years ago was resistant to aphids but after years of development it had to be scrapped when it was found to contain toxic levels (to humans) of solanine (an alkaloid). Genetic engineering can now allow the gene for the insecticidal toxin produced by *Bacillus thuringiensis* to be transferred to and expressed in cotton plants. Whether such developments are good or bad needs to be evaluated carefully.

- **Crop rotation**

Farmers have always practised crop rotation (growing different crops on the same land each year) but the range of crops in the rotation is generally smaller nowadays than formerly, because of the specialisation of land use for maximum yields. Crop rotation reduces the build up of certain pests and diseases.

Pesticides are important in crop protection

INSECTICIDES (insects) and ACARICIDES (mites)

FUNGICIDES (fungal diseases, e.g. rusts)

MOLLUSCICIDES (slugs and snails)

RODENTICIDES (rats and mice)

HERBICIDES (plant competitors, e.g. wild oats)

GRAIN STORE

... and for wildlife!

- There is an increasing demand for food
- Pesticides increase yield
- Demand can be satisfied by smaller land area.

MORE LAND FOR WILDLIFE!

As with all conservation issues, there must be a ***compromise.***

Replanting of forests is good for wildlife – but uses herbicides and insecticides to allow trees to become established!

WITHOUT PESTICIDES

LAND TO BE CULTIVATED

LAND FOR WILDLIFE

WITH PESTICIDES

LAND TO BE CULTIVATED

LAND FOR WILDLIFE

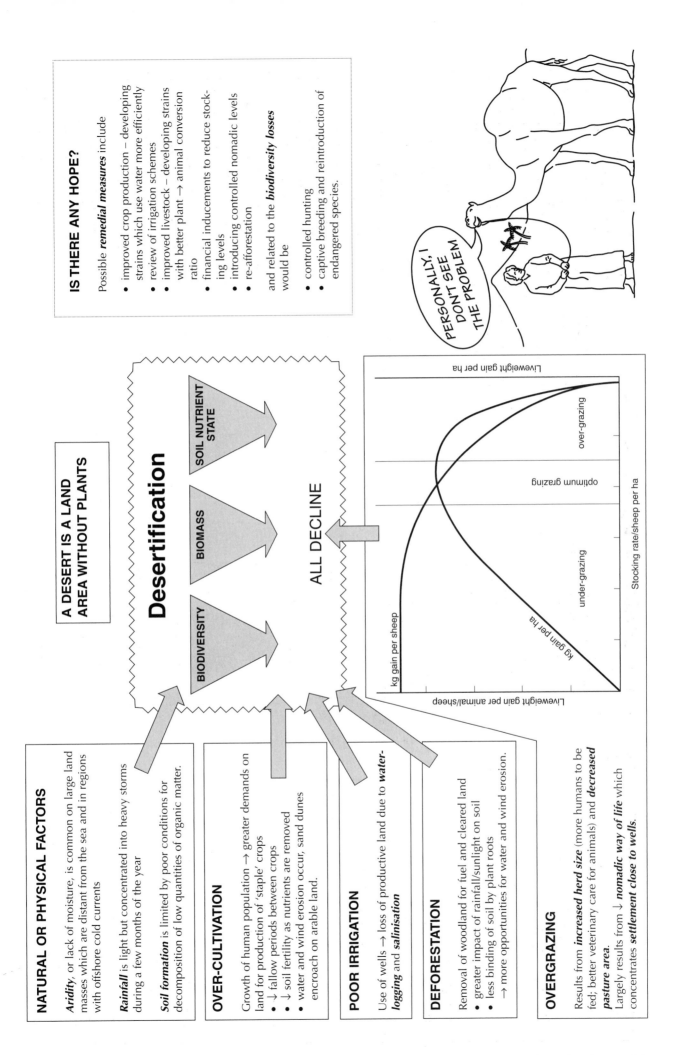

IS THERE ANY HOPE?

Possible *remedial measures* include

- improved crop production – developing strains which use water more efficiently
- review of irrigation schemes
- improved livestock – developing strains with better plant → animal conversion ratio
- financial inducements to reduce stocking levels
- introducing controlled nomadic levels
- re-afforestation

and related to the *biodiversity losses* would be

- controlled hunting
- captive breeding and reintroduction of endangered species.

A DESERT IS A LAND AREA WITHOUT PLANTS

Desertification

BIODIVERSITY · BIOMASS · SOIL NUTRIENT STATE

ALL DECLINE

NATURAL OR PHYSICAL FACTORS

Aridity, or lack of moisture, is common on large land masses which are distant from the sea and in regions with offshore cold currents

Rainfall is light but concentrated into heavy storms during a few months of the year

Soil formation is limited by poor conditions for decomposition of low quantities of organic matter.

OVER-CULTIVATION

Growth of human population → greater demands on land for production of 'staple' crops

- ↓ fallow periods between crops
- ↓ soil fertility as nutrients are removed
- water and wind erosion occur, sand dunes encroach on arable land.

POOR IRRIGATION

Use of wells → loss of productive land due to *water-logging* and *salinisation*

DEFORESTATION

Removal of woodland for fuel and cleared land

- greater impact of rainfall/sunlight on soil
- less binding of soil by plant roots
- → more opportunities for water and wind erosion.

OVERGRAZING

Results from *increased herd size* (more humans to be fed; better veterinary care for animals) and *decreased pasture area*.
Largely results from ↓ *nomadic way of life* which concentrates *settlement close to wells*.

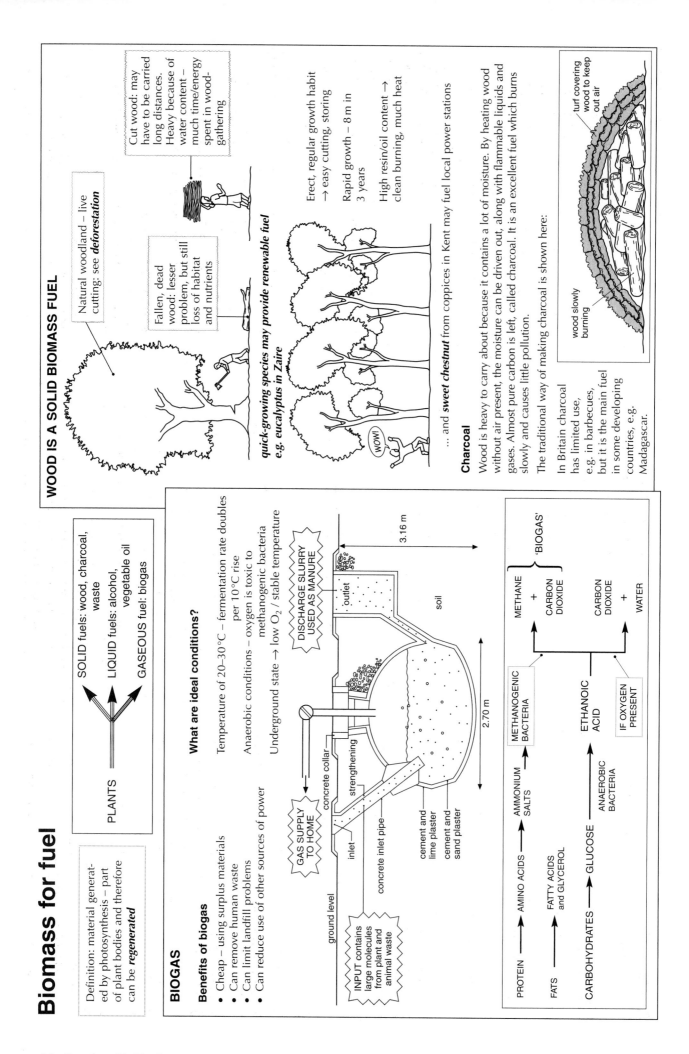

Ethanol – is it the fuel of the future?

It is generated by anaerobic respiration (fermentation) of sugar.

FEEDSTOCK : where does the glucose come from?

USA uses CORN   Brazil uses SUGAR CANE

CORN   WASH and
                                        CRUSH
SOAK: soak
liquid →   CANE PULP
CATTLEFEED

        SEED   SEPARATION: waste
                                 is BAGASSE – burn
GRIND: separated   → energy for
germ → corn oil   distillation*

        STARCH   SYRUP

ENZYMATIC   PURIFICATION:
HYDROLYSIS   impure sucrose solution
uses amylase   (MOLASSES) → CATTLEFEED

ENZYMES:
amylases and
cellulases

FUEL VALUE

Hydrated ethanol
(95% ethanol : 5% water) burns
very well but very difficult to
'start' the engine.

Gasohol
(90% petrol : 10% ethanol) burns
well **and** starts well even in
unmodified engines.

ENVIRONMENTAL IMPACT

- Ethanol fuels do not require 'anti-knock' additives
- Ethanol fuels produce fewer emissions

Emission	Pump (exhaust)	'Gasohol' petrol	Straight hydrated ethanol
Hydrocarbons	3.32	2.44	2.50
Carbon monoxide	49.37	20.62	17.00
Nitrogen oxides, NO_X	1.86	2.86	1.66
Aldehydes:			
☐ methanal	0.05	0.07	0.04
☐ ethanal	0.05	0.16	0.52
Ethanol	–	0.19	1.19

Emissions (in grams per mile) from a 1.4 litre Ford engine tested in Brazil

- Like all biomass fuels, ethanol does not increase atmospheric CO_2 levels (photosynthesis removes CO_2, combustion produces it).

WHAT ARE THE PLUSES AND MINUSES?

+
- Reduced consumption of oil reserves, since gasohol can be regenerated
- Reduced expenditure for developing, oil-poor countries
- Only valuable if there is a plentiful supply of cheap feedstock.

But

–
- Not yet very efficient
- Reduced availability of land for crops

FERMENTATION AND DISTILLATION STAGES ARE ENERGY DEMANDING*

BUBBLE CO_2: ENCOURAGE
ANAEROBIC CONDITIONS

SUGAR

OTHER
NUTRIENTS

DRIED YEAST
→ Vitamins
'Marmite'

ALCOHOL
SOLUTION
(15–20%)

DISTILLATION → PURE ETHANOL

$$C_6H_{12}O_6$$

$$2C_2H_5OH + 2CO_2$$

FERMENTATION AGENT has traditionally been
the yeast *Saccharomyces*. Useful mutants/engineered
strains can

- cope with high temperatures generated during fermentation
- survive at higher alcohol concentrations (up to 15% compared with 8–9% for 'normal' strains)
- respire anaerobically even when O_2 present due to absence of some electron transfer chain components

Recently *Zygomonas mobilis* ('Tequila') has attracted
attention:

- produces fewer impurities such as glycerol, thus has *higher ethanol yield*
- smaller than yeast, therefore higher SA : vol. ratio, more rapid uptake/loss of sugar/ethanol, *more rapid fermentation*

CROPS OR CARS: IS THIS THE BEST WAY TO USE THE LAND?

From the figures you can then go on to estimate how
many people could be fed using the grain which would
be used by a typical car in Europe or in the USA.

Consumer	Grain/pounds	Cropland/acres
Subsistence diet	400	0.2
Affluent diet	1 600	0.9
Typical European car	6 200	3.3
☐ 7000 miles/yr (37.5 m.p.g.)		
Typical US car	14 600	7.8
☐ 10000 miles/yr (15 m.p.g.)		

Section F Animal nutrition

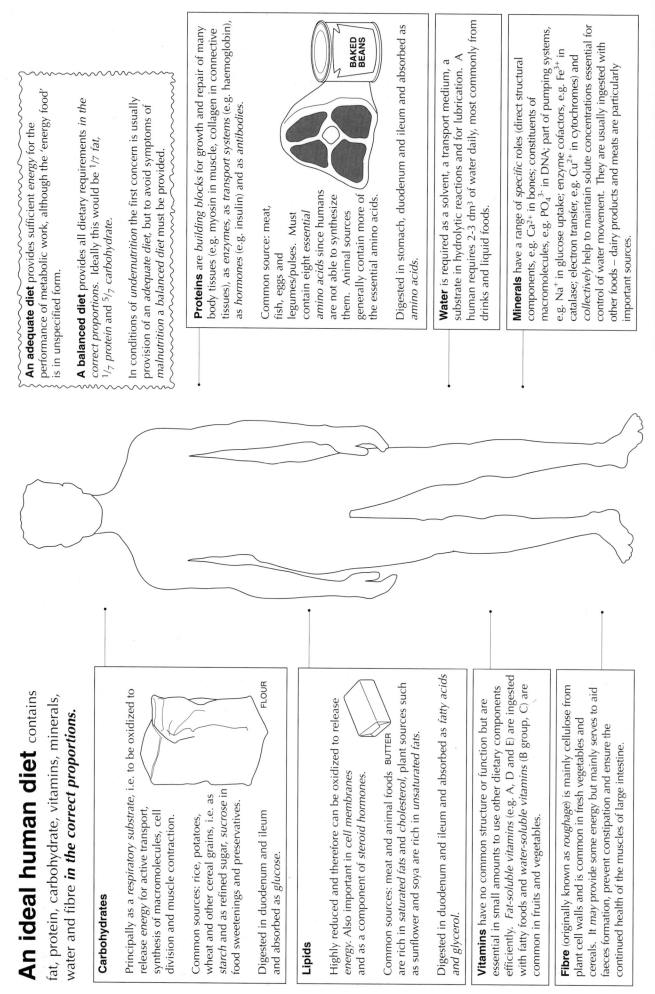

An adequate diet provides sufficient *energy* for the performance of metabolic work, although the 'energy food' is in unspecified form.

A balanced diet provides all dietary requirements *in the correct proportions*. Ideally this would be $^1/_7$ *fat*, $^1/_7$ *protein* and $^5/_7$ *carbohydrate*.

In conditions of *undernutrition* the first concern is usually provision of an *adequate* diet, but to avoid symptoms of *malnutrition* a balanced diet must be provided.

Proteins are *building blocks* for growth and repair of many body tissues (e.g. myosin in muscle, collagen in connective tissues), as *enzymes*, as *transport systems* (e.g. haemoglobin), as *hormones* (e.g. insulin) and as *antibodies*.

Common source: meat, fish, eggs and legumes/pulses. Must contain eight *essential amino acids* since humans are not able to synthesize them. Animal sources generally contain more of the essential amino acids.

Digested in stomach, duodenum and ileum and absorbed as *amino acids*.

Water is required as a solvent, a transport medium, a substrate in hydrolytic reactions and for lubrication. A human requires 2-3 dm^3 of water daily, most commonly from drinks and liquid foods.

Minerals have a range of *specific* roles (direct structural components, e.g. Ca^{2+} in bones; constituents of macromolecules, e.g. PO_4^{3-} in DNA; part of pumping systems, e.g. Na^+ in glucose uptake; enzyme cofactors, e.g. Fe^{3+} in catalase; electron transfer, e.g. Cu^{2+} in cytochromes) and *collectively* help to maintain solute concentrations essential for control of water movement. They are usually ingested with other foods – dairy products and meats are particularly important sources.

An ideal human diet contains fat, protein, carbohydrate, vitamins, minerals, water and fibre *in the correct proportions.*

Carbohydrates

Principally as a *respiratory substrate*, i.e. to be oxidized to release *energy* for active transport, synthesis of macromolecules, cell division and muscle contraction.

Common sources: rice, potatoes, wheat and other cereal grains, i.e. as *starch* and as refined sugar, *sucrose* in food sweetenings and preservatives.

Digested in duodenum and ileum and absorbed as *glucose*.

Lipids

Highly reduced and therefore can be oxidized to release *energy*. Also important in *cell membranes* and as a component of *steroid hormones*.

Common sources: meat and animal foods are rich in *saturated fats* and *cholesterol*, plant sources such as sunflower and soya are rich in *unsaturated fats*.

Digested in duodenum and ileum and absorbed as *fatty acids and glycerol*.

Vitamins

Vitamins have no common structure or function but are essential in small amounts to use other dietary components efficiently. *Fat-soluble vitamins* (e.g. A, D and E) are ingested with fatty foods and *water-soluble vitamins* (B group, C) are common in fruits and vegetables.

Fibre

Fibre (originally known as *roughage*) is mainly cellulose from plant cell walls and is common in fresh vegetables and cereals. It *may* provide some energy but mainly serves to aid faeces formation, prevent constipation and ensure the continued health of the muscles of large intestine.

Human digestive system I

Uvula: extension of soft palate which separates nasal chamber from pharynx.

Epiglottis: muscular flap which reflexly closes the trachea during swallowing to prevent food entry to respiratory tree.

Oesophagus: muscular tube which is dorsal to the trachea and connects the buccal cavity to the stomach. Muscular to generate peristaltic waves, which drive bolus of food downwards, and glandular to lubricate bolus with mucus. Semi-solid food passes to stomach in 4–8 seconds, very soft foods and liquids take only 1 second.

Cardiac sphincter: allows entry of food to stomach. Helps to retain food in stomach.

Stomach: a muscular bag which is distensible to permit storage of large quantities of food. Mucosal lining is glandular, with numerous gastric pits which secrete digestive juices. Three muscle layers including an oblique layer churn the stomach contents to ensure thorough mixing and eventual transfer of chyme to the duodenum.

Pyloric sphincter: opens to permit passage of chyme into duodenum and closes to prevent backflow of food from duodenum to stomach.

Colon (large intestine): absorbs water from faeces. Some B vitamins and vitamin K are synthesized by colonic bacteria. Mucus glands lubricate faeces.

Rectum: stores faeces before expulsion.

There are only *radial* muscle fibres in sphincters (including the bladder sphincter).
When the fibres are *relaxed* the sphincter is *closed.*
When the fibres are *contracted* the sphincter is *open.*
Since the sphincter is closed for most of the time, the muscle fibres are relaxed and there is no fatigue.

Palate: separates breathing and feeding pathways, allowing both processes to go on simultaneously so different types of teeth evolved.

Teeth: cut, tear and grind food so that solid foods are reduced to smaller particles for swallowing, and the food has a larger surface area for enzyme action.

Salivary glands: produce *saliva,* which is 99% water plus mucin, chloride ions (activate amylase), hydrogen carbonate and phosphate (maintain pH about 6.5), lysozyme and salivary amylase.

Parotoid
Sublingual
Submandibular

Tongue: manoeuvres food for chewing and rolls food into a bolus for swallowing. Mixes food with saliva.

Diaphragm: a muscular 'sheet' separating the thorax and abdomen.

Liver: an accessory organ which produces bile and stores it in the gall bladder.

Bile duct: carries bile from gall bladder to duodenum.

Pancreas: an accessory organ producing a wide range of digestive secretions, as well as hormones.

Duodenum: the first 30 cm of the small intestine. Receives pancreatic secretions and bile and produces an alkaline mucus for protection, lubrication and chyme neutralization.

Ileum: up to 6 m in length – main site for absorption of soluble products of digestion.

Appendix: no function in humans. It is a vestige of the caecum in other mammals (herbivores).

Anal sphincter: regulates release of faeces (defecation).

Human digestive system II

Digestion of protein, fat and carbohydrate

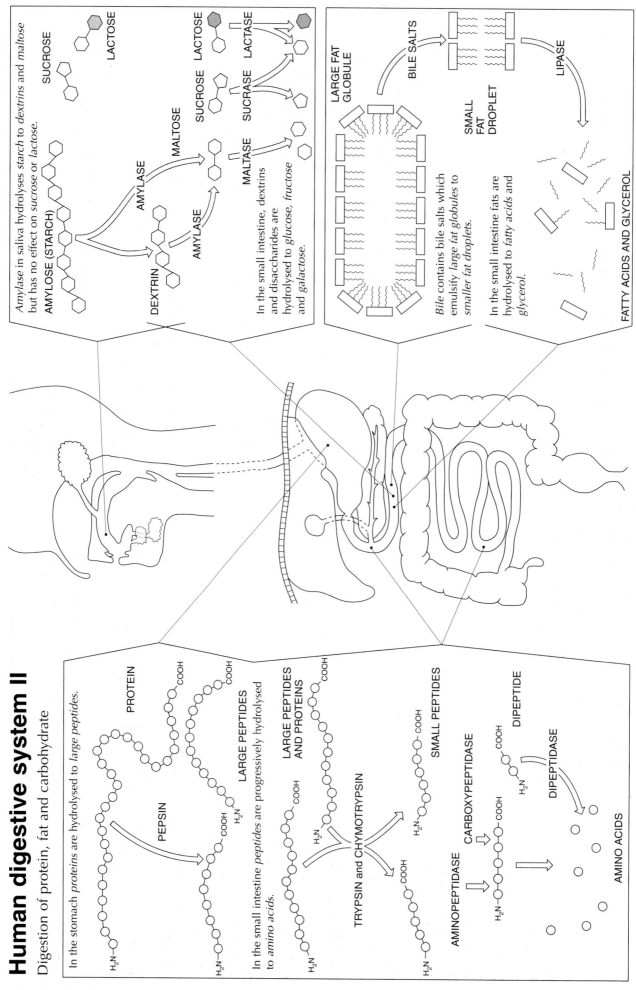

In the stomach *proteins* are hydrolysed to *large peptides*.

PROTEIN

PEPSIN

LARGE PEPTIDES

In the small intestine *peptides* are progressively hydrolysed to *amino acids*.

LARGE PEPTIDES AND PROTEINS

TRYPSIN and CHYMOTRYPSIN

SMALL PEPTIDES

AMINOPEPTIDASE

CARBOXYPEPTIDASE

DIPEPTIDE

DIPEPTIDASE

AMINO ACIDS

Amylase in saliva hydrolyses starch to *dextrins* and *maltose* but has no effect on *sucrose* or *lactose*.

SUCROSE

LACTOSE

AMYLOSE (STARCH)

AMYLASE

MALTOSE

DEXTRIN

AMYLASE

AMYLASE

MALTASE

MALTOSE

SUCROSE

SUCRASE

LACTASE

SUCROSE

LACTOSE

LACTASE

In the small intestine, dextrins and disaccharides are hydrolysed to glucose, fructose and galactose.

LARGE FAT GLOBULE

BILE SALTS

SMALL FAT DROPLET

LIPASE

Bile contains bile salts which emulsify *large fat globules* to *smaller fat droplets*.

In the small intestine fats are hydrolysed to *fatty acids and glycerol*.

FATTY ACIDS AND GLYCEROL

Production of saliva is under **nervous control** at both **conscious** and **unconscious** levels

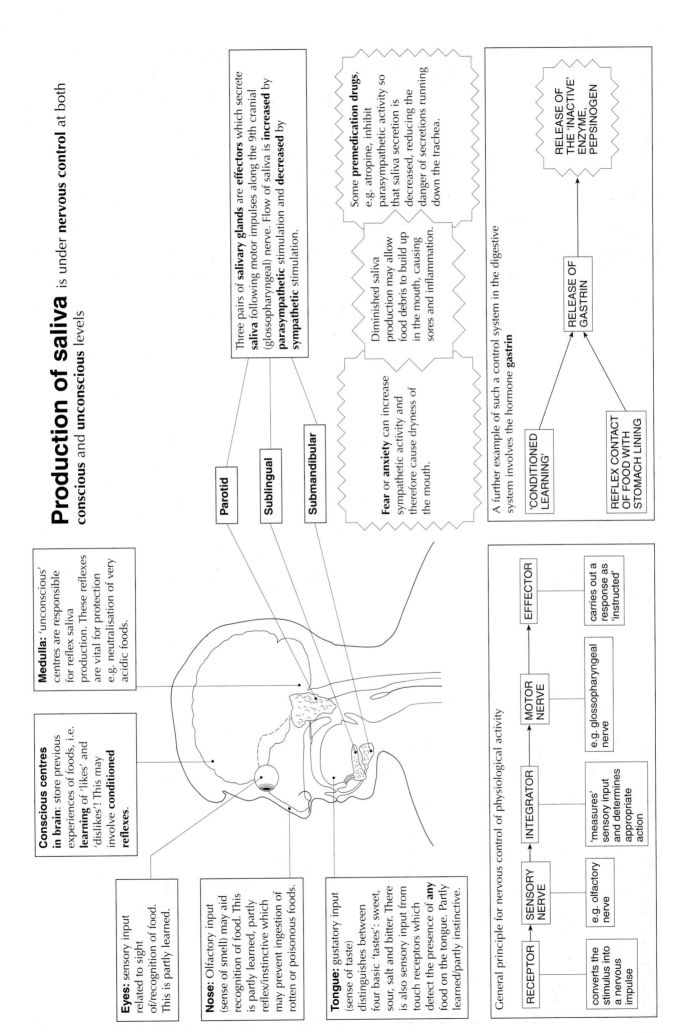

Eyes: sensory input related to sight of/recognition of food. This is partly learned.

Nose: Olfactory input (sense of smell) may aid recognition of food. This is partly learned, partly reflex/instinctive which may prevent ingestion of rotten or poisonous foods.

Tongue: gustatory input (sense of taste) distinguishes between four basic 'tastes': sweet, sour, salt and bitter. There is also sensory input from touch receptors which detect the presence of **any** food on the tongue. Partly learned/partly instinctive.

Conscious centres in brain: store previous experiences of foods, i.e. **learning** of 'likes' and 'dislikes'! This may involve **conditioned reflexes.**

Medulla: 'unconscious' centres are responsible for reflex saliva production. These reflexes are vital for protection e.g. neutralisation of very acidic foods.

Parotid

Sublingual

Submandibular

Three pairs of **salivary glands** are **effectors** which secrete **saliva** following motor impulses along the 9th cranial (glossopharyngeal) nerve. Flow of saliva is **increased** by **parasympathetic** stimulation and **decreased** by **sympathetic** stimulation.

Some **premedication drugs**, e.g. atropine, inhibit parasympathetic activity so that saliva secretion is decreased, reducing the danger of secretions running down the trachea.

Diminished saliva production may allow food debris to build up in the mouth, causing sores and inflammation.

Fear or **anxiety** can increase sympathetic activity and therefore cause dryness of the mouth.

A further example of such a control system in the digestive system involves the hormone **gastrin**

'CONDITIONED LEARNING'

REFLEX CONTACT OF FOOD WITH STOMACH LINING

→ RELEASE OF GASTRIN → RELEASE OF THE 'INACTIVE' ENZYME, PEPSINOGEN

General principle for nervous control of physiological activity

RECEPTOR → SENSORY NERVE → INTEGRATOR → MOTOR NERVE → EFFECTOR

- converts the stimulus into a nervous impulse
- e.g. olfactory nerve
- 'measures' sensory input and determines appropriate action
- e.g. glossopharyngeal nerve
- carries out a response as 'instructed'

Transverse section of a generalised gut

Serosa (serous coat): composed of connective tissue and epithelium. Protects inner layers from abrasion, and provides support for lymphatic and blood vessels. Also called the **visceral peritoneum.**

Lymph nodule (Peyer's patch): aggregate of lymphocytes to help prevent infection, since mucosa is very exposed to microbes in food.

Mucosa: surface epithelium with secretary cells (type depends on site – typically some mucus-secreting goblet cells are present).

lamina propria (loose fibrous tissue with mucosal glands, and blood and lymphatic vessels)

Lumen: the gut cavity, and the site of digestion. It is from the lumen that the products of digestion are absorbed.

Muscularis mucosa: a thin muscle layer which can help local movements of areas of villi.

Submucosa: contains dense connective tissue and some adipose tissue which supports the nervous and blood supply for the mucosa.

Submucosal gland: only in the duodenum, when the **Brunner's glands** produce an alkaline, mucus secretion.

Mesentery: connective tissue sheet which suspends the gut in the abdomen.

Longitudinal muscle and **circular muscle** is smooth muscle (except for oesophagus) which helps to break down food physically, mixes it with secretions and propels food through the digestive tract by *peristalsis.*

Auerbach's plexus: the major nerve supply to the gut. Consists of both sympathetic and para-sympathetic fibres and mostly controls gut motility.

Meissner's plexus: part of the autonomic supply to the muscularis mucosa and also has a role in controlling secretions by the gut.

External (accessory) gland: salivary glands, pancreas and liver all produce secretions which are poured into the lumen via ducts.

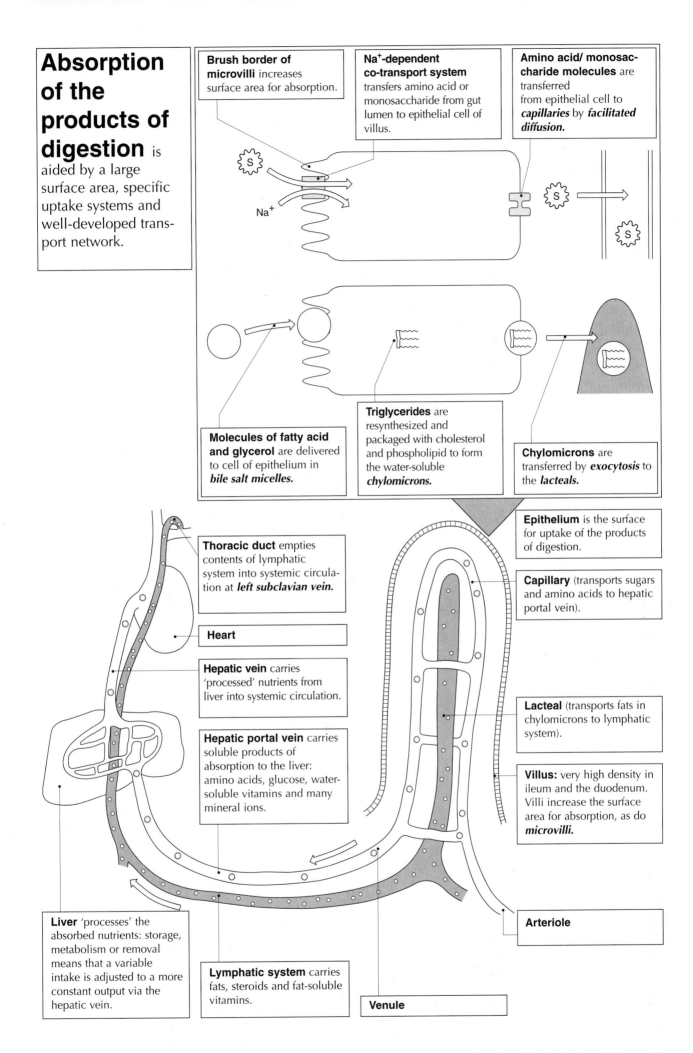

Absorption of the products of digestion is aided by a large surface area, specific uptake systems and well-developed transport network.

Brush border of microvilli increases surface area for absorption.

Na⁺-dependent co-transport system transfers amino acid or monosaccharide from gut lumen to epithelial cell of villus.

Amino acid/ monosaccharide molecules are transferred from epithelial cell to *capillaries* by *facilitated diffusion*.

Molecules of fatty acid and glycerol are delivered to cell of epithelium in *bile salt micelles*.

Triglycerides are resynthesized and packaged with cholesterol and phospholipid to form the water-soluble *chylomicrons*.

Chylomicrons are transferred by *exocytosis* to the *lacteals*.

Thoracic duct empties contents of lymphatic system into systemic circulation at *left subclavian vein.*

Heart

Hepatic vein carries 'processed' nutrients from liver into systemic circulation.

Hepatic portal vein carries soluble products of absorption to the liver: amino acids, glucose, water-soluble vitamins and many mineral ions.

Epithelium is the surface for uptake of the products of digestion.

Capillary (transports sugars and amino acids to hepatic portal vein).

Lacteal (transports fats in chylomicrons to lymphatic system).

Villus: very high density in ileum and the duodenum. Villi increase the surface area for absorption, as do *microvilli.*

Arteriole

Liver 'processes' the absorbed nutrients: storage, metabolism or removal means that a variable intake is adjusted to a more constant output via the hepatic vein.

Lymphatic system carries fats, steroids and fat-soluble vitamins.

Venule

Undernutrition

can be **general** (causing starvation) or **specific** (lack of a single nutrient). There are several dominant conditions worldwide.

ANAEMIA is caused by a low haemoglobin nutritional deficiency of **iron, vitamin B$_{12}$** or of the **amino acids** required for haemoglobin synthesis. Vitamin C promotes the absorption of iron, and high fibre diets which contain phytate may reduce iron bioavailability.

There may be 150 000 000 sufferers worldwide.

XEROPHTHALMIA, which is also called **dry eye**, is caused by a **lack of vitamin A**. The absence of this precursor leads to low rhodopsin synthesis (causing poor dark adaptation). It also causes keratinisation of epithelia, and is often first seen as a roughening of the skin. Prolonged deficiency causes blindness (true xerophthalmia) in about 20 000 people per annum.

Good nourishment allows the liver to store sufficient vitamin A to last for several years, but there is little retinol in the typical vegetarian diet of the Third World person.

GOITRE is characterised by a swelling of the neck and is caused by a **lack of iodine**. The condition is exacerbated by **goitrogens**, which inhibit iodine uptake by the thyroid gland. Goitre in pregnant females may lead to the birth of a **cretin** – a child with a conspicuously low Intelligence Quotient (I.Q.).

There are about 200 000 000 sufferers worldwide.

RICKETS, which causes poor bone formation, is caused by a **lack of calcium** or of **vitamin D**. It is very widespread in Asia and in Africa and, as well as causing locomotory difficulties, it leads to childbirth problems if the deficiency occurs in young women.

THE VALUE OF A MIXED DIET

To avoid malnutrition it is important to eat an adequate amount of a wide variety of foods. This is particularly the case if the diet is largely of vegetable origin.

A mixed diet:
- increases food acceptability since a monotonous diet can lead to undereating and energy deficiencies. This may initiate protein catabolism in an attempt to compensate;
- offers a better balance of vitamins and minerals; and
- dilutes toxins or pollutants which might be present in any particular food.

Ideally the diet will be balanced, i.e. will provide all of the essential nutrients in the correct proportions. For example, no more than 14% of the diet should be composed of fats, and it is recommended that at least 50% of kilojoules are supplied by complex polysaccharides.

PROTEIN ENERGY MALNUTRITION (PEM)

Marasmus and **Kwashiorkor** are the two extreme manifestations of PEM, although there are many intermediate cases.

Marasmus occurs most frequently in children aged between 9 and 12 months, and is the result of dietary deficiency of both *kilojoules (energy)* and *growth foods (protein)*.

A typical sequence of development of the disease is

| early weaning/end of breast feeding |
→
| weight loss and weakness |
→
| frequent infections and diarrhoea |
→
| 'starvation' therapy in an attempt to control diarrhoea |
→
| emaciation, wrinkled skin, misery |

Marasmus

Normal hair

Old person's face

Thin muscles, thin fat

No oedema

Very underweight

There are few biochemical changes noted, i.e. the sufferer shows adaptation. As a result recovery may be impossible, and a return to the 'normal' diet can be dangerous. The adaptation may include atrophy of the pancreas and intestinal mucosa, which means that the sufferer can produce very few digestive enzymes and has little absorptive area to take up any soluble foods which are available.

Kwashiorkor in African means 'the rejected one', since the sufferer is commonly a child which, after a prolonged period of breast feeding, is displaced from the mother's breast by the arrival of a new baby.

A typical pattern of disease development is

Kwashiorkor

Hair changes

Misery

Moon face

Thin muscles, fat present

Oedema

Underweight

| replacement of milk with 'starchy' family diet |
→
| protein deficiency in diet |
→
| oedema (retention of fluid in tissues) |
→
| limited development of skin (turns 'flaky'), hair (loses pigmentation and falls out) and muscle |
→
| lethargy and little interest in surroundings |

A child suffering from Kwashiorkor shows certain **biochemical changes** which include the accumulation of fat in the liver (causes cirrhosis) and a severe reduction in blood albumin level (responsible for the oedema).

Children who survive PEM are often retarded mentally, and recovery is poor since during the first three years of life the brain grows to 80% of its full adult size (and is composed of up to 50% dry weight of protein).

DEFICIENCIES OF THE VEGETARIAN DIET

Iron: is very much more common in meats, especially liver, and anaemia is three times more common in vegetarians than in non-vegetarians. Green leafy vegetables are a good source, but menstruating women may need an iron supplement.

Calcium: often very low in vegetarian diets, especially if the hull of grains and legumes are removed. Calcium can easily be provided if milk products are part of the diet, but the problem can be made worse by the high fibre content of vegetarian diets since phytate in fibre lowers the bioavailability of calcium.

Vitamin B$_{12}$ is made by micro-organisms and, apart from wheatgerm and beer, is not present in vegetable products. Deficiency leads to pernicious anaemia, since this vitamin is essential for the production of erythrocytes.

Vitamin D: not found in plant foods, but very common in dairy products so not a problem for ovolactovegetarians. Vitamin D is synthesised in the skin if exposure to sunlight is adequate.

Protein: the most quoted failing of vegetarian diets. Vegetable protein does not contain all amino acids in the correct relative proportions for the synthesis of animal proteins – for example corn is deficient in lysine and beans in methionine. The key is to select a combination of plant foods which complement one another.

The eight essential amino acids for humans

Trytophan
Methionine
Valine
Threonine
Phenylalanine
Leucine
Isoleucine
Lysine

Corn and other grains

Beans and other legumes

Third World countries may have a vegetarian diet dictated not by choice but by sheer economics. Since the diet is likely to be dominated by a single 'energy' food, protein deficiency diseases such as Kwashiokor and Marasmus are far more likely to occur in these areas.

BENEFITS OF THE VEGETARIAN DIET

Current research suggests a reduction in

- hypertension
- colon cancer
- breast cancer
- mature-onset diabetes
- diverticulitis
- dental caries
- cardiovascular disease

in vegetarians, especially those who neither drink nor smoke. These benefits are thought to be largely due to

- increased content of dietary fibre
- higher ratio of unsaturated : saturated fatty acids

Vegetarians may choose a diet of grains, fruits, vegetables and nuts for ethical or nutritional reasons.

A vegetarian diet must, of course, satisfy the basic nutritional demands of an animal, i.e. the provision of fat, protein, carbohydrate, vitamins, minerals, fibre and water.

This can be relatively straightforward if the diet contains a small quantity of animal products, thus

Ovolactovegetarians (who eat milk products and eggs) and

Pescovegetarians (who eat fish) have few problems maintaining a balanced diet.

Cereals (grains) such as rice and wheat, contain carbohydrate (energy), protein, dietary fibre, vitamin B complex, iron and calcium. The vitamins and minerals are largely confined to the husk, and poor cooking or preparative techniques may considerably reduce cereal food value.

Oilseeds (sunflower, for example) provide energy (both carbohydrate and fat), some protein and a range of essential fatty acids. There is a high proportion of unsaturated fatty acids in these oils. The fat-soluble vitamins A and E may be present.

Those at greater risk on a vegetarian diet:

- **children,** since the abundance of fibrous, bulky foods may suppress appetite (that "full up" feeling) and thus limit 'energy' and protein intake.
- **expectant mothers,** with a greater dietary demand for protein, 'energy' and calcium.
- **vegans** (who eat neither fish, dairy products nor eggs) may need to supplement their diet with calcium, vitamin B$_{12}$ and vitamin D. There may also be deficiencies in some essential fatty acids which are found almost exclusively in fish.

Fruits may be rich in sugars (glucose, fructose and sucrose), pectins and gums (soluble dietary fibre), vitamin C (particularly citrus fruits, blackcurrants, tomatoes and peppers), and provide small quantities of vitamin K.

Leaves provide dietary fibre, vitamins C, A, E and K. Some (spinach, watercress) are rich in iron.

Legumes provide protein in large amounts, carbohydrate (energy), fibre, iron, calcium and vitamin B complex. Any vegetarian diet should contain legume material such as peas or beans.

Roots such as carrot, beet and cassava are rich in complex polysaccharides (energy), and provide small quantities of protein and vitamin C. Potato is not a root (it is a stem tuber) but as it grows underground is often treated as one.

Diets and dieting

are usually associated with attempts to control body mass, although there may be other reasons for their use

MICRODIETS MAY SERVE SPECIAL FUNCTIONS

Some diets are designed not for weight loss but for reduction of some other medical problem. For example,

Gluten-free diets (gluten is a protein found in cereals, including wheat) are prescribed for sufferers which coeliac disease in which an immune response destroys the villi and causes starvation through the inability to absorb nutrients.

Reduced-sodium diets for sufferers from hypertension

Reduced-sugar diets for diabetics

Iron-enriched diets for sufferers of anaemia, particularly women at menstruation

Evening primrose oil for individuals with multiple sclerosis

Extra carrots for nocturnal bunnies!

REMEMBER – An Adequate Diet must provide **fats**, **proteins**, **carbohydrates** in the correct proportions, and **vitamins**, **minerals**, **water** and **fibre**.

Typical weight reduction diets might

* limit alcohol intake
* reduce fat intake, which includes the elimination of 'hidden fat' foods such as cakes and biscuits
* reduce the intake of refined, simple sugars
* increase the intake of dietary fibre (the F-plan diet) so that food is passed through the alimentary canal at a greater rate to limit the absorption of 'potential energy'

DIETS FOR WEIGHT CONTROL

The most successful programme for weight control is one which includes

controlled diet and exercise

limit the intake of kjoules | increase the consumption of kjoules

The reasons for employing a weight-reducing diet may be

cosmetic: a desire to fit into a socially-acceptable body shape

health: problems associated with obesity include increased incidence of diabetes, cardiovascular accidents and arthritis

Weight reduction programmes which include an exercise regime are more likely to be successful, since

a. exercise may promote metabolism, thus increasing the kJ demand rather than allowing an adaptation of the type described opposite.

b. exercise may itself improve body form (muscle tone and posture) so that the dieter is encouraged to continue.

Statistics suggest that a dieter who is able to record a weight loss of 4–6 kg over a four week period is likely to continue with such a weight reduction programme.

'CRASH' DIETS (those which attempt to lose 2 - 3 kg per week) include

The 'Milk' Diet: comprises 1800 cm³ (3 pints) of milk with supplementary iron, vitamins and inert bulk laxative such as bran.It provides 4900 kJ and 59g protein per day.

It is cheap and uncomplicated but very monotonous.

The VLCD (very low calorie diet: also known as the 'Cambridge' or 'protein sparing modified diet'. Provides only 1200–1800 kJ and about 40–50g of protein per day.

Sometimes recommended for the very obese, since weight loss can be rapid, but there is a loss of protein from the blood and skin, and the mass of the heart is reduced.

Considered unsuitable for infants, children, pregnant or breast-feeding women, or people with porphyria or gout. Should not be used for more than 3 or 4 weeks as the sole source of nourishment.

'Crash' diets are seldom successful in the long term. Great motivation is necessary to continue with them, they make social contacts difficult, and the body can readjust metabolic rate so that even a reduced kJ intake can lead to fat storage.

stabilisation – reduced metabolism demands fewer kJ

further reduction in metabolic rate as body stores available kJ rather than using them

initial weight loss as metabolic demand for kJ exeeeds intake

weight/kg

desired weight

10000   8000

Energy intake / kJ per day

Metabolism and metabolic rate

BASAL METABOLIC RATE (B.M.R.) is the energy consumed by the body at rest.

This energy is required to perform vital functions such as breathing movements, heartbeat, maintenance of ion gradients across membranes and the synthesis of molecules such as proteins

B.M.R. is measured lying down (no activity) in a warm (no thermal stress) quiet (minimal mental stress) room 12–18 hours after the last meal (no digestive activity).

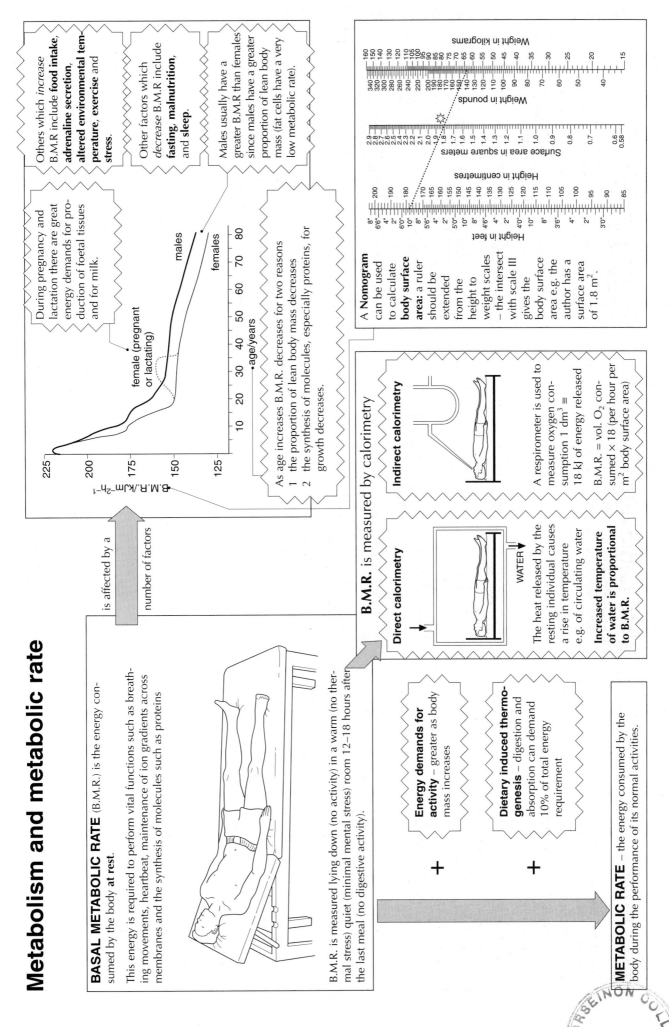

is affected by a number of factors

Others which *increase* B.M.R include **food intake, adrenaline secretion, altered environmental temperature, exercise** and **stress.**

Other factors which *decrease* B.M.R include **fasting, malnutrition,** and **sleep.**

Males usually have a greater B.M.R than females since males have a greater proportion of lean body mass (fat cells have a very low metabolic rate).

During pregnancy and lactation there are great energy demands for production of foetal tissues and for milk.

As age increases B.M.R. decreases for two reasons
1 the proportion of lean body mass decreases
2 the synthesis of molecules, especially proteins, for growth decreases.

B.M.R./kJm⁻²h⁻¹ — $B.M.R./kJ\,m^{-2}\,h^{-1}$

males

females

female (pregnant or lactating)

age/years

A **Nomogram** can be used to calculate **body surface area:** a ruler should be extended from the height to weight scales – the intersect with scale III gives the body surface area e.g. the author has a surface area of 1.8 m².

Weight in kilograms

Weight in pounds

Surface area in square meters

Height in centimetres

Height in feet

B.M.R. is measured by calorimetry

Direct calorimetry

WATER

The heat released by the resting individual causes a rise in temperature e.g. of circulating water

Increased temperature of water is proportional to B.M.R.

Indirect calorimetry

A respirometer is used to measure oxygen consumption 1 dm³ ≡ 18 kJ of energy released

$B.M.R. = $ vol. O_2 consumed $\times 18$ (per hour per m² body surface area)

Energy demands for activity – greater as body mass increases

+

Dietary induced thermogenesis – digestion and absorption can demand 10% of total energy requirement

+

METABOLIC RATE – the energy consumed by the body during the performance of its normal activities.

Food additives

can be defined as 'non-nutritive substances added to food to improve appearance, flavour, texture and/or storage properties'.

They may be

natural extracted directly from natural products

nature identical synthetic but of the same formula as the natural product

synthetic non-natural, although often related formulas

COLOURING (E100 NUMBERS)

Added to improve colour which may have been lost during cooking (e.g. peas become paler) or dilution.

Natural e.g. anthocyanin (E163) from grape skins
β-carotene (E160) from tomatoes

Synthetic – many are derived from coal tar
e.g. Tartrazine (E102)

PREVENTION OF SPOILAGE – PRESERVATIVES (E200 NUMBERS)

These inhibit the growth of fungi or bacteria which might otherwise change the taste or nutritional value of a food

e.g. sulphur dioxide/sulphites (E220–E227) control browning of potatoes
extend life of soft drinks
nitrates (E251) – used in ham and sausages

PREVENTION OF SPOILAGE – ANTI-OXIDANTS (E300 NUMBERS)

These inhibit food deterioration caused by atmospheric oxidation

e.g. vitamin C (ascorbic acid – E300) – added to fruit drinks
vitamin E (tocopherol – E306) – added to margarines/oils

These compounds remove free radicals which otherwise accelerate rancidity.

SOME ADDITIVES MAY BE A LEGAL REQUIREMENT

e.g. calcium/vitamin D added to milk
vitamin E to margarine
vitamin C to fruit juices
vitamin D/protein to bread

NO artificial colourings are allowed to be added to baby foods.

Legal controls
- All must be identified on food packaging, all must be given an 'E number'.
- All must be pre-tested on animals.

ADDITIVES MAY BE DANGEROUS

Tartrazine – linked to hyperactivity / poor concentration in children

Sulphur dioxide – may cause sensitivity in asthmatics

Nitrates – may have carcinogenic (cancer-inducing) or teratogenic (developmental abnormality inducing) properties

Monosodium glutamate – may induce vomiting, migraine and nausea.

** N.B. Some natural compounds e.g. 'colour' in strawberries may cause allergic reactions.

ADDITIVES SHOULD NOT BE CONFUSED WITH RESIDUES

Additives are used to improve food quality, in the ways described above. Residues, on the other hand, are used to increase food production. As a result of their extensive use, or poor food production techniques, these agents may remain in food presented to humans.

Some chemicals which may remain as residues, and which cause concern for human health are

Pesticides – may become concentrated in food chains, causing harm to the top consumers (including humans)

Fertilisers – nitrates and phosphates may leach into water supplies; some problems for children fed on reconstituted powdered milk

Antibiotics – used to prevent infection of cattle/poultry; may lead to selection of antibiotic-resistant strains of microbes

Hormones – oestrogen (promotes chicken growth) may lead to feminisation of males.

IMPROVEMENT OF TEXTURE/CONSISTENCY

Emulsifiers: combine normally immiscible substances
e.g. Lecithin E322

Stabilisers: prevent separation of the components of emulsions
e.g. Xanthan E415

Thickeners: alter consistency of food
e.g. cellulose E460–466.

FLAVOURINGS AND FLAVOUR ENHANCERS

(some E600 numbers, some not numbered)

Flavourings are the commonest food additives – some 3000 are in use – many are natural, but synthetic alternatives are usually cheaper and easier to obtain

e.g. vanilla
propyl pentanoate (pineapple flavour).

Flavour enhancers are not flavourings, but make existing flavours taste stronger by stimulation of the taste buds.

e.g. monosodium glutamate (MSG – E621).

NEW NATU-LOW Guaranteed to contain no food **BUT HIGH IN ADDITIVES**

Food spoilage is caused by enzymes – it can be dangerous.

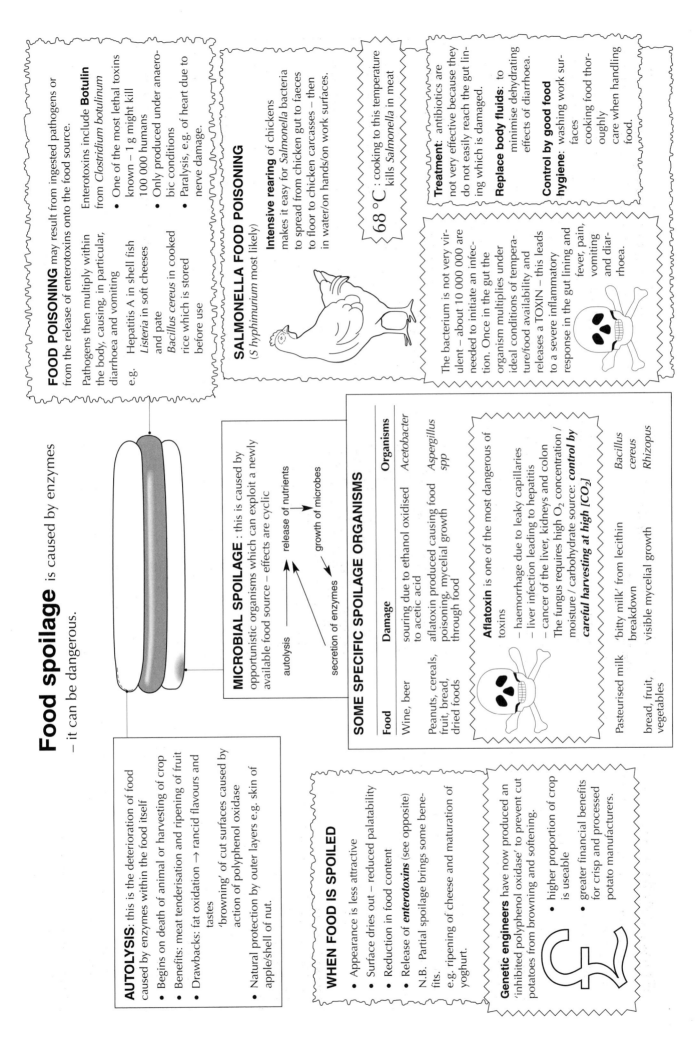

FOOD POISONING

FOOD POISONING may result from ingested pathogens or from the release of enterotoxins onto the food source.

Pathogens then multiply within the body, causing, in particular, diarrhoea and vomiting

e.g. Hepatitis A in shell fish
Listeria in soft cheeses and pate
Bacillus cereus in cooked rice which is stored before use

Enterotoxins include **Botulin** from *Clostridium botulinum*
- One of the most lethal toxins known – 1 g might kill 100 000 humans
- Only produced under anaerobic conditions
- Paralysis, e.g. of heart due to nerve damage.

SALMONELLA FOOD POISONING
(*S hyphimurium* most likely)

Intensive rearing of chickens makes it easy for *Salmonella* bacteria to spread from chicken gut to faeces to floor to chicken carcasses – then in water/on hands/on work surfaces.

$68\,°C$: cooking to this temperature kills *Salmonella* in meat

The bacterium is not very virulent – about 10 000 000 are needed to initiate an infection. Once in the gut the organism multiplies under ideal conditions of temperature/food availability and releases a TOXIN – this leads to a severe inflammatory response in the gut lining and fever, pain, vomiting and diarrhoea.

Treatment: antibiotics are not very effective because they do not easily reach the gut lining which is damaged.

Replace body fluids: to minimise dehydrating effects of diarrhoea.

Control by good food hygiene: washing work surfaces
cooking food thoroughly
care when handling food.

AUTOLYSIS

AUTOLYSIS: this is the deterioration of food caused by enzymes within the food itself
- Begins on death of animal or harvesting of crop
- Benefits: meat tenderisation and ripening of fruit
- Drawbacks: fat oxidation → rancid flavours and tastes
'browning' of cut surfaces caused by action of polyphenol oxidase
- Natural protection by outer layers e.g. skin of apple/shell of nut.

WHEN FOOD IS SPOILED

- Appearance is less attractive
- Surface dries out – reduced palatability
- Reduction in food content
- Release of *enterotoxins* (see opposite)

N.B. Partial spoilage brings some benefits.
e.g. ripening of cheese and maturation of yoghurt.

Genetic engineers have now produced an 'inhibited polyphenol oxidase' to prevent cut potatoes from browning and softening.
- higher proportion of crop is useable
- greater financial benefits for crisp and processed potato manufacturers.

MICROBIAL SPOILAGE

MICROBIAL SPOILAGE : this is caused by opportunistic organisms which can exploit a newly available food source – effects are cyclic

autolysis → release of nutrients → growth of microbes → secretion of enzymes → (cyclic)

SOME SPECIFIC SPOILAGE ORGANISMS

Food	Damage	Organisms
Wine, beer	souring due to ethanol oxidised to acetic acid	*Acetobacter*
Peanuts, cereals, fruit, bread, dried foods	aflatoxin produced causing food poisoning, mycelial growth through food	*Aspergillus spp*
Pasteurised milk	'bitty milk' from lecithin breakdown	*Bacillus cereus*
bread, fruit, vegetables	visible mycelial growth	*Rhizopus*

Aflatoxin is one of the most dangerous of toxins
– haemorrhage due to leaky capillaries
– liver infection leading to hepatitis
– cancer of the liver, kidneys and colon
The fungus requires high O_2 concentration / moisture / carbohydrate source: *control by careful harvesting at high* $[CO_2]$

Food preservation

involves inhibition of microbial activity.

HIGH OSMOLARITY – salt or sugar can be used to generate high solute concentrations in foods. Such conditions inhibit the growth of micro-organisms which inevitably lose water by exosmosis.

FREEZING – most fresh foods contain over 60% water. If water is frozen it cannot be used by micro-organisms.

There are **positive** points

very little loss of nutritive value – 'quick frozen' food may have more nutrients than food which is 'fresh' but takes 2–3 days to reach point of sale

... but some **negative** points too

expansion of water → cell damage → 'mush' (e.g. strawberries)

'drip' from frozen food → loss of soluble nutrients (e.g. vitamin C) on thawing.

DEHYDRATION – one of the most efficient methods involves *freeze drying*: freeze rapidly then rewarm under reduced pressure → ice sublimation → **porous structure**

Benefits: better for rehydration allied to an N_2-containing atmosphere offers 2–3 years storage.

Drawback: any fats present are easily oxidised (**rancidity**) because of open structure.

IRRADIATION – although sterilisation by radiation is permitted for medical supplies and drugs it is not permitted for food preservation in the UK.

^{60}cobalt

^{137}caesium

γ-radiation (dose required to kill microbes is usually greater than does permitted for humans, but should decay before consumption)

... but
- May be some induced radiation
- Very expensive, so use tends to be confined to high-cost foods e.g. prawns
- some loss of vitamins A, B, C and E.

N.B. not yet allowed in UK (although currently the only effective method for reducing *Salmonella* in frozen meat).

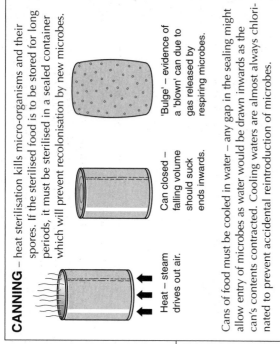

CANNING – heat sterilisation kills micro-organisms and their spores. If the sterilised food is to be stored for long periods, it must be sterilised in a sealed container which will prevent recolonisation by new microbes.

'Bulge' – evidence of a 'blown' can due to gas released by respiring microbes.

Can closed – falling volume should suck ends inwards.

Heat – steam drives out air.

Cans of food must be cooled in water – any gap in the sealing might allow entry of microbes as water would be drawn inwards as the can's contents contracted. Cooling waters are almost always chlorinated to prevent accidental reintroduction of microbes.

The **botulinum cook** – the use of pressure cookers allows boiling water to exceed 100 °C. The modern HTST (high temperature, short time) technique allows a temperature of 121 °C for three minutes. These conditions will even kill the spores of *Clostridium botulinum* – this organism produces a toxin – botulin – which is so toxic that it is estimated that 500 g would kill almost the whole population of the UK.

CONTROL OF pH – microbes may grow more slowly under acidic conditions.

Yoghurt — Lactic acid is produced by lactose fermentation

Pickles — Preserved by added vinegar (3% aqueous solution of ethanoic acid).

CHEMICAL PRESERVATIVES – include sulphur dioxide and sulphites, benzoic acid and benzoates and nitrates/nitrites.

- Typically act as **anti-oxidants** and inhibit auto-oxidation of fats
- Removal of O_2 gas and ↓ pH both limit growth of microbes
- SO_2/SO_2^-: widely used in sausages/dried fruits/soft drinks
- Benzoates: soft drinks
- NO_3^-/NO_2^-: meats e.g. ham/bacon are cured this way; some danger from carcinogenic **nitrosamines** ($NO_2^- + -NH_2$ in food → $-NONH_2$).

Before vegetable foods are dehydrated (or, indeed, stored in any other way) they are **blanched** in boiling water or steam. This denatures enzymes such as catalase and ascorbic acid oxidase and improves both appearance and storage life of the food.

Food packaging serves a number of purposes:

- provision of information e.g. energy content
- advertising/attraction of consumer
- prevention of spoilage by microbes.

Before vegetable foods are dehydrated (or, indeed, stored in any other way) they are **blanched** in boiling water or steam. This denatures enzymes such as **catalase** and **ascorbic acid oxidase** and improves both appearance and storage life of the food.

LABELLING should include
- energy content
- salt and saturated fat content
- additives (with E numbers)
- allergens (e.g. peanuts).

Cartons may be lined with a **formaldehyde resin** to prevent deterioration of the paper/cardboard. This resin may migrate into the food.

WAXED PAPERS
- limit water movement between food and atmosphere so control hydration/dehydration and regulate weight of product
- limit water availability so prevent the multiplication of microbes.

MEAT
has a colour which is attractive to consumers. The red colour is due primarily to myoglobin in the muscle, rather than haemoglobin of the blood – the colour of the meat surface depends on the type of myoglobin molecule and also on its chemical state.

$$Fe^{II} \rightarrow Fe^{III}$$

metamyoglobin (brown)

$+ O_2$ → oxymyoglobin (red)

Packaging must allow O_2 in (consumers prefer 'red' meat) but prevent water loss.

The bright pink colour of **cured** meat is due to **nitrosomyoglobin**, formed by a combination of myoglobin with nitrogen(II) oxide (nitric acid) produced from nitrates and nitrites used in the curing process. Oxygen is *not* required to maintain this colour, so packaging materials for cured meat products can be impermeable to oxygen.

PLASTICS
are very malleable (can be easily shaped) and so are valuable packaging material for a variety of food products, particularly liquids.
There are a number of concerns about the use of stabilisers and flow-inducing compounds in the manufacture of plastics:

- compounds added to plastics to improve 'flow' into moulding equipment may be **oestrogenic** (promote production of female sex hormones)
- oestrogenic compounds may leach into contents → reduced sperm count and diminished masculinity!

POLYETHYLENE
film may offer controlled humidity and gas exchange:

allows CO_2 out/limits O_2 in → controlled rate of respiration

N.B. may have leaching of fat-soluble compounds from film into fatty foods (e.g. cheese).

Cans are made from a **tin-based steel**. They must be lined with a wax-like layer if they are used to store acidic foods such as citrus fruits: 'colour' such as flavenoids may react with tin. Canned foods may become pale.

VACUUM PACKING:
- excludes O_2/H_2O
- no germination of seeds in foods
- no multiplication of microbes
- combined with dehydration → lightweight foods (e.g. for astronauts).

VACU-SAVE

Dehydrated water

CONTROLLED ATMOSPHERE:
the relative concentration of carbon dioxide and oxygen can control rates of respiration and ripening.

↑CO_2/↓O_2 to slow rate of **ripening** (N.B. do **not** allow to become anaerobic → nasty, unpleasant tastes and smells).

High [CO_2] → pectinase activity → firmer fruits

Also regulated by **ethene** which acts as a 'switch' for enzymes involved in ripening.

Maxi-Flakes

GRAPEFRUIT

NUTS

Microbes and milk

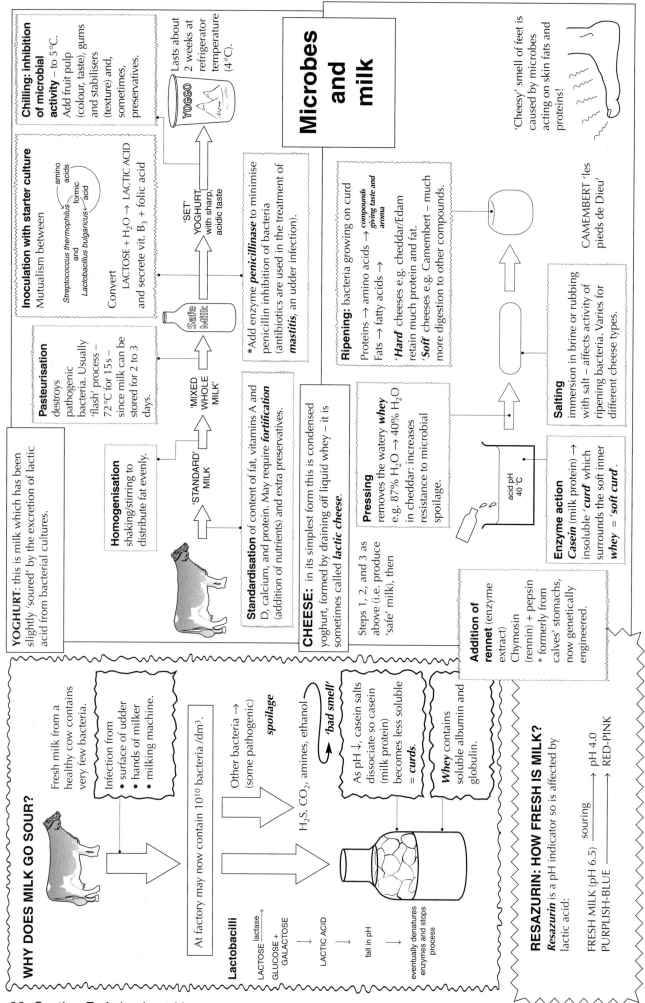

Chilling: inhibition of microbial activity – to 5°C. Add fruit pulp (colour, taste), gums and stabilisers (texture) and, sometimes, preservatives.

Lasts about 2 weeks at refrigerator temperature (4°C).

YOGGO

Inoculation with starter culture

Mutualism between

Streptococcus thermophilus and *Lactobacillus bulgaricus*

amino acids → formic acid

Convert

LACTOSE + H_2O → LACTIC ACID and secrete vit. B_3 + folic acid

'SET' YOGHURT, with sharp, acidic taste

Pasteurisation destroys pathogenic bacteria. Usually 'flash' process – 72°C for 15 s – since milk can be stored for 2 to 3 days.

Safe Milk

*Add enzyme **penicillinase** to minimise penicillin inhibition of bacteria (antibiotics are used in the treatment of **mastitis**, an udder infection).

Homogenisation shaking/stirring to distribute fat evenly.

'STANDARD' MILK

Standardisation of content of fat, vitamins A and D, calcium, and protein. May require **fortification** (addition of nutrients) and extra preservatives.

'MIXED WHOLE MILK'

Ripening: bacteria growing on curd

Proteins → amino acids → *compounds giving taste and aroma*
Fats → fatty acids →

'**Hard**' cheeses e.g. cheddar/Edam retain much protein and fat.
'**Soft**' cheeses e.g. Camembert – much more digestion to other compounds.

Salting immersion in brine or rubbing with salt – affects activity of ripening bacteria. Varies for different cheese types.

CAMEMBERT 'les pieds de Dieu'

'Cheesy' smell of feet is caused by microbes acting on skin fats and proteins!

YOGHURT: this is milk which has been slightly 'soured' by the excretion of lactic acid from bacterial cultures.

CHEESE: in its simplest form this is condensed yoghurt, formed by draining off liquid whey – it is sometimes called *lactic cheese.*

Steps 1, 2, and 3 as above (i.e. produce 'safe' milk), then

Pressing removes the watery *whey* e.g. 87% H_2O → 40% H_2O in cheddar: increases resistance to microbial spoilage.

acid pH 40 °C

Addition of rennet (enzyme extract)
Chymosin (rennin) + pepsin
* formerly from calves' stomachs, now genetically engineered.

Enzyme action *Casein* (milk protein) → insoluble '*curd*' which surrounds the soft inner whey = '*soft curd*'.

WHY DOES MILK GO SOUR?

Fresh milk from a healthy cow contains very few bacteria.

Infection from
• surface of udder
• hands of milker
• milking machine.

At factory may now contain 10^{10} bacteria /dm³.

Other bacteria → (some pathogenic) *spoilage*

H_2S, CO_2, amines, ethanol → '*bad smell*'

As pH ↓, casein salts dissociate so casein (milk protein) becomes less soluble = *curds*.

Whey contains soluble albumin and globulin.

Lactobacilli

LACTOSE → lactase →

GLUCOSE + GALACTOSE →

LACTIC ACID →

fall in pH →

eventually denatures enzymes and stops process

RESAZURIN: HOW FRESH IS MILK?

Resazurin is a pH indicator so is affected by lactic acid:

FRESH MILK (pH 6.5) —souring→ pH 4.0

PURPLISH-BLUE —————→ RED-PINK

PRODUCTION OF SOY SAUCE INVOLVES BOTH AEROBIC AND ANAEROBIC FERMENTATION

AEROBIC

Soy beans → Soaking and cooking → 'SOY' MIXTURE or 'MASH' 'PULP'

Wheat flour and *Aspergillus* added

Salt added

Spread into thin layers – kept at 30 °C for 2–3 days. Breaks down starch and protein to sugars and amino acids.

ANAEROBIC

Fermented with yeast and *Lactobacillus* for 3–6 months.

'crude' SAUCE

Filtered to remove solids

Pasteurised

Sealed into sterile bottles

Soy Sauce / Soy Sauce

WINE MAKING FROM GRAPE SUGAR

GRAPES → CRUSHED 'MUST' → FILTERED JUICE

Some skin remains in red wine: colour and healthy antioxidants.

CLOUDY SOLUTION → CLEAR SOLUTION

May **add sulphite** or **pasteurise** to stop action of bacteria (or wine may turn to vinegar!)

Adsorbents (e.g. crushed shells) may be added to remove suspended particles.

Fermentation may include conversion of *malic acid* to *lactic acid* by *Lactobacillus*.

N.B. Alcohol, like lactic acid, is a poison. It eventually kills the yeast cells which produce it, and does the same to human cells if taken in too large a quantity!

Hic!

If a culture of yeast is supplied with glucose and water, at a temperature of around 28 °C it will reproduce by *budding*.

Bud forming on parent cell.

Rapid budding produces chains of cells.

Different strains of the yeast *Saccharomyces cerevisiae* are specialised for *baking* and for different forms of *brewing*.

AMYLOSE and AMYLOPECTIN

α and β amylase

MALTOSE + GLUCOSE

GLUCOSE $\xrightarrow{\text{ANAEROBIC CONDITIONS}}$ ENERGY + CARBON DIOXIDE + ETHANOL

CARBON DIOXIDE IS PARTICULARLY IMPORTANT IN BREAD MAKING

Yeast + sugar warmed together = RAISING AGENT

FLOUR, SALT, WARM WATER → DOUGH → (FERMENTATION) RISEN DOUGH → BAKING

(Carbon dioxide 'bubbles' cause the dough to swell.)

BAKING • kills yeast • evaporates alcohol.

(Carbon dioxide 'bubbles' which traps CO_2 → texture

MIXING (KNEADING) DOUGH creates a protein framework which traps CO_2 → texture of bread. Framework is created when **gluten** molecules form **disulphide bridge** cross-links.

FLOUR IMPROVERS e.g. *ascorbic acid* speed up this process by oxidising –SH groups.

SH SH SH SH SH SH → S–S S–S S–S S–S

Food from fermentation

includes bread, wine, mycoprotein and soy sauce.

MYCOPROTEIN is the bodies of a filamentous fungus.

OXYGEN maintains aerobic conditions

Culture of *Fusarium graminareum* grows at 30 °C and pH6

Glucose syrup (carbon source for energy and organic molecules)

Ammonia (nitrogen source → amino acids)

Mineral salts

Choline (stimulates growth of long fibres)

Biotin (vitamin required for respiration)

Harvested fungus can be textured and flavoured.

GOOD POINTS	BAD POINTS
Rapid production	High RNA content can cause gout and kidney damage
High protein content	Cell walls are indigestible by humans
Low fat and salt content	
High fibre content	

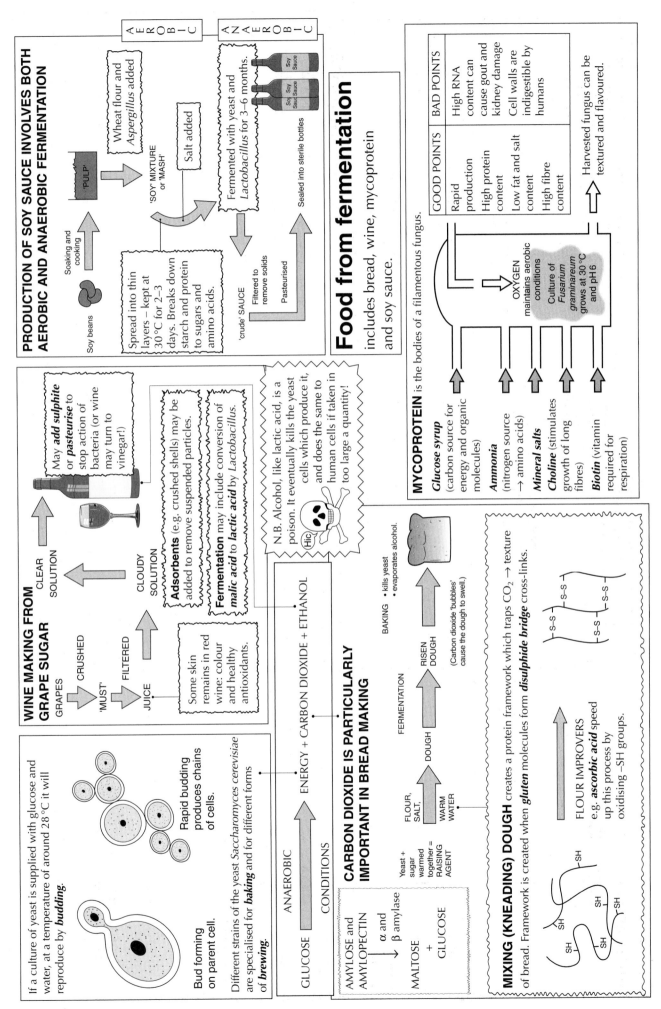

Obesity: possible health problems associated with being overweight.

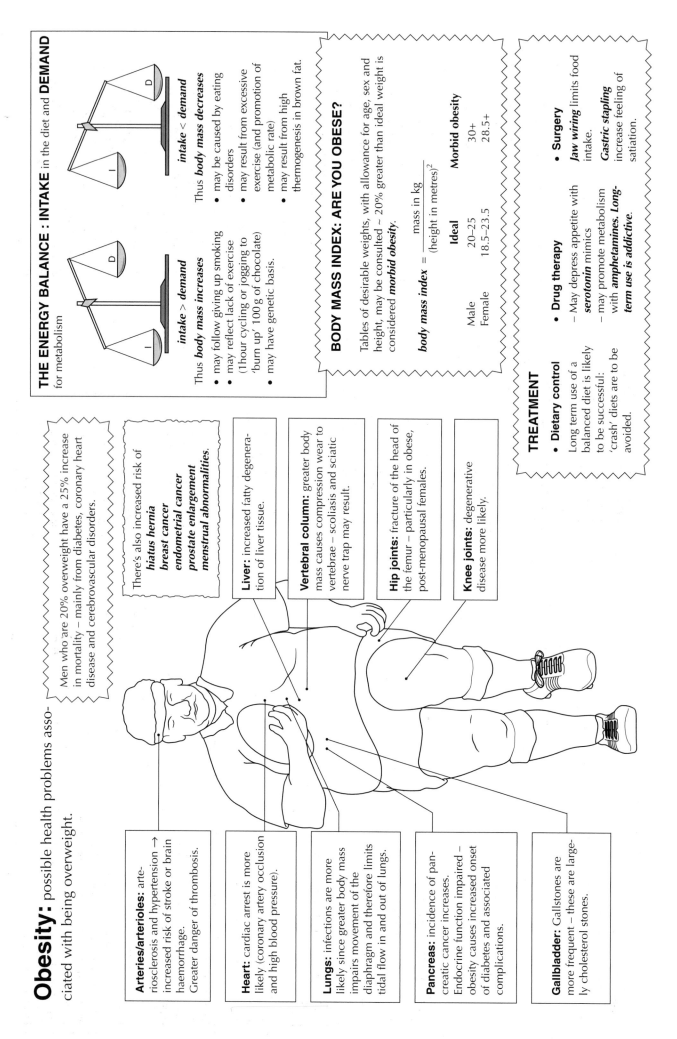

Men who are 20% overweight have a 25% increase in mortality – mainly from diabetes, coronary heart disease and cerebrovascular disorders.

There's also increased risk of
**hiatus hernia
breast cancer
endometrial cancer
prostate enlargement
menstrual abnormalities.**

Arteries/arterioles: arteriosclerosis and hypertension → increased risk of stroke or brain haemorrhage. Greater danger of thrombosis.

Heart: cardiac arrest is more likely (coronary artery occlusion and high blood pressure).

Lungs: infections are more likely since greater body mass impairs movement of the diaphragm and therefore limits tidal flow in and out of lungs.

Pancreas: incidence of pancreatic cancer increases. Endocrine function impaired – obesity causes increased onset of diabetes and associated complications.

Gallbladder: Gallstones are more frequent – these are largely cholesterol stones.

Liver: increased fatty degeneration of liver tissue.

Vertebral column: greater body mass causes compression wear to vertebrae – scoliasis and sciatic nerve trap may result.

Hip joints: fracture of the head of the femur – particularly in obese, post-menopausal females.

Knee joints: degenerative disease more likely.

THE ENERGY BALANCE: INTAKE in the diet and DEMAND for metabolism

intake > demand

Thus *body mass increases*

- may follow giving up smoking
- may reflect lack of exercise (1 hour cycling or jogging to 'burn up' 100 g of chocolate)
- may have genetic basis.

intake < demand

Thus *body mass decreases*

- may be caused by eating disorders
- may result from excessive exercise (and promotion of metabolic rate)
- may result from high thermogenesis in brown fat.

BODY MASS INDEX: ARE YOU OBESE?

Tables of desirable weights, with allowance for age, sex and height, may be consulted ~ 20% greater than ideal weight is considered *morbid obesity.*

$$body\ mass\ index = \frac{mass\ in\ kg}{(height\ in\ metres)^2}$$

	Ideal	Morbid obesity
Male	20–25	30+
Female	18.5–23.5	28.5+

TREATMENT

- **Dietary control**
  Long term use of a balanced diet is likely to be successful: 'crash' diets are to be avoided.

- **Drug therapy**
  – May depress appetite with *serotonin* mimics
  – may promote metabolism with **amphetamines. Long-term use is addictive.**

- **Surgery**
  Jaw wiring limits food intake.
  Gastric stapling increase feeling of satiation.

The liver is the largest gland,

weighing between 1.0 and 2.3 kg. It is situated in the upper abdomen, just beneath the diaphragm, and has four lobes. On the posterior surface is the **portal fissure** where various structures – hepatic portal vein, hepatic artery, hepatic vein, bile duct, lymph vessels – together with sympathetic and parasympathetic nerve fibres, enter or leave the gland.

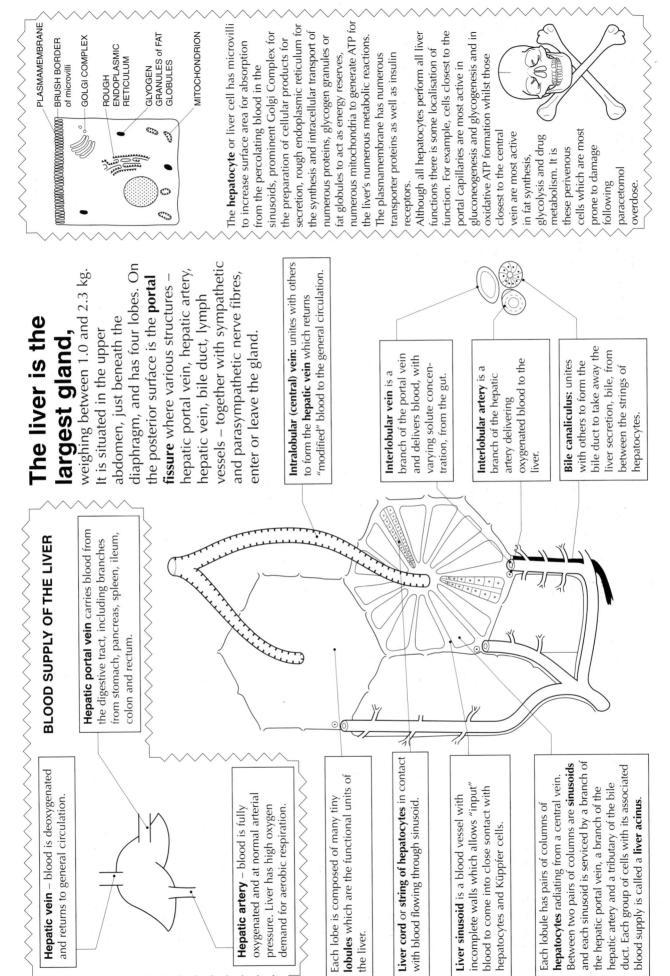

PLASMAMEMBRANE

BRUSH BORDER of microvilli

GOLGI COMPLEX

ROUGH ENDOPLASMIC RETICULUM

GLYOGEN GRANULES of FAT GLOBULES

MITOCHONDRION

The **hepatocyte** or liver cell has microvilli to increase surface area for absorption from the percolating blood in the sinusoids, prominent Golgi Complex for the preparation of cellular products for secretion, rough endoplasmic reticulum for the synthesis and intracellular transport of numerous proteins, glycogen granules or fat globules to act as energy reserves, numerous mitochondria to generate ATP for the liver's numerous metabolic reactions.

The plasmamembrane has numerous transporter proteins as well as insulin receptors.

Although all hepatocytes perform all liver functions there is some localisation of function. For example, cells closest to the portal capillaries are most active in gluconeogenesis and glycogenesis and in oxidative ATP formation whilst those closest to the central vein are most active in fat synthesis, glycolysis and drug metabolism. It is these perivenous cells which are most prone to damage following paracetomol overdose.

BLOOD SUPPLY OF THE LIVER

Hepatic portal vein carries blood from the digestive tract, including branches from stomach, pancreas, spleen, ileum, colon and rectum.

Intralobular (central) vein: unites with others to form the **hepatic vein** which returns "modified" blood to the general circulation.

Interlobular vein is a branch of the portal vein and delivers blood, with varying solute concentration, from the gut.

Interlobular artery is a branch of the hepatic artery delivering oxygenated blood to the liver.

Bile canaliculus: unites with others to form the bile duct to take away the liver secretion, bile, from between the strings of hepatocytes.

Hepatic vein – blood is deoxygenated and returns to general circulation.

Hepatic artery – blood is fully oxygenated and at normal arterial pressure. Liver has high oxygen demand for aerobic respiration.

Each lobe is composed of many tiny **lobules** which are the functional units of the liver.

Liver cord or **string of hepatocytes** in contact with blood flowing through sinusoid.

Liver sinusoid is a blood vessel with incomplete walls which allows "input" blood to come into close contact with hepatocytes and Küpffer cells.

Each lobule has pairs of columns of **hepatocytes** radiating from a central vein. Between two pairs of columns are **sinusoids** and each sinusoid is serviced by a branch of the hepatic portal vein, a branch of the hepatic artery and a tributary of the bile duct. Each group of cells with its associated blood supply is called a **liver acinus.**

Liver structure and function is illustrated by a single sinusoid.

Protein metabolism

Deamination of excess amino acids

Amino acid + oxygen → oxo acid + ammonia

and subsequent formation of *urea*

$$2NH_3 + CO_2 \longrightarrow CO(NH_2)_2 + H_2O$$

Transamination involves the transfer of an -NH_2 group from an amino acid to a different carbon skeleton

Amino acid I + oxo acid → Amino acid II + oxo acid I

The eight 'essential' amino acids cannot be synthesized in this way.
Synthesis of plasma proteins, including albumin, globulins, heparin, fibrinogen, prothrombin, factor VIII.

Branch of the hepatic portal vein (interlobular vein)
Delivers blood from the absorptive regions of the gut carrying a *variable concentration* of the soluble products of digestion.

Branch of hepatic artery (interlobular artery)
Delivers oxygenated blood at high pressure – hepatocytes have a high oxygen demand and are very vulnerable to hypoxia. Also delivers lactate from anaerobic respiration in skeletal muscle.

Bile ductile
Takes away *bile* to be stored in the gall bladder. Bile is 90% water + *bile salts* (aid emulsification of fats) + *bile pigments* (an excretory product) + *cholesterol + salts.* Release is triggered by CCK-PZ from wall of duodenum.

Heat production
High metabolic rate and considerable energy consumption make the liver the main heat-producing organ in the body. Metabolic rate and hence heat production is under the control of thyroxine.

Küppfer cell
A fixed cell of the reticulo-endothelial system which removes old ('effete') red blood cells. Iron is stored as ferritin, globin → amino acid pool and pyrrole rings are excreted as bile pigments.

Lipid metabolism
1. Excess carbohydrate is converted to fat.
2. Desaturation of stored fats prior to oxidation.
3. Synthesis/degradation of phospholipids and cholesterol.
4. Synthesis of lipid-transporting globulins.

Storage of minerals and vitamins
principally *iron* (as ferritin), some potassium, copper and trace elements.

mainly the *fat-soluble* vitamins A, D, E. *Small* amounts of B_{12}, C. Synthesis of vitamin A from carotene.

Branch of hepatic vein (intralobular vein)
Delivers blood with a *constant concentration* of products to the

Macrophage
Wandering cells of the reticulo-endothelial system which identify and remove pathogens by phagocytosis.

Detoxification
Often by *oxidation* (e.g. alcohol → ethanal), often by the protein P450, sometimes by *reduction* or *methylation.* Many drugs damage the liver, in original or intermediate forms, e.g. chloroform, paracetamol, tetracyclines, alcohol, anabolic steroids. Detoxified products are usually excreted but sometimes stored (e.g. DDT).

Carbohydrate metabolism
Glycogenesis:
– promoted by *insulin*
Glycogenolysis:
– promoted by *glucagon*
Lactate metabolism:
– initiated by *lactate dehydrogenase*
Gluconeogenesis:
– promoted by *cortisone* and *adrenaline*
Non-carbohydrate sources → glucose (e.g. glycerol, amino acids)

glucose ⇌ gluc. 6 (P) ⇌ gluc. 1 (P) ⇌ glycogen

glycogen ⇌ gluc. 1 (P) ⇌ gluc. 6 (P) ⇌ glucose

lactate → pyruvate → glucose → glycogen

Hormone removal
Rapid inactivation of testosterone and aldosterone. More gradual breakdown of insulin, glucagon, thyroxine, cortisone, oestrogen and progesterone.

Alternative methods of feeding

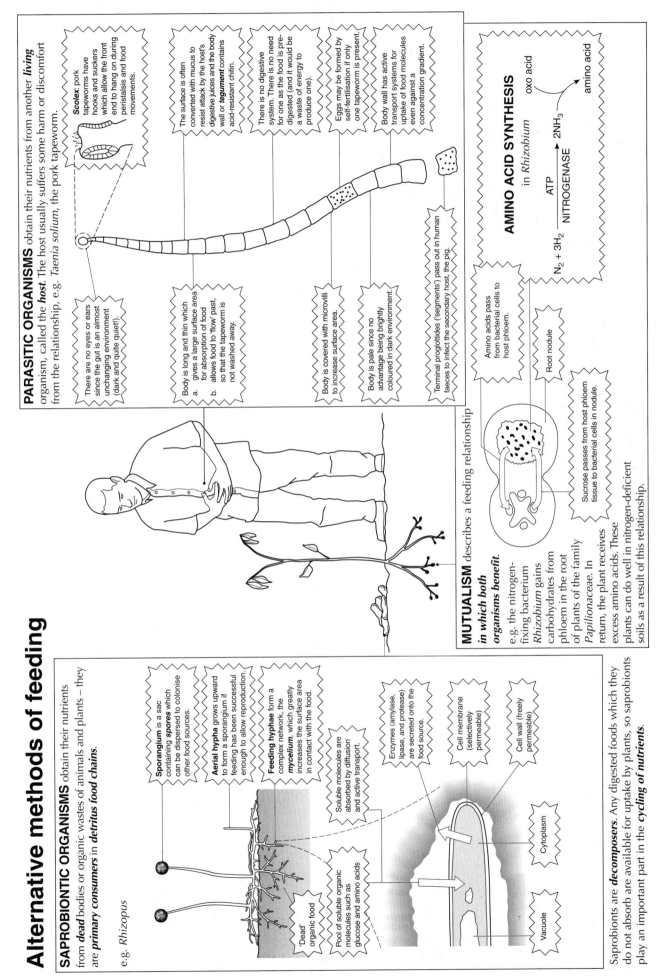

PARASITIC ORGANISMS obtain their nutrients from another *living* organism, called the *host*. The host usually suffers some harm or discomfort from the relationship, e.g. *Taenia solium*, the pork tapeworm.

Scolex: pork tapeworms have hooks and suckers which allow the front end to hang on during peristalsis and food movements.

There are no eyes or ears since the gut is an almost unchanging environment (dark and quite quiet!).

The surface is often converted with mucus to resist attack by the host's digestive juices and the body wall or *tegument* contains acid-resistant chitin.

Body is long and thin which
a. gives a large surface area for absorption of food
b. allows food to 'flow' past, so that the tapeworm is not washed away.

There is no digestive system. There is no need for one as the food is pre-digested (and it would be a waste of energy to produce one).

Body is covered with microvilli to increase surface area.

Eggs may be formed by self-fertilisation if only one tapeworm is present.

Body is pale since no advantage being brightly coloured in dark environment.

Body wall has active transport systems for uptake of food molecules even against a concentration gradient.

Terminal proglottides ('segments') pass out in human faeces to infect the secondary host, the pig.

SAPROBIONTIC ORGANISMS obtain their nutrients

from *dead* bodies or organic wastes of animals and plants – they are *primary consumers* in *detritus food chains*.

e.g. *Rhizopus*

Sporangium is a sac containing *spores* which can be dispersed to colonise other food sources.

Aerial hypha grows upward to form a sporangium if feeding has been successful enough to allow reproduction.

Feeding hyphae form a complex network, the *mycelium*, which greatly increases the surface area in contact with the food.

Soluble molecules are absorbed by diffusion and active transport.

'Dead' organic food

Pool of soluble organic molecules such as glucose and amino acids

Enzymes (amylase, lipase, and protease) are secreted onto the food source.

Cell membrane (selectively permeable)

Cell wall (freely permeable)

Cytoplasm

Vacuole

Saprobionts are *decomposers*. Any digested foods which they do not absorb are available for uptake by plants, so saprobionts play an important part in the *cycling of nutrients*.

MUTUALISM describes a feeding relationship *in which both organisms benefit*.

e.g. the nitrogen-fixing bacterium *Rhizobium* gains carbohydrates from phloem in the root of plants of the family *Papilionaceae*. In return, the plant receives excess amino acids. These plants can do well in nitrogen-deficient soils as a result of this relationship.

Amino acids pass from bacterial cells to host phloem.

Root nodule

Sucrose passes from host phloem tissue to bacterial cells in nodule.

AMINO ACID SYNTHESIS

in *Rhizobium*

$$N_2 + 3H_2 \xrightarrow[\text{NITROGENASE}]{\text{ATP}} 2NH_3$$

oxo acid → amino acid

Ruginants

Ruminants

have a four-chambered stomach to obtain maximum value from low-quality foods.

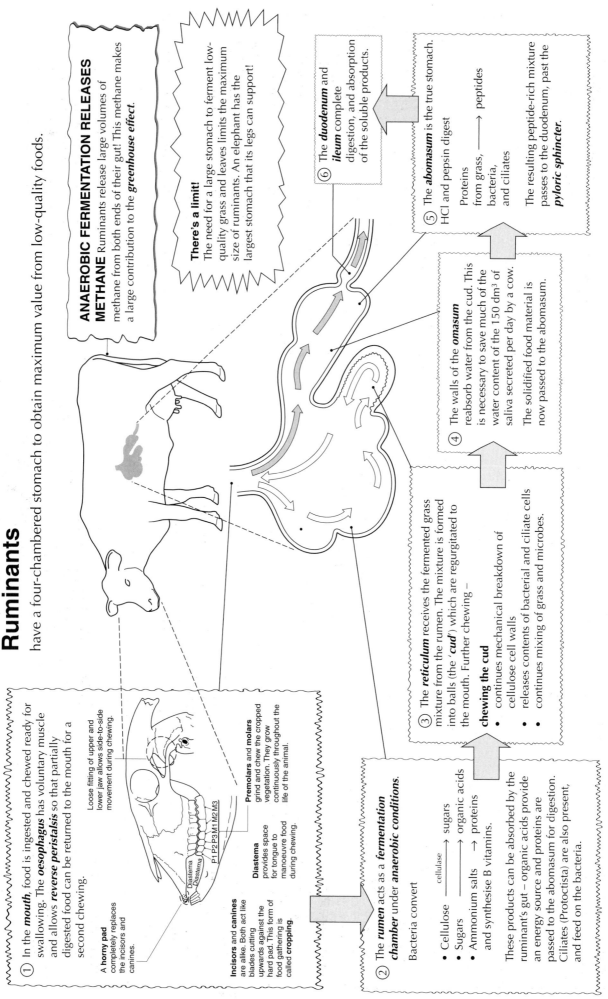

ANAEROBIC FERMENTATION RELEASES METHANE Ruminants release large volumes of methane from both ends of their gut! This methane makes a large contribution to the *greenhouse effect*.

There's a limit! The need for a large stomach to ferment low-quality grass and leaves limits the maximum size of ruminants. An elephant has the largest stomach that its legs can support!

① In the **mouth**, food is ingested and chewed ready for swallowing. The **oesophagus** has voluntary muscle and allows *reverse peristalsis* so that partially digested food can be returned to the mouth for a second chewing.

Loose fitting of upper and lower jaw allows side-to-side movement during chewing.

A **horny pad** completely replaces the incisors and canines.

Premolars and **molars** grind and chew the cropped vegetation. They grow continuously throughout the life of the animal.

Diastema provides space for tongue to manoeuvre food during chewing.

Incisors and **canines** are alike. Both act like blades cutting upwards against the hard pad. This form of food gathering is called **cropping**.

Diastema Diastema
P1 P2 P3 M1 M2 M3

② The **rumen** acts as a *fermentation chamber* under *anaerobic conditions*.

Bacteria convert

- Cellulose $\xrightarrow{\text{cellulase}}$ sugars
- Sugars $\longrightarrow$ organic acids
- Ammonium salts $\longrightarrow$ proteins and synthesise B vitamins.

These products can be absorbed by the ruminant's gut – organic acids provide an energy source and proteins are passed to the abomasum for digestion. Ciliates (Protoctista) are also present, and feed on the bacteria.

③ The **reticulum** receives the fermented grass mixture from the rumen. The mixture is formed into balls (the 'cud') which are regurgitated to the mouth. Further chewing –

chewing the cud
- continues mechanical breakdown of cellulose cell walls
- releases contents of bacterial and ciliate cells
- continues mixing of grass and microbes.

④ The walls of the **omasum** reabsorb water from the cud. This is necessary to save much of the water content of the 150 dm³ of saliva secreted per day by a cow. The solidified food material is now passed to the abomasum.

⑤ The **abomasum** is the true stomach. HCl and pepsin digest

Proteins from grass, bacteria, and ciliates $\longrightarrow$ peptides

The resulting peptide-rich mixture passes to the duodenum, past the *pyloric sphincter*.

⑥ The **duodenum** and **ileum** complete digestion, and absorption of the soluble products.

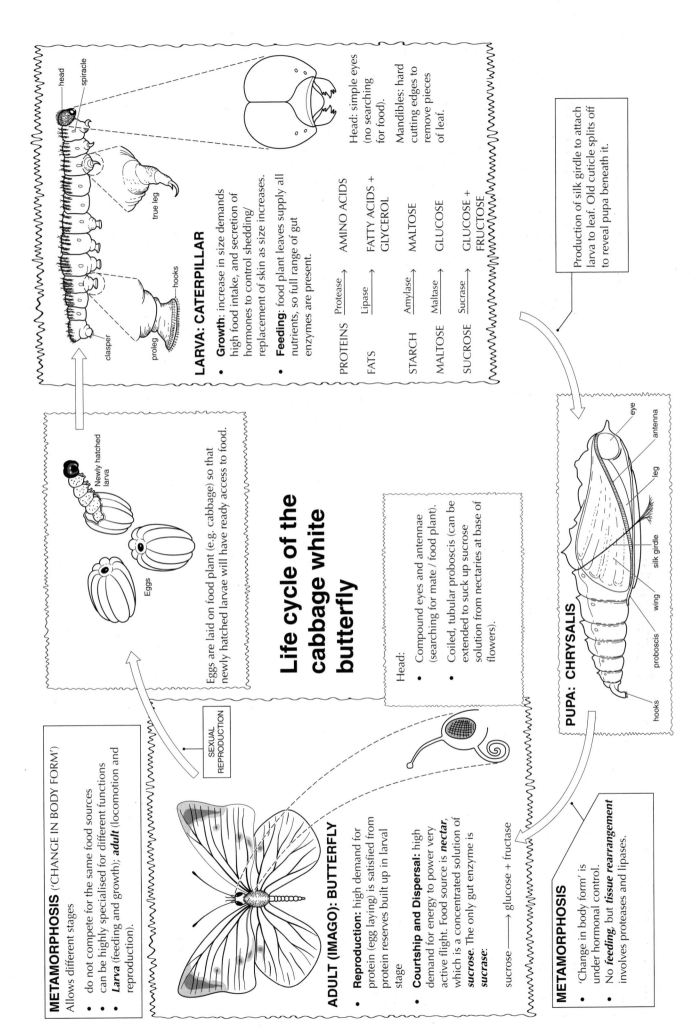

head
spiracle

Head: simple eyes (no searching for food).

Mandibles: hard cutting edges to remove pieces of leaf.

true leg

clasper
proleg
hooks

LARVA: CATERPILLAR

- **Growth:** increase in size demands high food intake, and secretion of hormones to control shedding/replacement of skin as size increases.
- **Feeding:** food plant leaves supply all nutrients, so full range of gut enzymes are present.

PROTEINS	$\xrightarrow{\text{Protease}}$	AMINO ACIDS
FATS	$\xrightarrow{\text{Lipase}}$	FATTY ACIDS + GLYCEROL
STARCH	$\xrightarrow{\text{Amylase}}$	MALTOSE
MALTOSE	$\xrightarrow{\text{Maltase}}$	GLUCOSE
SUCROSE	$\xrightarrow{\text{Sucrase}}$	GLUCOSE + FRUCTOSE

Production of silk girdle to attach larva to leaf. Old cuticle splits off to reveal pupa beneath it.

METAMORPHOSIS ('CHANGE IN BODY FORM')
Allows different stages

- do not compete for the same food sources
- can be highly specialised for different functions
- **Larva** (feeding and growth); **adult** (locomotion and reproduction).

Newly hatched larva

Eggs

Eggs are laid on food plant (e.g. cabbage) so that newly hatched larvae will have ready access to food.

Life cycle of the cabbage white butterfly

SEXUAL REPRODUCTION

Head:
- Compound eyes and antennae (searching for mate / food plant).
- Coiled, tubular proboscis (can be extended to suck up sucrose solution from nectaries at base of flowers).

eye
antenna
leg
silk girdle
wing
proboscis
hooks

PUPA: CHRYSALIS

ADULT (IMAGO): BUTTERFLY

- **Reproduction:** high demand for protein (egg laying) is satisfied from protein reserves built up in larval stage
- **Courtship and Dispersal:** high demand for energy to power very active flight. Food source is **nectar**, which is a concentrated solution of **sucrose**. The only gut enzyme is **sucrase:**

sucrose $\longrightarrow$ glucose + fructase

METAMORPHOSIS

- 'Change in body form' is under hormonal control.
- No **feeding**, but **tissue rearrangement** involves proteases and lipases.

Principles of respiration: a
number of processes are involved in the provision/consumption of **oxygen** and the excretion/production of **carbon dioxide.**

- **Pulmonary ventilation** moves gases between atmosphere and respiratory surface.
- **External respiration** occurs when gases diffuse across the respiratory surface.
- **Internal respiration** occurs when gases diffuse between circulating blood and respiring cells.
- **Tissue/cell respiration** occurs when oxygen is consumed and carbon dioxide is produced during the oxidation of foods to release energy.

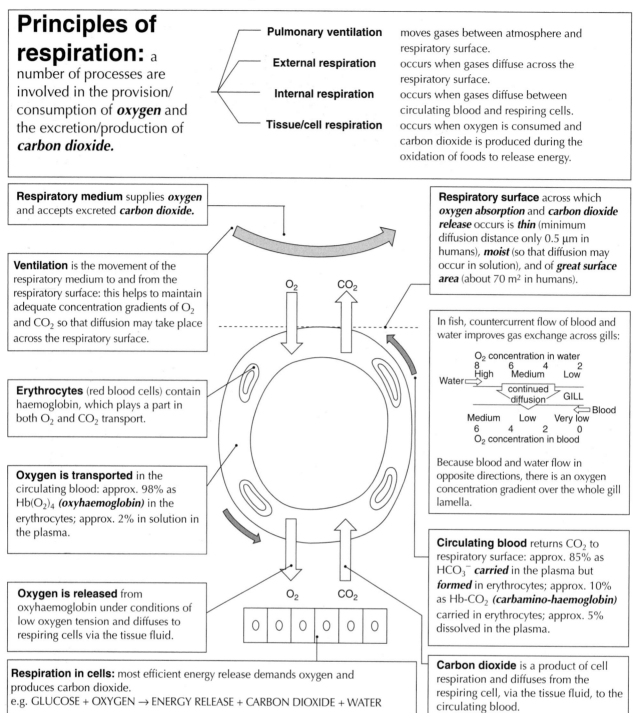

Respiratory medium supplies *oxygen* and accepts excreted *carbon dioxide.*

Ventilation is the movement of the respiratory medium to and from the respiratory surface: this helps to maintain adequate concentration gradients of O_2 and CO_2 so that diffusion may take place across the respiratory surface.

Erythrocytes (red blood cells) contain haemoglobin, which plays a part in both O_2 and CO_2 transport.

Oxygen is transported in the circulating blood: approx. 98% as $Hb(O_2)_4$ *(oxyhaemoglobin)* in the erythrocytes; approx. 2% in solution in the plasma.

Oxygen is released from oxyhaemoglobin under conditions of low oxygen tension and diffuses to respiring cells via the tissue fluid.

Respiratory surface across which *oxygen absorption* and *carbon dioxide release* occurs is *thin* (minimum diffusion distance only 0.5 μm in humans), *moist* (so that diffusion may occur in solution), and of *great surface area* (about 70 m² in humans).

In fish, countercurrent flow of blood and water improves gas exchange across gills:

O_2 concentration in water
8   6   4   2
High   Medium   Low
Water ⇒
continued diffusion   GILL
⇐ Blood
Medium   Low   Very low
6   4   2   0
O_2 concentration in blood

Because blood and water flow in opposite directions, there is an oxygen concentration gradient over the whole gill lamella.

Circulating blood returns CO_2 to respiratory surface: approx. 85% as HCO_3^- *carried* in the plasma but *formed* in erythrocytes; approx. 10% as $Hb-CO_2$ *(carbamino-haemoglobin)* carried in erythrocytes; approx. 5% dissolved in the plasma.

Carbon dioxide is a product of cell respiration and diffuses from the respiring cell, via the tissue fluid, to the circulating blood.

Respiration in cells: most efficient energy release demands oxygen and produces carbon dioxide.
e.g. GLUCOSE + OXYGEN → ENERGY RELEASE + CARBON DIOXIDE + WATER

FEATURE	AMOEBA	EARTHWORM	INSECT	MAMMAL	FISH
RESPIRATORY MEDIUM	WATER	AIR	AIR	AIR	WATER
SURFACE	CELL MEMBRANE	MOIST BODY SURFACE	TRACHEOLES contact cells directly	ALVEOLI of lungs	LAMELLAE of gills
TRANSPORT SYSTEM	NONE	BLOOD VESSELS and SIMPLE 'HEART' AS PUMP	NONE	Blood in DOUBLE CIRCULATION	Blood in SINGLE CIRCULATION
VENTILATION	NONE	NONE	Little – some ABDOMINAL MOVEMENT	Negative pressure system initiates TIDAL LOW of air	Muscular movements drive ONE-WAY flow of water

Lung structure and function may be affected by a variety of disease conditions.

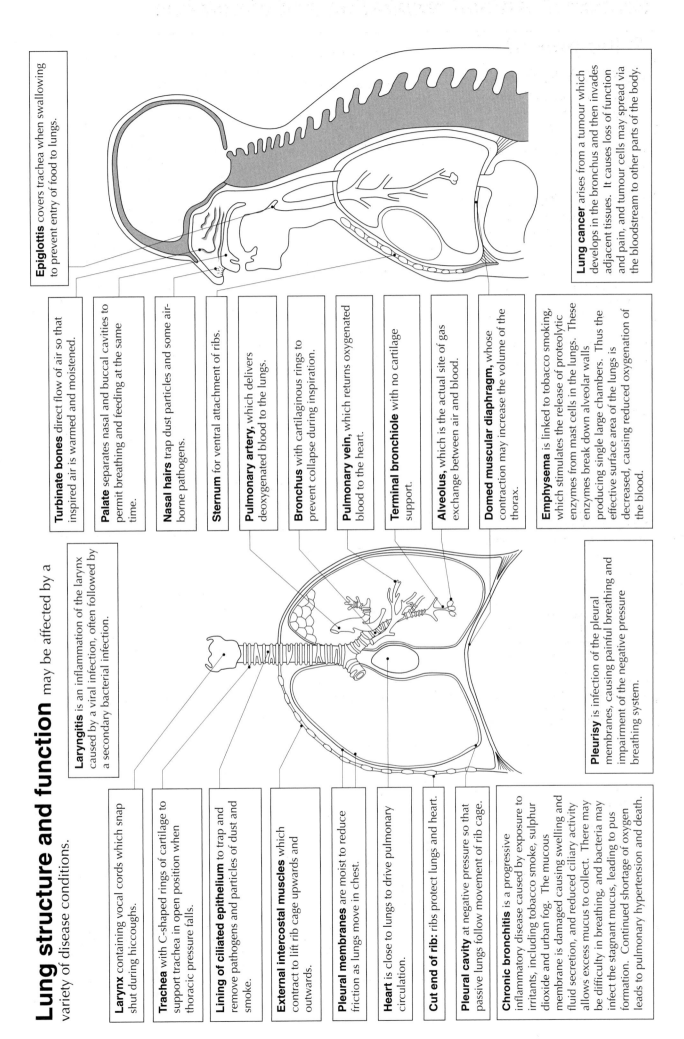

Epiglottis covers trachea when swallowing to prevent entry of food to lungs.

Lung cancer arises from a tumour which develops in the bronchus and then invades adjacent tissues. It causes loss of function and pain, and tumour cells may spread via the bloodstream to other parts of the body.

Turbinate bones direct flow of air so that inspired air is warmed and moistened.

Palate separates nasal and buccal cavities to permit breathing and feeding at the same time.

Nasal hairs trap dust particles and some air-borne pathogens.

Sternum for ventral attachment of ribs.

Pulmonary artery, which delivers deoxygenated blood to the lungs.

Bronchus with cartilaginous rings to prevent collapse during inspiration.

Pulmonary vein, which returns oxygenated blood to the heart.

Terminal bronchiole with no cartilage support.

Alveolus, which is the actual site of gas exchange between air and blood.

Domed muscular diaphragm, whose contraction may increase the volume of the thorax.

Emphysema is linked to tobacco smoking, which stimulates the release of proteolytic enzymes from mast cells in the lungs. These enzymes break down alveolar walls producing single large chambers. Thus the effective surface area of the lungs is decreased, causing reduced oxygenation of the blood.

Laryngitis is an inflammation of the larynx caused by a viral infection, often followed by a secondary bacterial infection.

Pleurisy is infection of the pleural membranes, causing painful breathing and impairment of the negative pressure breathing system.

Larynx containing vocal cords which snap shut during hiccoughs.

Trachea with C-shaped rings of cartilage to support trachea in open position when thoracic pressure falls.

Lining of ciliated epithelium to trap and remove pathogens and particles of dust and smoke.

External intercostal muscles which contract to lift rib cage upwards and outwards.

Pleural membranes are moist to reduce friction as lungs move in chest.

Heart is close to lungs to drive pulmonary circulation.

Cut end of rib: ribs protect lungs and heart.

Pleural cavity at negative pressure so that passive lungs follow movement of rib cage.

Chronic bronchitis is a progressive inflammatory disease caused by exposure to irritants, including tobacco smoke, sulphur dioxide and urban fog. The mucous membrane is damaged causing swelling and fluid secretion, and reduced ciliary activity allows excess mucus to collect. There may be difficulty in breathing, and bacteria may infect the stagnant mucus, leading to pus formation. Continued shortage of oxygen leads to pulmonary hypertension and death.

Fine structure of the lung: exchange of gases in the alveolus requires

1. A tube for the movement of gases to and from the atmosphere (e.g. *bronchiole*).
2. A surface across which gases may be transported between air and blood (i.e. *alveolar membrane*).
3. A vessel which can take away oxygenated blood or deliver carboxylated blood (i.e. a *branch of the pulmonary circulation*).

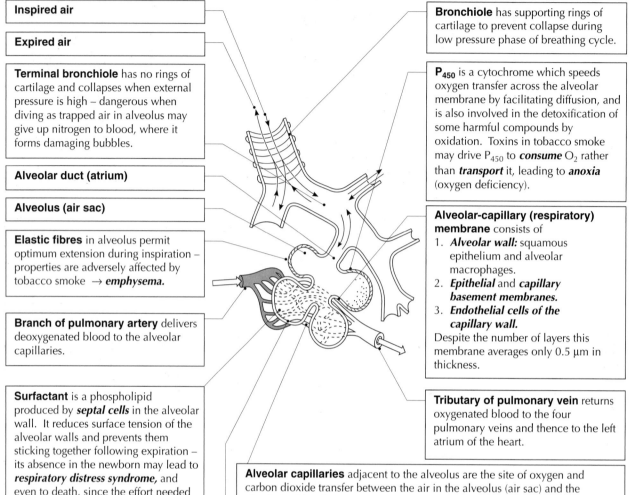

Inspired air

Expired air

Terminal bronchiole has no rings of cartilage and collapses when external pressure is high – dangerous when diving as trapped air in alveolus may give up nitrogen to blood, where it forms damaging bubbles.

Alveolar duct (atrium)

Alveolus (air sac)

Elastic fibres in alveolus permit optimum extension during inspiration – properties are adversely affected by tobacco smoke → *emphysema*.

Branch of pulmonary artery delivers deoxygenated blood to the alveolar capillaries.

Surfactant is a phospholipid produced by *septal cells* in the alveolar wall. It reduces surface tension of the alveolar walls and prevents them sticking together following expiration – its absence in the newborn may lead to *respiratory distress syndrome*, and even to death, since the effort needed to breathe is 7–10 × normal.

Bronchiole has supporting rings of cartilage to prevent collapse during low pressure phase of breathing cycle.

P$_{450}$ is a cytochrome which speeds oxygen transfer across the alveolar membrane by facilitating diffusion, and is also involved in the detoxification of some harmful compounds by oxidation. Toxins in tobacco smoke may drive P$_{450}$ to *consume* O_2 rather than *transport* it, leading to *anoxia* (oxygen deficiency).

Alveolar-capillary (respiratory) membrane consists of
1. *Alveolar wall:* squamous epithelium and alveolar macrophages.
2. *Epithelial* and *capillary basement membranes*.
3. *Endothelial cells of the capillary wall*.
Despite the number of layers this membrane averages only 0.5 μm in thickness.

Tributary of pulmonary vein returns oxygenated blood to the four pulmonary veins and thence to the left atrium of the heart.

Alveolar capillaries adjacent to the alveolus are the site of oxygen and carbon dioxide transfer between the air in the alveolus (air sac) and the circulating blood.

Stretch receptors provide sensory input, which initiates the *Hering-Breuer reflex* control of the breathing cycle.

Changes in composition of inspired and expired air

	INSPIRED	ALVEOLAR	EXPIRED	
O_2	20.95	13.80	16.40	Oxygen diffuses from alveoli into blood: expired air has an increased proportion of oxygen due to additional oxygen added from the anatomical dead space.
CO_2	0.04	5.50	4.00	Carbon dioxide concentration in alveoli is high because CO_2 diffuses from blood: the apparent fall in CO_2 concentration in expired air is due to dilution in the anatomical dead space.
N_2	79.01	80.70	79.60	The apparent increase in the concentration of nitrogen, a metabolically inert gas, is due to a *relative* decrease in the proportion of oxygen rather than an *absolute* increase in nitrogen.
$H_2O(g)$	VARI–ABLE	SATURATED		The moisture lining the alveoli evaporates into the alveolar air and is then expired unless the animal has anatomical adaptations to prevent this (e.g. the extensive nasal hairs in desert rats).
Temp.	ATMOS–PHERIC	BODY		Heat lost from the blood in the pulmonary circulation raises the temperature of the alveolar air.

Smoking and lung disease

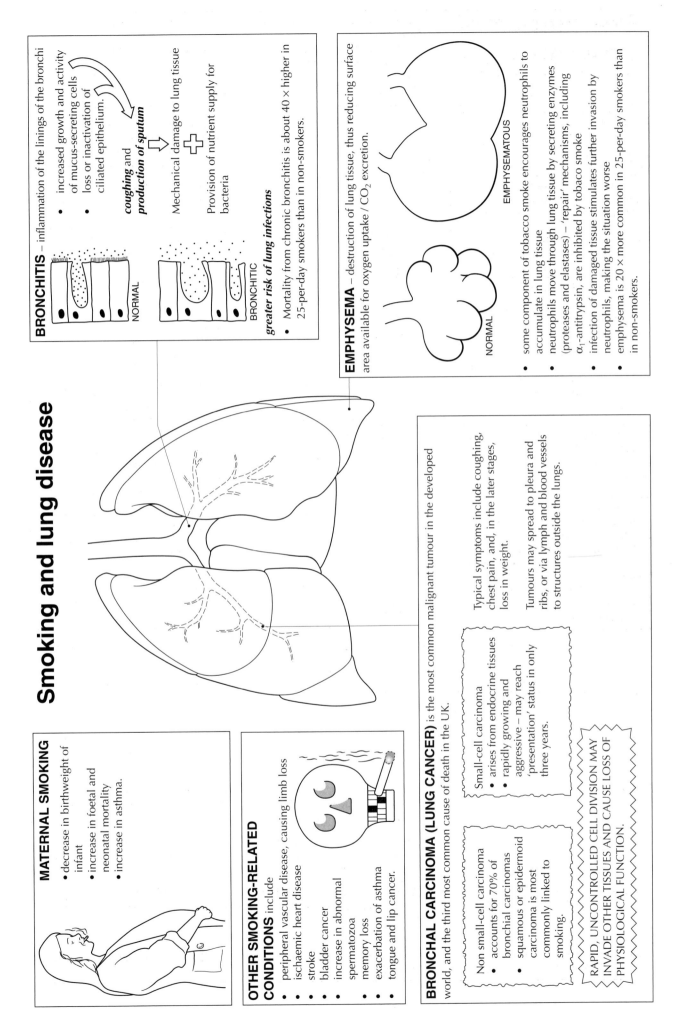

BRONCHITIS – inflammation of the linings of the bronchi

- increased growth and activity of mucus-secreting cells
- loss or inactivation of ciliated epithelium.

coughing and *production of sputum*

Mechanical damage to lung tissue

Provision of nutrient supply for bacteria

NORMAL

BRONCHITIC

greater risk of lung infections

- Mortality from chronic bronchitis is about 40 × higher in 25-per-day smokers than in non-smokers.

EMPHYSEMA – destruction of lung tissue, thus reducing surface area available for oxygen uptake / CO_2 excretion.

NORMAL

EMPHYSEMATOUS

- some component of tobacco smoke encourages neutrophils to accumulate in lung tissue
- neutrophils move through lung tissue by secreting enzymes (proteases and elastases) – 'repair' mechanisms, including α_1-antitrypsin, are inhibited by tobacco smoke
- infection of damaged tissue stimulates further invasion by neutrophils, making the situation worse
- emphysema is 20 × more common in 25-per-day smokers than in non-smokers.

MATERNAL SMOKING

- decrease in birthweight of infant
- increase in foetal and neonatal mortality
- increase in asthma.

OTHER SMOKING-RELATED CONDITIONS include

- peripheral vascular disease, causing limb loss
- ischaemic heart disease
- stroke
- bladder cancer
- increase in abnormal spermatozoa
- memory loss
- exacerbation of asthma
- tongue and lip cancer.

BRONCHAL CARCINOMA (LUNG CANCER) is the most common malignant tumour in the developed world, and the third most common cause of death in the UK.

Non small-cell carcinoma
- accounts for 70% of bronchial carcinomas
- squamous or epidermoid carcinoma is most commonly linked to smoking.

Small-cell carcinoma
- arises from endocrine tissues
- rapidly growing and aggressive – may reach 'presentation' status in only three years.

Typical symptoms include coughing, chest pain, and, in the later stages, loss in weight.

Tumours may spread to pleura and ribs, or via lymph and blood vessels to structures outside the lungs.

RAPID, UNCONTROLLED CELL DIVISION MAY INVADE OTHER TISSUES AND CAUSE LOSS OF PHYSIOLOGICAL FUNCTION.

Pulmonary ventilation is a result of changes in pressure within the thorax.

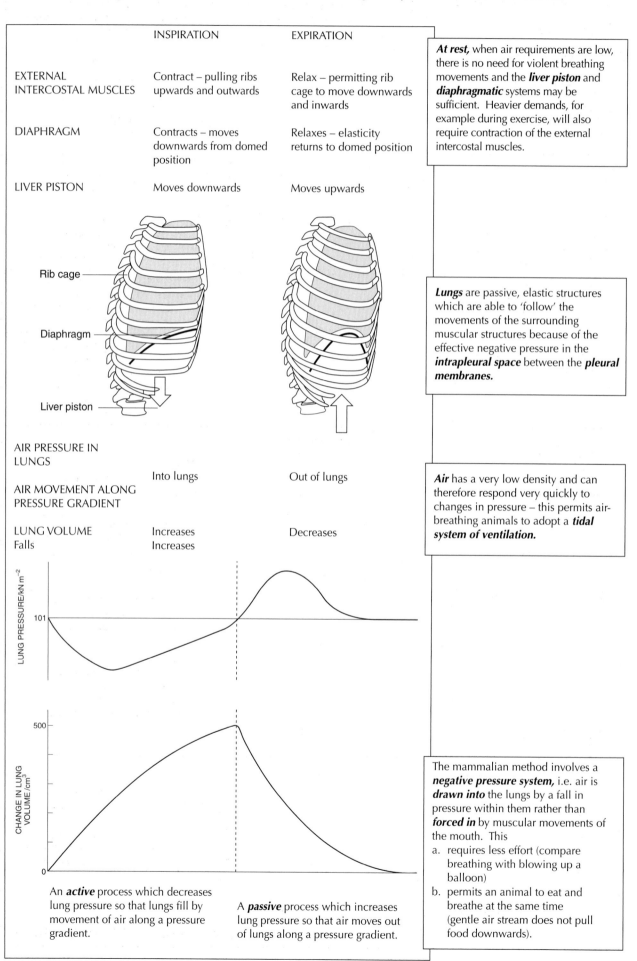

	INSPIRATION	EXPIRATION
EXTERNAL INTERCOSTAL MUSCLES	Contract – pulling ribs upwards and outwards	Relax – permitting rib cage to move downwards and inwards
DIAPHRAGM	Contracts – moves downwards from domed position	Relaxes – elasticity returns to domed position
LIVER PISTON	Moves downwards	Moves upwards

Rib cage

Diaphragm

Liver piston

AIR PRESSURE IN LUNGS | Falls | Increases

AIR MOVEMENT ALONG PRESSURE GRADIENT | Into lungs | Out of lungs

LUNG VOLUME | Increases | Decreases

LUNG PRESSURE/kN m⁻²
101

CHANGE IN LUNG VOLUME/cm³
500
0

An **active** process which decreases lung pressure so that lungs fill by movement of air along a pressure gradient.

A **passive** process which increases lung pressure so that air moves out of lungs along a pressure gradient.

At rest, when air requirements are low, there is no need for violent breathing movements and the **liver piston** and **diaphragmatic** systems may be sufficient. Heavier demands, for example during exercise, will also require contraction of the external intercostal muscles.

Lungs are passive, elastic structures which are able to 'follow' the movements of the surrounding muscular structures because of the effective negative pressure in the **intrapleural space** between the **pleural membranes.**

Air has a very low density and can therefore respond very quickly to changes in pressure – this permits air-breathing animals to adopt a **tidal system of ventilation.**

The mammalian method involves a **negative pressure system,** i.e. air is **drawn into** the lungs by a fall in pressure within them rather than **forced in** by muscular movements of the mouth. This
a. requires less effort (compare breathing with blowing up a balloon)
b. permits an animal to eat and breathe at the same time (gentle air stream does not pull food downwards).

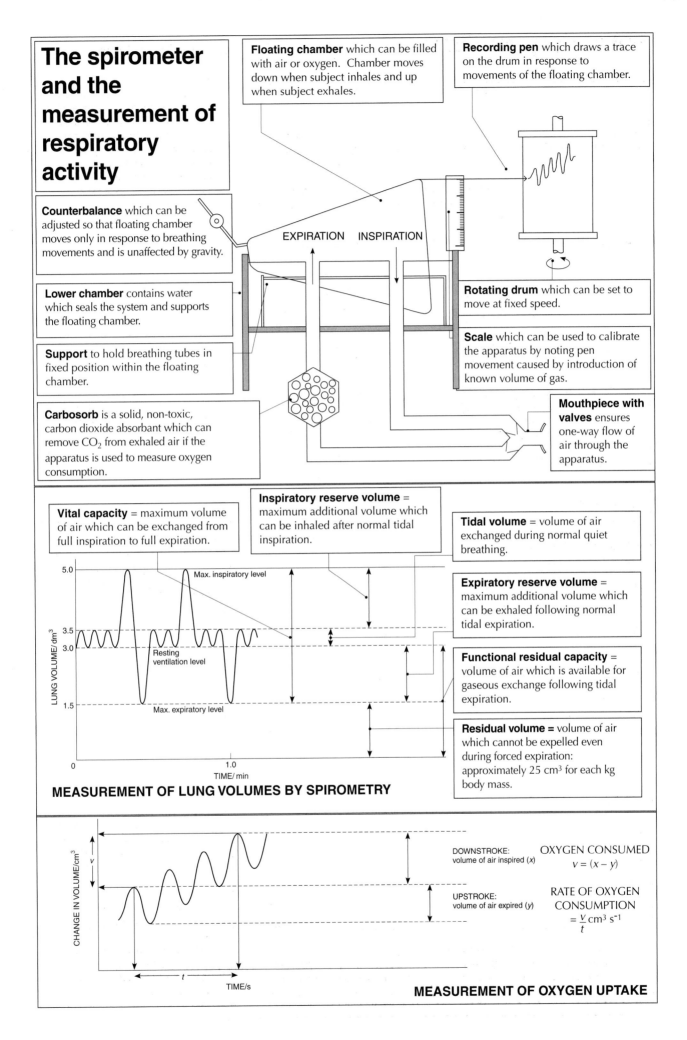

The spirometer and the measurement of respiratory activity

Floating chamber which can be filled with air or oxygen. Chamber moves down when subject inhales and up when subject exhales.

Recording pen which draws a trace on the drum in response to movements of the floating chamber.

Counterbalance which can be adjusted so that floating chamber moves only in response to breathing movements and is unaffected by gravity.

EXPIRATION   INSPIRATION

Lower chamber contains water which seals the system and supports the floating chamber.

Rotating drum which can be set to move at fixed speed.

Support to hold breathing tubes in fixed position within the floating chamber.

Scale which can be used to calibrate the apparatus by noting pen movement caused by introduction of known volume of gas.

Carbosorb is a solid, non-toxic, carbon dioxide absorbant which can remove CO_2 from exhaled air if the apparatus is used to measure oxygen consumption.

Mouthpiece with valves ensures one-way flow of air through the apparatus.

Vital capacity = maximum volume of air which can be exchanged from full inspiration to full expiration.

Inspiratory reserve volume = maximum additional volume which can be inhaled after normal tidal inspiration.

Tidal volume = volume of air exchanged during normal quiet breathing.

Expiratory reserve volume = maximum additional volume which can be exhaled following normal tidal expiration.

Functional residual capacity = volume of air which is available for gaseous exchange following tidal expiration.

Residual volume = volume of air which cannot be expelled even during forced expiration: approximately 25 cm^3 for each kg body mass.

Max. inspiratory level
Resting ventilation level
Max. expiratory level

LUNG VOLUME/dm^3

5.0
3.5
3.0
1.5
0

1.0
TIME/ min

MEASUREMENT OF LUNG VOLUMES BY SPIROMETRY

CHANGE IN VOLUME/cm^3

v

t

TIME/s

DOWNSTROKE: volume of air inspired (x)

OXYGEN CONSUMED
$v = (x - y)$

UPSTROKE: volume of air expired (y)

RATE OF OXYGEN CONSUMPTION
$= \dfrac{v}{t}\ cm^3\ s^{-1}$

MEASUREMENT OF OXYGEN UPTAKE

Blood cells differ in structure and function.

If blood is spun for a few minutes in a high speed centrifuge it separates into two layers.

Serum is the name given to plasma from which the soluble protein fibrinogen (a protein involved in blood clotting) has been removed.

PLASMA (55%)

CELLS (45%)

Erythrocytes (red blood cells) are the most numerous of blood cells – about 5 000 000 per mm^3 of adult blood. They function in the transport of O_2 and CO_2, and they contribute to the buffering capacity of the blood. The red colour is due to the presence of the pigment haemoglobin. There are several advantages in packing the haemoglobin into cells rather than leaving it free in the cytoplasm – it keeps the blood viscosity low, it allows the best arrangement of enzymes and solutes for functioning of haemoglobin and it prevents a dramatic reduction in blood water potential. The typical lifespan of a red cell is 90–120 days, before they are destroyed in the spleen.

Platelets (thrombocytes) are fragments of cells which are involved in blood clotting (they disintegrate to release thromboplasts).

Neutrophils are the most abundant of the leucocytes (white blood cells). They are very short-lived (12–72 h), contain non-staining granules, and are responsible for the phagocytosis of micro-organisms. They migrate from the blood to the tissues, and are so active in phagocytosis that they are replaced at the rate of about 100 000 000 000 per day.

Monocytes are the largest of the leucocytes. They are agranulocytes (have non-granular cytoplasm) and have a large, bean-shaped nucleus. They spend a short time (2–3 days) in the circulatory system before moving into the tissues where they mature into phagocytic macrophages.

Lymphocytes make up about 30% of the circulating leucocytes. Although they are produced in the bone marrow they continue to develop and mature in the lymph nodes, the thymus gland and the spleen. They are responsible for the specific immune response – the B-lymphocytes produce antibodies and the T-lymphocytes have a number of roles, including co-ordination of the immune response and direct cell destruction. They are best identified by the prominent, deeply staining nucleus and the thin 'halo' of clear cytoplasm.

Basophils have an S-shaped nucleus and granules which stain blue. They secrete large amounts of histamine (which increases inflammation) and heparin (which helps to keep a balance between blood clotting and not clotting).

SCALE
5 µm

Eosinophils have a double-lobed nucleus and granules which stain red with the acid dye eosin. They help control the allergic response – for example, they secrete enzymes which inactivate histamine. Their numbers increase during allergic reactions and in response to some parasitic infections (e.g. tapeworm and hookworm).

Blood cells originate from *stem cells* in the bone marrow by the process of *haemopoiesis*.

UNCOMMITTED STEM CELL

HAEMOCYTOBLAST

LYMPHOBLAST

MEGA KARYOCYTE

MYELOBLAST

PROERYTHROBLAST

MONOBLAST

ERYTHROCYTE

LEUCOCYTE (e.g. neutrophil)

PLATELETS

LYMPHOCYTE

MACROPHAGE

Tissue fluid

(interstitial or intercellular fluid) is the immediate environment of the cells, and represents the 'internal environment' described by Claude Bernard in his definition of homeostasis.

Plasma proteins do not move from plasma to tissue fluid (cannot cross capillary endothelium) - largely responsible for solute potential of plasma.

Movement from tissue fluid to plasma
- *Water*
- *Carbon dioxide*
- *Nitrogenous waste*
- *Hormones and other secretions*

ARTERIAL END OF CAPILLARY

VENOUS END OF CAPILLARY

Movement from plasma to tissue fluid
- *Water*
- *Oxygen*
- *Soluble products of digestion*
- *Hormones*

Living cells place demands on the tissue fluid.

FORCES WHICH REGULATE THE FORMATION AND RECLAMATION OF TISSUE FLUID

Pressure potential (hydrostatic potential) is the pressure exerted on a fluid by its surroundings, e.g. by *pumping action of heart* and *elastic recoil of arteries.*

Solute potential (osmotic potential) is the force of attraction towards water molecules caused by dissolved solutes, particularly *ions* and *plasma proteins*.

Net force driving fluid movement at any point = pressure potential gradient – solute potential gradient

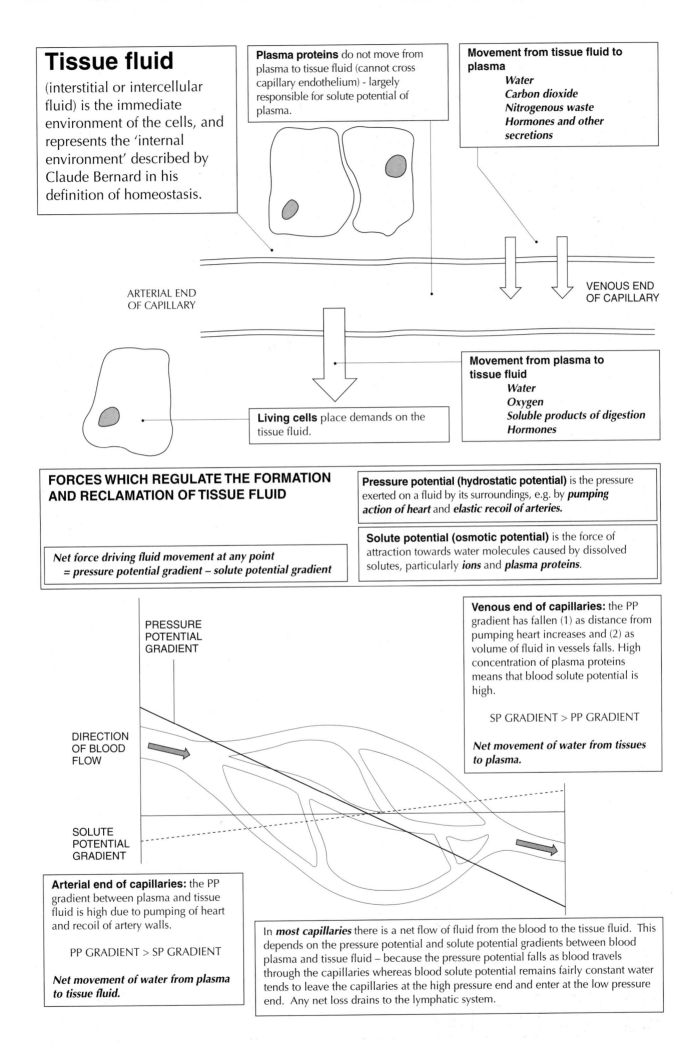

PRESSURE POTENTIAL GRADIENT

DIRECTION OF BLOOD FLOW

SOLUTE POTENTIAL GRADIENT

Venous end of capillaries: the PP gradient has fallen (1) as distance from pumping heart increases and (2) as volume of fluid in vessels falls. High concentration of plasma proteins means that blood solute potential is high.

SP GRADIENT > PP GRADIENT

Net movement of water from tissues to plasma.

Arterial end of capillaries: the PP gradient between plasma and tissue fluid is high due to pumping of heart and recoil of artery walls.

PP GRADIENT > SP GRADIENT

Net movement of water from plasma to tissue fluid.

In *most capillaries* there is a net flow of fluid from the blood to the tissue fluid. This depends on the pressure potential and solute potential gradients between blood plasma and tissue fluid – because the pressure potential falls as blood travels through the capillaries whereas blood solute potential remains fairly constant water tends to leave the capillaries at the high pressure end and enter at the low pressure end. Any net loss drains to the lymphatic system.

Functions of blood

TRANSPORT

Soluble products of digestion/absorption (such as glucose, amino acids, vitamins and minerals) from the gut to the liver and then to the general circulation. Fatty acids are transported from the gut to the lymph system and then to the general circulation.

Waste products of metabolism (such as urea, creatinine and lactate) from sites of production to sites of removal, such as the liver and kidney.

Hormones (such as insulin, a peptide, testosterone, a steroid, and adrenaline, a catecholamine) from their sites of production in the glands to the target organs where they exert their effects.

Respiratory gases (oxygen and carbon dioxide) from their sites of uptake or production to their sites of utilization or removal. Oxygen transport is more closely associated with red blood cells, and carbon dioxide transport with the plasma.

Plasma proteins secreted from the liver and present in the circulating blood include fibrinogen (a blood clotting agent), globulins (involved with specific transport functions, e.g. of thyroxine, iron and copper) and albumin (which binds plasma Ca^{2+} ions).

REGULATORY

Blood solutes affect the water potential of the blood, and thus the water potential gradient between the blood and the tissue fluid. The size of this water potential gradient, determined principally by plasma concentrations of Na^+ ions and plasma proteins, thus **regulates water movement** between blood and tissues.

The **water content** of the blood plays a part in **regulation of body temperature** since it may transfer heat between thermogenic (heat-generating) centres, such as the liver, skeletal muscle and brown fat, and heat sinks such as the skin, the brain and the kidney.

pH maintenance is an important function of **blood buffer systems** such as the hydrogencarbonate and phosphate equilibria, and is a secondary role of haemoglobin and some plasma proteins.

PROTECTIVE

Platelets, plasma proteins (e.g. fibrinogen) and many other plasma factors (e.g. Ca^{2+}) protect against **blood loss** and the **entry of pathogens** by the clotting mechanism.

Leucocytes protect against **toxins and potential pathogens** by both non-specific (e.g. phagocytosis) and specific (e.g. antibody production and secretion) immune responses.

Haemoglobin–oxygen association curve

shows the relationship between haemoglobin saturation with oxygen (i.e. the % of haemoglobin in the form of oxyhaemoglobin) and the partial pressure of oxygen in the environment (the pO_2 or oxygen tension). The curve may be defined by the LOADING and UNLOADING TENSION.

Loading tension
pO_2 at which the haemoglobin is 95% (100% saturation is very rare).

Uploading tension
pO_2 at which the haemoglobin is 50% saturated.

The S-shape of the curve is most significant. Simply put, it means that oxygen associates with haemoglobin, and remains associated with it, at oxygen tensions typical of the alveolar capillaries, the pulmonary vein, the aorta and the arteries, but that it very rapidly dissociates from haemoglobin at oxygen tensions typical of those found in respiring tissues. This dissociation is almost complete at the low oxygen tensions found in the most active tissues. In other words, oxygen release from oxyhaemoglobin is tailored to the tissues' demands for this gas.

The reasons for this S-shapedness is that haemoglobin and oxygen illustrate co-operative binding, that is, the binding of the first oxygen molecule to haemoglobin alters the shape of the haemoglobin molecule slightly so that the binding of a second molecule of oxygen is made easier, and so on until haemoglobin has its full complement of four molecules of oxygen. Conversely, when one molecule of oxygen dissociates from the oxyhaemoglobin, the haemoglobin, shape is adjusted to make release of successive molecules of oxygen increasing easy.

BOHR SHIFT

We have noted that the release of oxygen from oxyhaemoglobin is well suited to the oxygen demands of the tissues. The balance between supply and demand is even further adjusted by special local conditions in the tissues. For example, the curve is shifted to the right and made steeper by higher tensions of carbon dioxide, by falling pH, by increasing lactate concentration and by rising temperatures. All of these changes correspond to increased respiratory activity in a tissue, and the alteration in the oxygen association curve ensures that oxygen unloads from oxyhaemoglobin even more rapidly.

A similar Bohr shift of the Hb–O_2 association curve **to the right** (i.e. towards more O_2 released) results from **falling pH, increased [lactate], increased temperature.**

This phenomenon is called the Bohr shift.

Haemoglobin and DPG:
Diphosphoglycerate is found in erythrocytes during glycolysis, and is thus a good indicator of metabolic rate in cells. It binds reversibly to haemoglobin and promotes the release of oxygen from oxyhaemoglobin

---

The **'difference'** between the two curves represents the additional oxygen released from oxyhaemoglobin at any particular pO_2 (here shown as T mm Hg)

---

pO_2 typical of lungs, pulmonary veins and systemic arteries: haemoglobin is saturated with oxygen – the 'flatness' of the curve means that the Hb remains saturated (i.e. very little O_2 is released) despite a small reduction in O_2 – ideal for the **transport** of oxygen.

Steepness of curve corresponds to easy dissociation of oxyhaemoglobin (much release of O_2) as pO_2 falls to values typical of blood in capillaries of respiring tissues.

---

pO_2 typical of respiring tissues: almost complete dissociation of Hb–O_2 (most O_2 has been released)

Myoglobin and foetal haemoglobin

have molecular properties which promote oxygen transfer from circulating blood to skeletal muscle and to the foetus

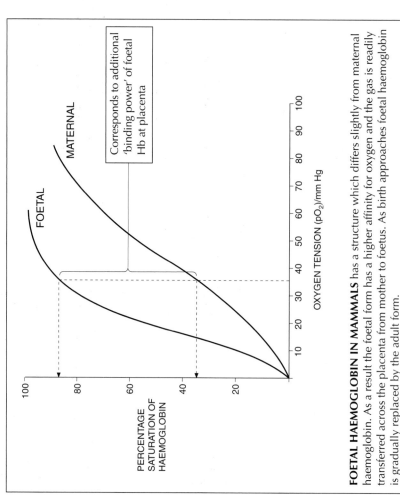

Corresponds to additional 'binding power' of foetal Hb at placenta

FOETAL HAEMOGLOBIN IN MAMMALS has a structure which differs slightly from maternal haemoglobin. As a result the foetal form has a higher affinity for oxygen and the gas is readily transferred across the placenta from mother to foetus. As birth approaches foetal haemoglobin is gradually replaced by the adult form.

MYOGLOBIN, MUSCLE AND MARATHONS

The muscles of mammals contains a red pigment called myoglobin which is structurally similar to one of the four sub-units of haemoglobin. This pigment may also bind to oxygen, but since there is only one haem group there can be no co-operative binding and the myoglobin–oxygen association curve is hyperbolic rather than sigmoidal.

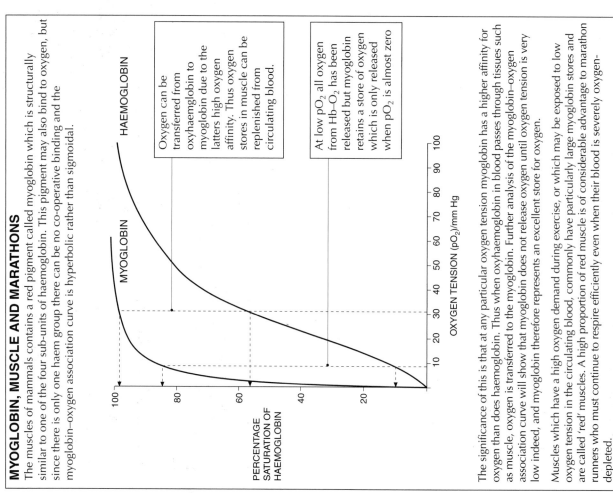

Oxygen can be transferred from oxyhaemglobin to myoglobin due to the latters high oxygen affinity. Thus oxygen stores in muscle can be replenished from circulating blood.

At low pO_2 all oxygen from Hb–O_2 has been released but myoglobin retains a store of oxygen which is only released when pO_2 is almost zero

The significance of this is that at any particular oxygen tension myoglobin has a higher affinity for oxygen than does haemoglobin. Thus when oxyhaemoglobin in blood passes through tissues such as muscle, oxygen is transferred to the myoglobin. Further analysis of the myoglobin–oxygen association curve will show that myoglobin does not release oxygen until oxygen tension is very low indeed, and myoglobin therefore represents an excellent store for oxygen.

Muscles which have a high oxygen demand during exercise, or which may be exposed to low oxygen tension in the circulating blood, commonly have particularly large myoglobin stores and are called 'red' muscles. A high proportion of red muscle is of considerable advantage to marathon runners who must continue to respire efficiently even when their blood is severely oxygen-depleted.

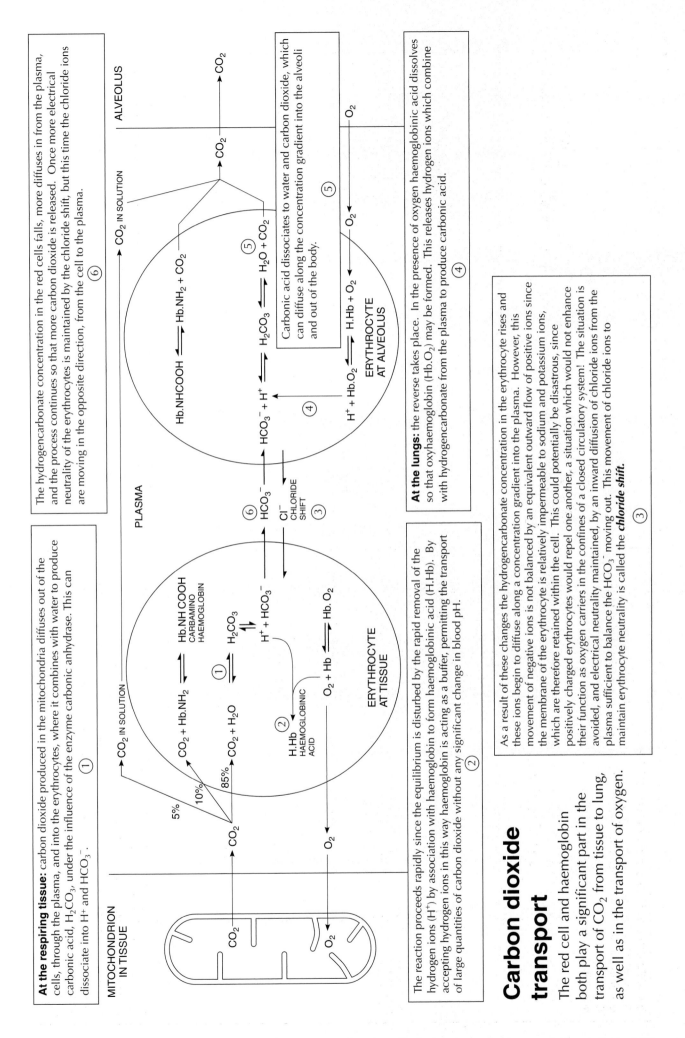

At the respiring tissue: carbon dioxide produced in the mitochondria diffuses out of the cells, through the plasma, and into the erythrocytes, where it combines with water to produce carbonic acid, H_2CO_3, under the influence of the enzyme carbonic anhydrase. This can dissociate into H^+ and HCO_3^-. ①

The hydrogencarbonate concentration in the red cells falls, more diffuses in from the plasma, and the process continues so that more carbon dioxide is released. Once more electrical neutrality of the erythrocytes is maintained by the chloride shift, but this time the chloride ions are moving in the opposite direction, from the cell to the plasma. ⑥

The reaction proceeds rapidly since the equilibrium is disturbed by the rapid removal of the hydrogen ions (H^+) by association with haemoglobin to form haemoglobinic acid (H.Hb). By accepting hydrogen ions in this way haemoglobin is acting as a buffer, permitting the transport of large quantities of carbon dioxide without any significant change in blood pH. ②

At the lungs: the reverse takes place. In the presence of oxygen haemoglobin (Hb.O_2) may be formed. This releases hydrogen ions which combine with hydrogencarbonate from the plasma to produce carbonic acid. ④

Carbonic acid dissociates to water and carbon dioxide, which can diffuse along the concentration gradient into the alveoli and out of the body. ⑤

As a result of these changes the hydrogencarbonate concentration in the erythrocyte rises and these ions begin to diffuse along a concentration gradient into the plasma. However, this movement of negative ions is not balanced by an equivalent outward flow of positive ions since the membrane of the erythrocyte is relatively impermeable to sodium and potassium ions, which are therefore retained within the cell. This could potentially be disastrous, since positively charged erythrocytes would repel one another, a situation which would not enhance their function as oxygen carriers in the confines of a closed circulatory system! The situation is avoided, and electrical neutrality maintained, by an inward diffusion of chloride ions from the plasma sufficient to balance the HCO_3^- moving out. This movement of chloride ions to maintain erythrocyte neutrality is called the **chloride shift.** ③

Carbon dioxide transport

The red cell and haemoglobin both play a significant part in the transport of CO_2 from tissue to lung, as well as in the transport of oxygen.

Mammalian double circulation comprises *pulmonary* (heart – lung – heart)
and *systemic* (heart – rest of body – heart) *circuits*. The complete separation of the two circuits permits rapid, high-pressure distribution of oxygenated blood essential in active, endothermic animals. There are many subdivisions of the systemic circuit, including **coronary, cerebral, hepatic portal** and, during fetal life only, **fetal circuits**. The circuits are typically named for the organ or system which they service – thus each kidney has a **renal** artery and vein. Each organ has an artery bringing oxygenated blood and nutrients, and a vein removing deoxygenated blood and waste.

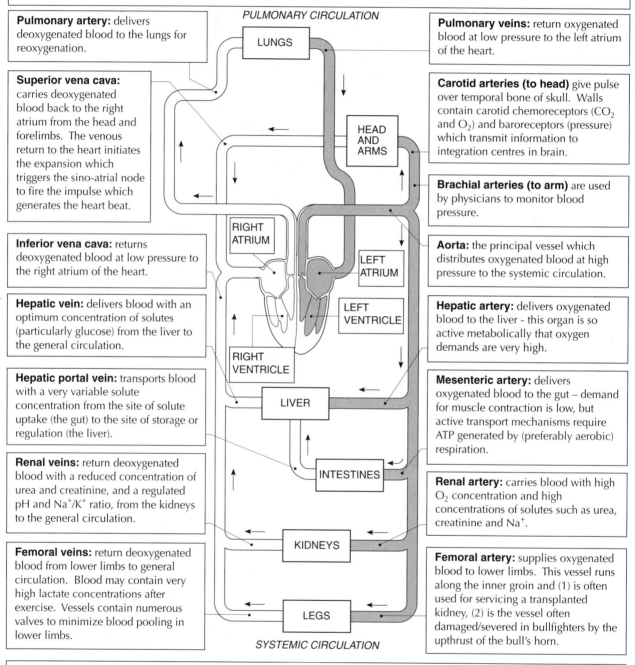

Pulmonary artery: delivers deoxygenated blood to the lungs for reoxygenation.

Superior vena cava: carries deoxygenated blood back to the right atrium from the head and forelimbs. The venous return to the heart initiates the expansion which triggers the sino-atrial node to fire the impulse which generates the heart beat.

Inferior vena cava: returns deoxygenated blood at low pressure to the right atrium of the heart.

Hepatic vein: delivers blood with an optimum concentration of solutes (particularly glucose) from the liver to the general circulation.

Hepatic portal vein: transports blood with a very variable solute concentration from the site of solute uptake (the gut) to the site of storage or regulation (the liver).

Renal veins: return deoxygenated blood with a reduced concentration of urea and creatinine, and a regulated pH and Na^+/K^+ ratio, from the kidneys to the general circulation.

Femoral veins: return deoxygenated blood from lower limbs to general circulation. Blood may contain very high lactate concentrations after exercise. Vessels contain numerous valves to minimize blood pooling in lower limbs.

PULMONARY CIRCULATION

LUNGS

HEAD AND ARMS

RIGHT ATRIUM

LEFT ATRIUM

LEFT VENTRICLE

RIGHT VENTRICLE

LIVER

INTESTINES

KIDNEYS

LEGS

SYSTEMIC CIRCULATION

Pulmonary veins: return oxygenated blood at low pressure to the left atrium of the heart.

Carotid arteries (to head) give pulse over temporal bone of skull. Walls contain carotid chemoreceptors (CO_2 and O_2) and baroreceptors (pressure) which transmit information to integration centres in brain.

Brachial arteries (to arm) are used by physicians to monitor blood pressure.

Aorta: the principal vessel which distributes oxygenated blood at high pressure to the systemic circulation.

Hepatic artery: delivers oxygenated blood to the liver - this organ is so active metabolically that oxygen demands are very high.

Mesenteric artery: delivers oxygenated blood to the gut – demand for muscle contraction is low, but active transport mechanisms require ATP generated by (preferably aerobic) respiration.

Renal artery: carries blood with high O_2 concentration and high concentrations of solutes such as urea, creatinine and Na^+.

Femoral artery: supplies oxygenated blood to lower limbs. This vessel runs along the inner groin and (1) is often used for servicing a transplanted kidney, (2) is the vessel often damaged/severed in bullfighters by the upthrust of the bull's horn.

The **flow of blood** is maintained in three ways.
1. *The pumping action of the heart:* the ventricles generate pressures great enough to drive blood through the arteries into the capillaries.

2. *Contraction of skeletal muscle:* the contraction of muscles during normal movements compress and relax the thin-walled veins causing pressure changes within them. Pocket valves in the veins ensure that this pressure directs the blood to the heart, without backflow.

3. *Inspiratory movements:* reducing thoracic pressure caused by chest and diaphragm movements during inspiration helps to draw blood back towards the heart.

Mammalian heart structure and function

The pressure generated by the left ventricle is greater than that generated by the right ventricle as the systemic circuit is more extensive than the pulmonary circuit.

The pressure generated by the atria is less than that generated by the ventricles since the distance from atria to ventricles is less than that from ventricles to circulatory system.

Volume: the same volume of blood passes through each side of the heart. Both ventricles pump the same volume of blood.

Aortic (semilunar) valve: prevents backflow from aorta to left ventricle.

Bicuspid (mitral, left atrioventricular) valve: ensures blood flow from left ventricle into aortic arch.

Left ventricle: generates pressure to force blood into the systemic circulation.

Chordae tendinae: short, inextensible fibres – mainly composed of collagen – which connect to free edges of atrioventricular valves to prevent 'blow-back' of valves when ventricular pressure rises during contraction of myocardium.

Papillary muscles: contract as wave of excitation spreads through ventricular myocardium and tighten the chordae tendineae just before the ventricles contract.

CARDIAC MUSCLE FIBRE

CROSS BRIDGE WITH GAP JUNCTION

MITOCHONDRION

INTERCALATED DISC

Aorta: carries oxygenated blood from the left ventricle to the systemic circulation. It is a typical elastic (conducting) artery with a wall that is relatively thick in comparison to the lumen, and with more elastic fibres than smooth muscle. This allows the wall of the aorta to accommodate the surges of blood associated with the alternative contraction and relaxation of the heart – as the ventricles contract the artery expands and as the ventricle relaxes the elastic recoil of the artery forces the blood onwards.

Left atrium

Pulmonary arteries

Right atrium

Right ventricle: generates pressure to pump deoxygenated blood to pulmonary circulation.

Myocardium is composed of cardiac muscle: intercalated discs separate muscle fibres, strengthen the muscle tissue and aid impulse conduction; cross-bridges promote rapid conduction throughout entire myocardium; numerous mitochondria permit rapid aerobic respiration. Cardiac muscle is myogenic (can generate its own excitatory impulse) and has a long refractory period (interval between two consecutive effective excitatory impulses), which eliminates danger of cardiac fatigue.

Pulmonary (semilunar) valve: is composed of three cusps or watchpocket flaps which are forced together when the pressure in the pulmonary artery exceeds that in the right ventricle, thus preventing backflow of blood into the relaxing chambers of the heart.

P ventricle > P artery

P ventricle < P artery

Tricuspid (right atrioventricular) valve: has three fibrous flaps with pointed ends which point into the ventricle. The flaps are pushed together when the ventricular pressure exceeds the atrial pressure so that blood is propelled past the inner edge of the valve through the pulmonary artery instead of through the valve and back into the atrium.

Superior (anterior) vena cava: carries deoxygenated blood back to the right atrium of the heart. As with other veins the wall is thin, with little elastic tissue or smooth muscle. In contrast to veins returning blood from below the heart there are no venous valves, since blood may return under the influence of gravity.

Control of heartbeat
1. The heartbeat is initiated in the **sino-atrial node** (particularly excitable myogenic tissue in the wall of the right atrium).
2. **Intrinsic** heart rate is about 78 beats per minute.
3. **External (extrinsic)** factors may modify basic heart rate:
   a. **vagus nerve** decreases heart rate;
   b. **accelerator (sympathetic) nerve** increases heart rate;
   c. **adrenaline** and **thyroxine** increase heart rate.
   Resting heart rate of 70 beats per minute indicates that heart has **vagal tone.**

Control of the heartbeat

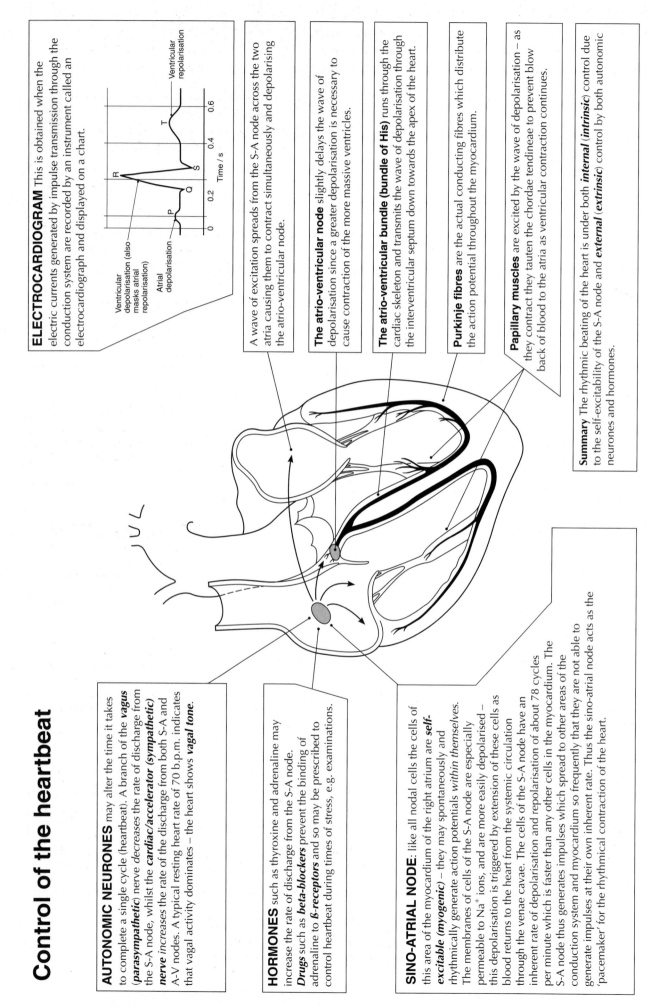

ELECTROCARDIOGRAM This is obtained when the electric currents generated by impulse transmission through the conduction system are recorded by an instrument called an electrocardiograph and displayed on a chart.

Ventricular depolarisation (also masks atrial repolarisation)

Atrial depolarisation

Ventricular repolarisation

Time / s

A wave of excitation spreads from the S-A node across the two atria causing them to contract simultaneously and depolarising the atrio-ventricular node.

The atrio-ventricular node slightly delays the wave of depolarisation since a greater depolarisation is necessary to cause contraction of the more massive ventricles.

The atrio-ventricular bundle (bundle of His) runs through the cardiac skeleton and transmits the wave of depolarisation through the interventricular septum down towards the apex of the heart.

Purkinje fibres are the actual conducting fibres which distribute the action potential throughout the myocardium.

Papillary muscles are excited by the wave of depolarisation – as they contract they tauten the chordae tendineae to prevent blow back of blood to the atria as ventricular contraction continues.

Summary The rhythmic beating of the heart is under both *internal* (*intrinsic*) control due to the self-excitability of the S-A node and *external* (*extrinsic*) control by both autonomic neurones and hormones.

AUTONOMIC NEURONES may alter the time it takes to complete a single cycle (heartbeat). A branch of the *vagus* (*parasympathetic*) nerve *decreases* the rate of discharge from the S-A node, whilst the *cardiac/accelerator* (*sympathetic*) *nerve increases* the rate of the discharge from both S-A and A-V nodes. A typical resting heart rate of 70 b.p.m. indicates that vagal activity dominates – the heart shows *vagal tone*.

HORMONES such as thyroxine and adrenaline may increase the rate of discharge from the S-A node. *Drugs* such as *beta-blockers* prevent the binding of adrenaline to *ß-receptors* and so may be prescribed to control heartbeat during times of stress, e.g. examinations.

SINO-ATRIAL NODE: like all nodal cells the cells of this area of the myocardium of the right atrium are *self-excitable* (*myogenic*) – they may spontaneously and rhythmically generate action potentials *within themselves*. The membranes of cells of the S-A node are especially permeable to Na^+ ions, and are more easily depolarised – this depolarisation is triggered by extension of these cells as blood returns to the heart from the systemic circulation through the venae cavae. The cells of the S-A node have an inherent rate of depolarisation and repolarisation of about 78 cycles per minute which is faster than any other cells in the myocardium. The S-A node thus generates impulses which spread to other areas of the conduction system and myocardium so frequently that they are not able to generate impulses at their own inherent rate. Thus the sino-atrial node acts as the 'pacemaker' for the rhythmical contraction of the heart.

Blood flow through the heart is controlled by two phenomena: the opening and closing of the valves and the contraction and relaxation of the myocardium. Both activities occur without direct stimulation from the nervous system; the valves are controlled by the pressure changes in each heart chamber, and the contraction of the cardiac muscle is stimulated by its conduction system.

Inpulse transmission through the heart's conduction system generates electrical currents which can be recorded using an electrocardiograph and presented as an **electrocardiogram** (ECG).

The QRS wave (complex) represents **ventricular depolarisation** and is strong enough to mask atrial repolarisation.

The P wave indicates **atrial depolarisation** – the spread of an impulse from the SA node through the two atria. It is less powerful than the QRS complex since the atria are less massive than the ventricles

The T wave represents **ventricular repolarisation**.

The dicrotic notch is the slight increase in pressure due to recoil of the arteries as the semilunar valve closes

As ventricular pressure exceeds atrial pressure the **semilunar values** open to allow blood from ventricles to arteries

As the ventricular pressure falls below the arterial pressure the **semilunar valves** snap shut to prevent the backflow of blood from arteries to ventricle.

Contraction of the ventricles without any emptying (note no reduction in volume) causes a sharp rise in ventricular pressure

Sharp rise in ventricular pressure above atrial pressure closes the **cuspid valves** to prevent backflow of blood from ventricles to atria

The arterial blood pressure falls as the ventricles relax (no blood entering arteries) and blood moves along systemic and pulmonary circuits.

Ventricular volume rises as blood moves (passively and by atrial contraction) from atria

About 70% of blood flows passively from atria to ventricles but this increase in pressure during atrial systole is necessary to force the remaining 30% through to the ventricles.

Ventricular volume falls as ventricles contract and semilunar valves open allowing blood flow through to arteries

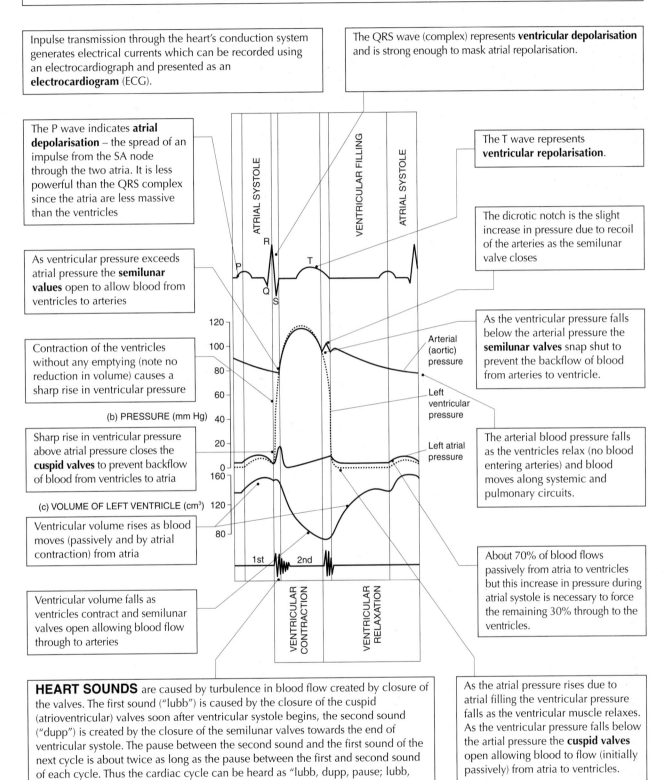

HEART SOUNDS are caused by turbulence in blood flow created by closure of the valves. The first sound ("lubb") is caused by the closure of the cuspid (atrioventricular) valves soon after ventricular systole begins, the second sound ("dupp") is created by the closure of the semilunar valves towards the end of ventricular systole. The pause between the second sound and the first sound of the next cycle is about twice as long as the pause between the first and second sound of each cycle. Thus the cardiac cycle can be heard as "lubb, dupp, pause; lubb, dupp, pause; lubb, dupp, pause".

As the atrial pressure rises due to atrial filling the ventricular pressure falls as the ventricular muscle relaxes. As the ventricular pressure falls below the artial pressure the **cuspid valves** open allowing blood to flow (initially passively) from atria to ventricles.

Blood flows from an area of higher pressure to one of lower pressure. The diagram relates pressure changes to heartbeat in the left side of the heart – the same pattern exists in the right side although pressures are somewhat lower there.

The lymphatic system

Tonsils are aggregations of large lymphatic nodules embedded in a mucous membrane. There is a single *pharyngeal* tonsil or *adenoid*, and two pairs (the *palatine* and the *lingual* tonsils), all arranged in a ring at the junction of the pharynx and the oral cavity. They contain phagocytes and lymphocytes which protect against the invasion of foreign substances from the mouth.

Right lymphatic duct receives lymph drainage from the right side of the head, the right upper trunk and the right arm, and empties the lymph into the systemic blood circulation at the junction of the right subclavian and right jugular veins.

Axillary nodes

The **thymus gland** is a paired organ most obvious in the pre-pubertal individual. It has a significant role in the immune response since it produces the T-lymphocytes which have a number of roles in the hierarchy of the immune system.

Peyer's patches are aggregated lymph nodes in the wall of the ileum, where they are ideally situated to offer protection against the invasion of potential pathogens or absorption of toxins from the gut contents.

The **lymphatic vessels** form a low-pressure return system for reclaimed tissue fluid. Since their contents are at low pressure the lymphatics are well-supplied with semi-lunar valves to ensure flow in one direction (towards the heart).

Thoracic duct receives lymph drained from all other areas of the body, and returns it to the blood circulation at the junction of the left jugular and left subclavian veins.

Cervical nodes

Inguinal nodes

The spleen is the largest mass of lymphatic tissue in the body, and has the role of filtration of *blood* (not lymph, like the other lymph nodes). In the spleen the blood, plus any potential pathogens, is exposed to lymphocytes which may then be triggered to produce appropriate antibodies. The spleen also phagocytoses bacterial and worn-out red blood cells, acts as a blood reservoir and, during fetal development, produces red blood cells.

Lymph nodes are located along the lymphatic vessels, usually in groups, some *superficial* (easily located during infection) and some *deep*. As lymph passes through these nodes it is filtered of foreign substances which are trapped within a network of fibres and then phagocytosed by *macrophages* or destroyed by products of *T-cells*. Lymph nodes also contain *B-lymphocytes*, which secrete antibodies, or may themselves leave the node and circulate to other parts of the body.

AFFERENT VESSEL

EFFERENT VESSEL

FIBROUS NETWORK WITH LYMPHOCYTES AND PHAGOCYTES

Relationship between lymphatic and blood vascular systems

Blood is forced into the systemic circulation at about 4.5 cm³ per minute.

HEART

BLOOD CAPILLARIES

SOLUTES DELIVERED TO CELLS

WATER AND SOME WASTES

LYMPHATIC CAPILLARIES

Lymph returns to low pressure vessels of the blood vascular system.

LYMPH NODE

Lymph is delivered to the lymph nodes at about 1.5 cm³ per minute.

At the lymph node, *phagocytes* remove toxins and pathogens, and *B-lymphocytes* secrete antibodies.

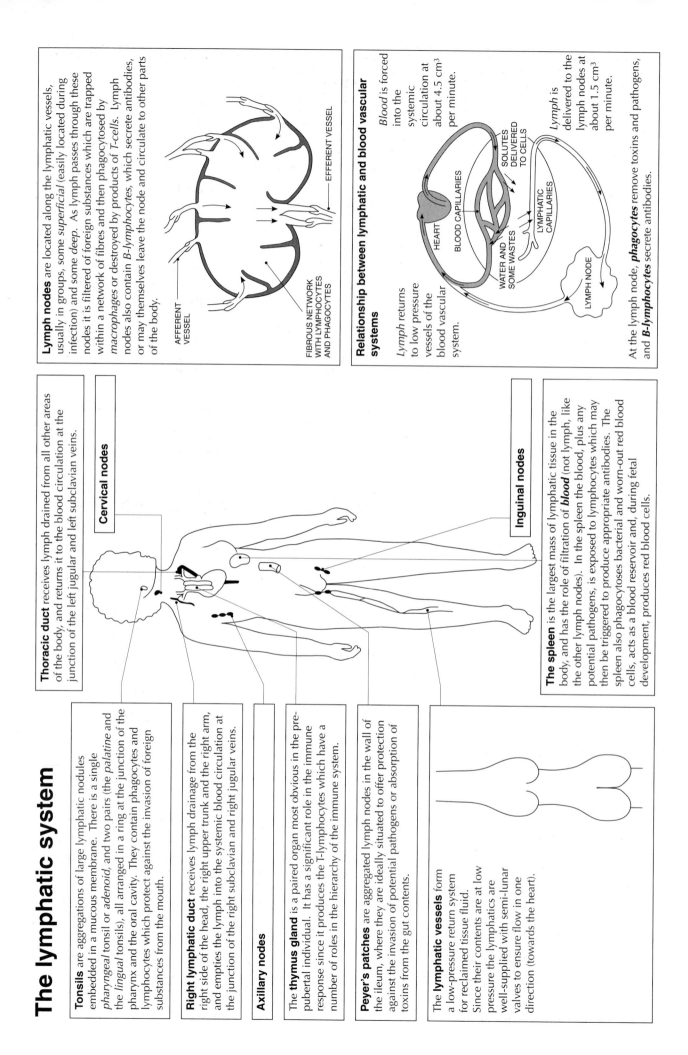

Vasomotor centre is the **integrator** in the control of circulatory changes. It is located in the **medulla** and has **sensory input** from cerebral cortex and baroreceptors and **motor output** to heart, precapillary sphincters and adrenal glands.

Changes in blood pressure: at the base of the *internal* carotid artery there is a swelling, the carotid sinus, that contains stretch receptors. A rise in blood pressure stimulates the stretch receptors and sensory impulses pass to the cardiovascular centre in the brain. Motor impulses pass from the brain to the heart and peripheral arterioles which cause a decrease in heart rate and vasoconstriction. Thus the blood pressure falls again.

Higher centres in the brain can also have an effect on the circulatory system. During periods of excitement (waiting for an examination) or shock (getting the exam results), the cerebral cortex passes nerve impulses to the adrenal glands which respond by secreting the hormone adrenaline into the bloodstream. Adrenaline increases cardiac frequency and constriction of the peripheral arterioles, an effect similar to increasing blood pressure by stimulation from the cardiovascular centre. Although the nervous and hormonal systems produce identical effects, there is an important difference. The nervous system (cardiovascular centre) adjusts the body to changes **that are taking place**. The hormonal system (cerebral cortex/adrenaline) adjusts the body to changes **that have yet to occur**. The hormonal system of control prepares the body for action in emergencies.

external carotid artery

carotid body

nerves to brain

internal carotid artery

carotid sinus

common carotid artery

Changes in cardiac output: the secretion of adrenaline from the adrenal medulla and the activity of the sympathetic nervous system both increase cardiac output. This is a result of both increased heart rate and increased stroke volume.

The circulatory system in action:
adjustment to exercise

The circulatory system must be able to respond to the changing requirements of the tissues which it supplies. The circulatory system does not work in isolation, and changes in circulation are often accompanied by adjustments to breathing rate and skeletal muscle tone.

The response to exercise involves
1. A change in cardiac output
2. A change in blood distribution
3. A change in blood pressure
4. A change in gas exchange

Note the typical homeostatic sequence

STIMULUS
↓
RECEPTOR
↓
INTEGRATOR
↓
EFFECTOR
↓
RESPONSE
↓
RETURN TO NORM

Changes in blood distribution: adrenaline and the sympathetic nervous system act upon smooth muscle fibres in the walls of arteries. As a result blood vessels in the skin and abdominal organs undergo vasoconstriction so that blood which is normally "stored" in these organs is put into more active circulation. At the same time vasodilation permits a greater blood flow through the coronary vessels and the skeletal muscles. There is not enough blood in the circulation to fill the whole of the circulatory system in the dilated state.

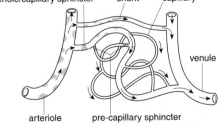

arteriole/capillary sphincter   shunt   capillary

venule

arteriole   pre-capillary sphincter

Local control of blood flow

There is also a local control of blood flow and distribution since the sphincters present at the entrances to capillary beds and arterio-venous shunts are sensitive to local changes in the concentration of CO_2 and lactate. Thus during exercise an increase in these waste products causes relaxation of the capillaries in the muscles and more blood flow through, thus removing the products. Once the level of these products returns to normal, the sphincter contracts again.

Capillaries can also respond to changes in body temperature. As the working skeletal muscle generates heat the capillaries dilate to permit the "heated" blood to pass more easily towards the body surface.

Heart disease

affects 1 in 4 persons between the ages of 30 and 60.
High risk factors include
1 High blood cholesterol level
2 Cigarette smoking
3 Obesity
4 Stress
5 Lack of regular exercise

Pericarditis and myocarditis are inflammation of the pericardium and myocardium respectively. Such inflammations may result from bacterial infection (e.g. following a tooth extraction or respiratory complaint) or a depressed immune response following chemotherapy or radiotherapy.

Valvular disease: Stenosis is caused by fibrosis of the valve following inflammation, **valvular incompetence** is a functional defect caused by failure of a valve to close completely, allowing blood to flow back into a ventricle as it relaxes.

Congenital defects:
Coarction of the aorta – one segment of the aorta is too narrow so that the flow of oxygenated blood to the body is reduced.

Persistent ductus arteriosus – the ductus arteriosus should close at birth – if it does not blood flows between the aorta and the pulmonary artery, reducing the volume entering the **systemic** circulation but causing congestion in the **pulmonary** circulation.
Septal defect ("Hole in the heart") is most commonly due to incomplete closure of the foramen ovale at birth, allowing blood to flow between the two atria. This may reduce the flow of oxygenated blood or more severely may raise right sided blood pressure to dangerous levels.

Ischaemic heart disease is due to narrowing or closure of one or more branches of the conorary arteries. The narrowing is caused by atheromatous plaques, closure either by plaques alone or complicated by thrombosis.

Angina pectoris is severe ischaemic pain following physical effort. The narrowing vessel may supply sufficient oxygen and nutrient laden blood to meet the demands of the cardiac muscle during rest or moderate exercise but is unable to dilate enough to allow for the increased blood flow needed by an active myocardium during severe exercise. Angina is treated by a combination of analgesics (painkillers) and vasodilators.

Rheumatic fever (rarely seen today) is an autoimmune disease which occurs 2–4 weeks after a throat infection, caused by *Streptococcus pyogenes*. The antibodies developed to combat the infection cause damage to the heart, in a way that is not yet understood.

Effects include
1 development of fibrous nodules on the mitral and aortic valves (rarely on the tricuspid and pulmonary valves)
2 fibrosis of the connective tissue of the mycardium
3 accumulation of fluid in the pericardial cavity – in severe cases the cavity is obliterated and heart expansion during diastole is severely affected

Myocardial infarction (commonly called "heart attack"): Infarction means the death of an area of tissue following an interrupted blood supply. The tissue beyond the obstruction dies and is replaced by non-contractile tissue – the heart loses some of its strength. In CHRONIC ISCHAEMIC HEART DISEASE many small infarcts form and collectively and gradually lead to myocardial weakness, but in ACUTE ISCHAEMIC HEART DISEASE the infarct is large since a large artery is closed – death may occur suddenly for a number of reasons:
1 Acute cardiac failure due to shock
2 Severe arrhythmia due to disruption of the conducting system
3 Rupture of the ventricle wall, usually within two weeks of the infarction

Recent studies suggest that much tissue damage occurs when blood flow is **restored** not when it is **interrupted.** When blood supply is interrupted the affected tissue releases an enzyme called XANTHOXIDASE – when blood supply is restored the enzyme reacts with oxygen in the blood to produce SUPEROXIDE FREE RADICALS which cause enormous tissue damage by protein oxidation. Drugs such as SUPEROXIDE DISMUTASE and ALLOPURINOL, are used to prevent the formation of the free radicals and limit tissue damage.

Disorders of the circulation

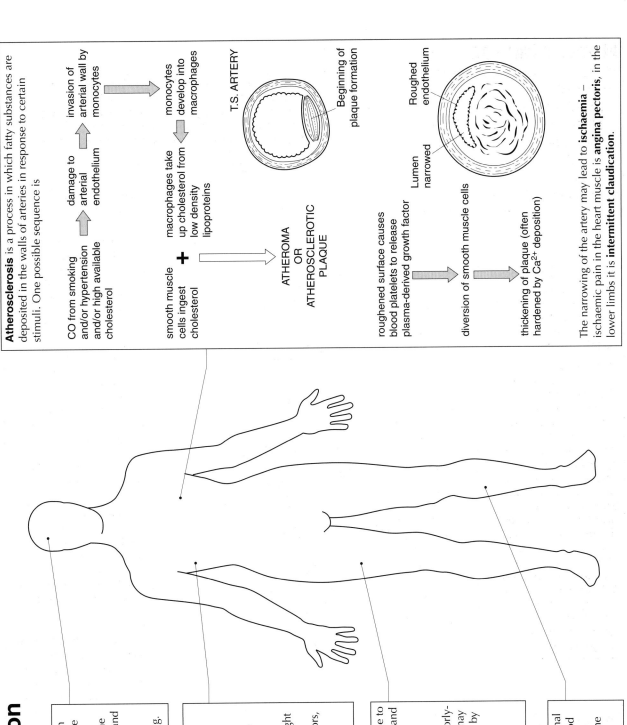

Atherosclerosis is a process in which fatty substances are deposited in the walls of arteries in response to certain stimuli. One possible sequence is

CO from smoking and/or hypertension and/or high available cholesterol → damage to arterial endothelium → invasion of arterial wall by monocytes → monocytes develop into macrophages → macrophages take up cholesterol from low density lipoproteins

smooth muscle cells ingest cholesterol

+ → ATHEROMA OR ATHEROSCLEROTIC PLAQUE

T.S. ARTERY

Beginning of plaque formation

Roughed endothelium

Lumen narrowed

roughened surface causes blood platelets to release plasma-derived growth factor → diversion of smooth muscle cells → thickening of plaque (often hardened by Ca²⁺ deposition)

The narrowing of the artery may lead to **ischaemia** – ischaemic pain in the heart muscle is **angina pectoris**, in the lower limbs it is **intermittent claudication**.

Aneurysms are thin, weakened sections of the wall of an artery or a vein that bulges outward forming a balloon-like sac. Aneurysms may be congenital or may be caused by atherosclerosis, syphilis or trauma. The vessel may become so thin that it bursts causing massive haemorrhage, pain and severe tissue damage. Unruptured aneurysms may exert pressure on adjacent tissues e.g. aortic aneurysm may compress the oesophagus, causing difficulty in swallowing.

Hypertension (high blood pressure) is the most common circulatory disease. Normal blood pressure is ¹²⁰/₈₀ – any value above ¹⁴⁰/₉₀ indicates hypertension and ¹⁶⁰/₉₅ is considered dangerous. Uncontrolled hypertension may damage heart (heart enlarges and demands more oxygen), brain (CVA, or stroke, which ruptures cerebral arteries supplying brain) and kidneys (glomoular arterioles are narrowed and deliver less blood). Treatment includes weight reduction in obese patients, restriction of Na²⁺ intake, increased Ca²⁺/K⁺ intake and stopping smoking. Vasodilators, beta blockers and diuretics are widely prescribed.

Varicose veins are so dilated that the valves do not close to prevent backflow of blood. The veins lose their elasticity and become congested. Predisposing factors include age, heredity, obesity, gravity and compression by adjacent structures. Common sites are the legs (where poor circulation may lead to haemorrhage or ulceration of poorly-nourished skin), the rectal-anal junction (haemorrhoids may bleed and cause mild anaemia), the oesophagus (caused by liver cirrhosis or right-sided heart failure – may lead to haemorrhage and death) and the spermatic cord (may reduce spermatogenesis).

Embolism is obstruction of a blood vessel by an abnormal mass of material (an embolus) such as a fragment of blood clot, atheromatous plaque, bone fragment or air bubble. Emboli may lodge in the coronary vessels, the lungs or the liver causing restricted circulation and, perhaps, death.

Section H Homeostasis in animals

Control systems in biology

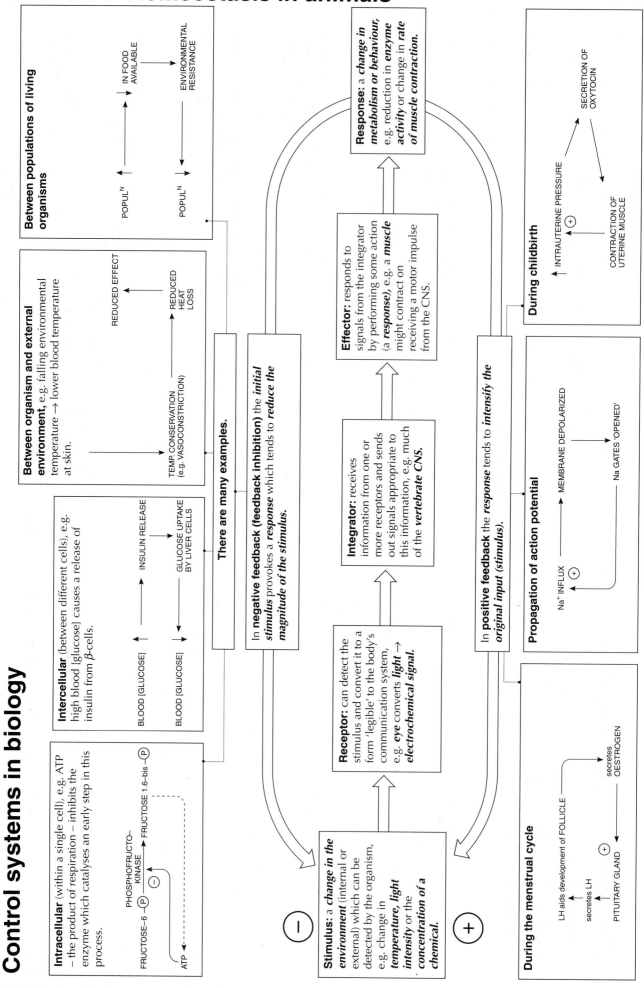

Intracellular (within a single cell), e.g. ATP – the product of respiration – inhibits the enzyme which catalyses an early step in this process.

PHOSPHOFRUCTO-
FRUCTOSE-6 –Ⓟ → KINASE → FRUCTOSE 1.6-bis –Ⓟ
—Ⓙ
ATP ┘

Intercellular (between different cells), e.g. high blood [glucose] causes a release of insulin from β-cells.

INSULIN RELEASE

BLOOD [GLUCOSE] ←

GLUCOSE UPTAKE
BY LIVER CELLS

BLOOD [GLUCOSE] →

Between populations of living organisms

POPUL.ᴺ ⇄ ─→ IN FOOD AVAILABLE ─→ ENVIRONMENTAL RESISTANCE

POPUL.ᴺ ⇄

Between organism and external environment, e.g. falling environmental temperature → lower blood temperature at skin.

─→ REDUCED EFFECT

TEMP. CONSERVATION
(e.g. VASOCONSTRICTION) ─→ REDUCED HEAT LOSS

There are many examples.

In **negative feedback (feedback inhibition)** the *initial stimulus* provokes a *response* which tends to *reduce the magnitude of the stimulus.*

Ⓘ

Stimulus: a *change in the environment* (internal or external) which can be detected by the organism, e.g. change in *temperature, light intensity* or the *concentration of a chemical.*

Receptor: can detect the stimulus and convert it to a form 'legible' to the body's communication system, e.g. *eye* converts *light → electrochemical signal.*

Integrator: receives information from one or more receptors and sends out signals appropriate to this information, e.g. much of the *vertebrate CNS.*

Effector: responds to signals from the integrator by performing some action (a *response*), e.g. a *muscle might contract* on receiving a motor impulse from the CNS.

Response: a *change in metabolism or behaviour,* e.g. reduction in *enzyme activity* or change in *rate of muscle contraction.*

In **positive feedback** the *response* tends to *intensify the original input (stimulus).*

Ⓙ

Propagation of action potential

Na⁺ INFLUX ⊕ → MEMBRANE DEPOLARIZED → Na GATES 'OPENED'

During childbirth

↑ INTRAUTERINE PRESSURE ⊕ → SECRETION OF OXYTOCIN → CONTRACTION OF UTERINE MUSCLE

During the menstrual cycle

LH aids development of FOLLICLE → secretes OESTROGEN

secretes LH ⊕

PITUITARY GLAND

Blood glucose concentration

is regulated by negative feedback

Insulin a. increases uptake of glucose by cells of liver, muscle and adipose tissue
b. increases conversion of glucose to **glycogen** and **fats**
c. decreases hydrolysis of glycogen to glucose, and synthesis of glucose from amino acids.

WHAT IS DIABETES?

Diabetes is a condition in which there are higher than normal blood glucose concentrations

- usually the result of the failure of the pancreas to secrete enough insulin

- symptoms include
  excessive thirst, hunger or urine production
  sweet smelling breath
  high 'overflow' of glucose into urine (test with **clinistix**)

- long-term effects if untreated include
  premature ageing
  cataract formation
  hardening of arteries
  heart disease

- treatment is by regular injection of pure insulin - much of this is now manufactured by **genetic engineering.**

- mild diabetes can be controlled by adjusting diet to reduce sugar intake.

Glucose tolerance test measures the body's response to high glucose intake (sugar solution).

SEVERE DIABETES

MILD DIABETES

NORMAL

Time

Glucose solution swallowed

Blood glucose concen-tration

TOO HIGH
= HYPERGLYCAEMIA

TOO LOW
= HYPOGLYCAEMIA

PANCREAS produces more of the hormone INSULIN

Blood sugar level measured as blood passes through the pancreas

PANCREAS produces more of the hormone GLUCAGON

Liver converts GLUCOSE → GLYCOGEN

NORMAL BLOOD SUGAR LEVEL

Liver converts GLYCOGEN → GLUCOSE

This conversion is also stimulated by the hormone **adrenaline**, secreted at times of stress from the adrenal medulla.

WHAT IS BLOOD SUGAR LEVEL?

Glucose is the cells' main source of energy, and it must always be available to them...

GLUCOSE + OXYGEN

CARBON DIOXIDE + WATER
+
ENERGY

Cells may carry out WORK

...so that the body keeps a constant amount of glucose in the blood. This is the **blood sugar level** and is usually maintained at about 1 mg of glucose per cm³ of blood.

STIMULUS

DETECTOR SYSTEM

REGULATOR

EFFECTORS

RESPONSE

CANCELS OUT STIMULUS

NOTE THE CLASSIC HOMEOSTATIC PRINCIPLE OF NEGATIVE FEEDBACK

Excretion involves removal of waste products of metabolism

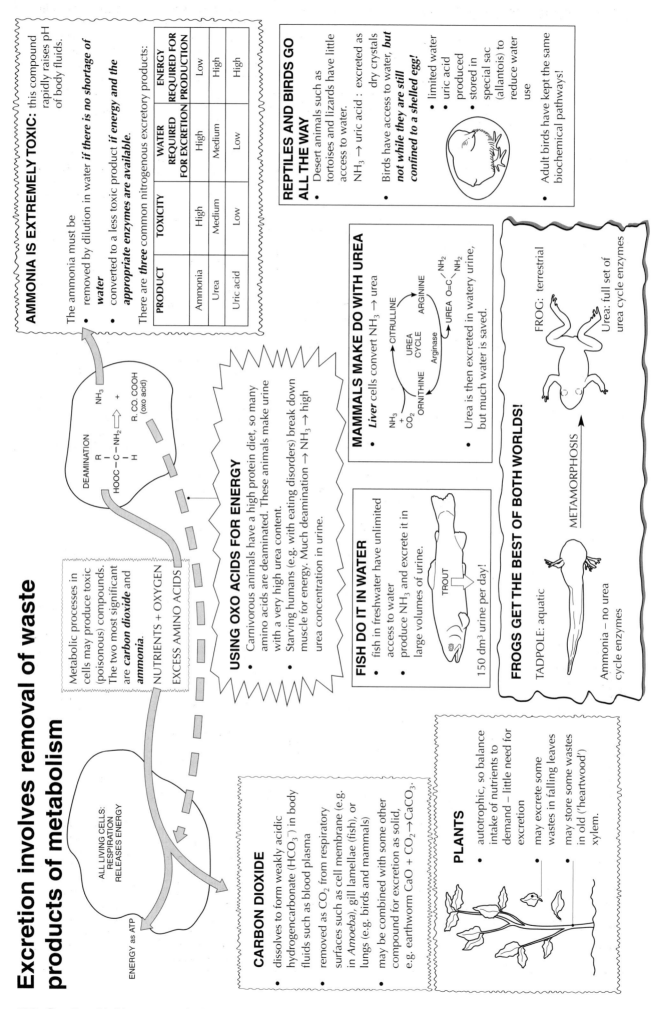

ALL LIVING CELLS: RESPIRATION RELEASES ENERGY

ENERGY as ATP

Metabolic processes in cells may produce toxic (poisonous) compounds. The two most significant are **carbon dioxide** and **ammonia.**

NUTRIENTS + OXYGEN

EXCESS AMINO ACIDS

DEAMINATION

$$HOOC-\underset{\underset{H}{|}}{\overset{\overset{R}{|}}{C}}-NH_2 \rightarrow NH_3 + R.CO.COOH \text{ (oxo acid)}$$

AMMONIA IS EXTREMELY TOXIC:
this compound rapidly raises pH of body fluids.

The ammonia must be
- removed by dilution in water *if there is no shortage of water*
- converted to a less toxic product *if energy and the appropriate enzymes are available.*

There are *three* common nitrogenous excretory products:

PRODUCT	TOXICITY	WATER REQUIRED FOR EXCRETION	ENERGY REQUIRED FOR PRODUCTION
Ammonia	High	High	Low
Urea	Medium	Medium	High
Uric acid	Low	Low	High

USING OXO ACIDS FOR ENERGY
- Carnivorous animals have a high protein diet, so many amino acids are deaminated. These animals make urine with a very high urea content.
- Starving humans (e.g. with eating disorders) break down muscle for energy. Much deamination → NH_3 → high urea concentration in urine.

CARBON DIOXIDE
- dissolves to form weakly acidic hydrogencarbonate (HCO_3^-) in body fluids such as blood plasma
- removed as CO_2 from respiratory surfaces such as cell membrane (e.g. in *Amoeba*), gill lamellae (fish), or lungs (e.g. birds and mammals)
- may be combined with some other compound for excretion as solid, e.g. earthworm $CaO + CO_2 \rightarrow CaCO_3$.

REPTILES AND BIRDS GO ALL THE WAY
- Desert animals such as tortoises and lizards have little access to water.
  NH_3 → uric acid : excreted as dry crystals
- Birds have access to water, **but not while they are still confined to a shelled egg!**
  - limited water
  - uric acid produced stored in special sac (allantois) to reduce water use
- Adult birds have kept the same biochemical pathways!

MAMMALS MAKE DO WITH UREA
- *Liver* cells convert NH_3 → urea

UREA CYCLE

CITRULLINE → ARGININE

NH_3 + CO_2 ORNITHINE

Arginase

UREA $O=C\begin{smallmatrix}NH_2\\NH_2\end{smallmatrix}$

- Urea is then excreted in watery urine, but much water is saved.

FISH DO IT IN WATER
- fish in freshwater have unlimited access to water
- produce NH_3 and excrete it in large volumes of urine.

TROUT

150 dm^3 urine per day!

FROGS GET THE BEST OF BOTH WORLDS!

TADPOLE: aquatic

Ammonia – no urea cycle enzymes

METAMORPHOSIS

FROG: terrestrial

Urea: full set of urea cycle enzymes

PLANTS
- autotrophic, so balance intake of nutrients to demand – little need for excretion
- may excrete some wastes in falling leaves
- may store some wastes in old ('heartwood') xylem.

The urinary system

INFERIOR VENA CAVA

ADRENAL GLAND

ADRENAL ARTERY AND VEIN

RENAL ARTERY

RIGHT KIDNEY

RENAL VEIN

DORSAL AORTA

URETER

BLADDER

URETHRA

Renal circulation: handles approx. 1200 cm³ of blood per minute ~ about 25% of cardiac output.

Renal papilla: apex of the pyramid from which the ends of the collecting ducts deliver the urine.

Ureter: propels urine from the pelvis to the bladder.

Interlobular artery: delivers blood at high pressure to the glomerular capillaries.

Interlobular vein: drains blood away from the glomerular filtration units and from the Loops of Henle.

Medulla: has a striated appearance due to the presence of the Loop of Henle, the collecting duct and the vasa rectae.

Renal pyramid: one of 5–12 triangular structures which make up the medullary region.

Kidney cortex: has a smooth texture and is the site of the Bowman's Capsules and the glomeruli, together with the proximal and distal convoluted tubules and their associated blood supply.

Pelvis: urine collects here from tips of renal papillae.

MEDULLA

CORTEX

Single nephron: (much enlarged) to show position in medulla and cortex.

Glomerular filtration occurs in the renal capsule: the selective structure is the **basement membrane of the glomerulus.** Water and solutes of relative molecular mass less than 68 000 form the filtrate.

Glomerular filtrate is formed at above 125 cm³ per minute in humans (this represents about 20% of the plasma delivered during that time). It contains all the materials present in the blood except blood cells and most proteins, which are too large to cross the basement membrane of the glomerulus.

Active controlled reabsorption of Na⁺ occurs in the distal convoluted tubule. This is followed by the osmotic movement of an equivalent volume of water down the water potential gradient.

Tubular reabsorption occurs in the proximal tubule. Solutes are selectively moved from the filtrate to the plasma by **active transport** and water follows by osmosis. Almost all glucose and amino acids, and high but variable amounts of ions, are reabsorbed here. Since water follows by osmosis (about 80% of the filtrate volume), and is not controlled by the proximal tubule, this is referred to as **obligatory water reabsorption.**

The different permeability properties of the two limbs of the Loop of Henle, together with their counterflow arrangement, allows a **countercurrent multiplication** to generate a high solute concentration in the tissue fluid of the medulla. The highest solute concentrations are generated deep in the medulla.

Variable (facultative) movement of water from urine collecting duct to medullary tissue fluid. This is dependent on (a) maintenance of high solute concentrations in tissue fluid and (b) controlled permeability of wall of the collecting duct.

Tubular secretion of H⁺ and NH₄⁺ from blood to urine helps to keep blood pH at its normal 7.4. These ion movements also help to conserve NaHCO₃.

Urine trickles into kidney pelvis - now only 1% of filtered volume, high concentrations of urea, creatinine and variable ion concentration. Typically about 1.5 dm³ per day.

The **countercurrent arrangement** of the loops of the vasa rectae and the sluggish movement of blood through them means that few ions are removed and the high solute concentration generated by the Loop of Henle is maintained.

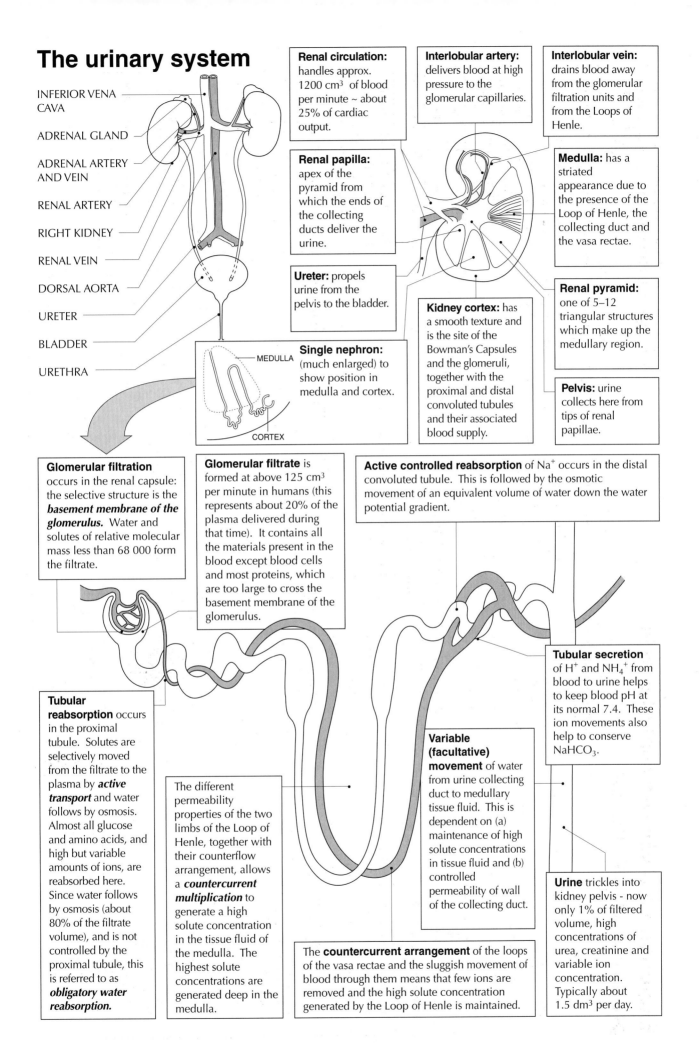

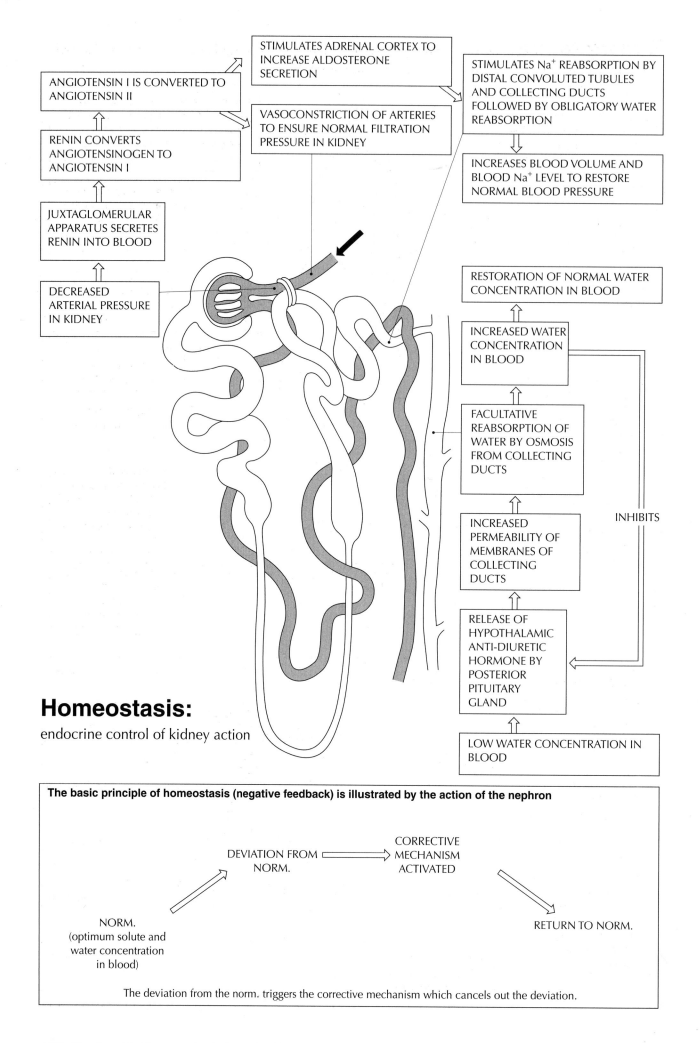

ANGIOTENSIN I IS CONVERTED TO
ANGIOTENSIN II

STIMULATES ADRENAL CORTEX TO
INCREASE ALDOSTERONE
SECRETION

STIMULATES Na⁺ REABSORPTION BY
DISTAL CONVOLUTED TUBULES
AND COLLECTING DUCTS
FOLLOWED BY OBLIGATORY WATER
REABSORPTION

RENIN CONVERTS
ANGIOTENSINOGEN TO
ANGIOTENSIN I

VASOCONSTRICTION OF ARTERIES
TO ENSURE NORMAL FILTRATION
PRESSURE IN KIDNEY

INCREASES BLOOD VOLUME AND
BLOOD Na⁺ LEVEL TO RESTORE
NORMAL BLOOD PRESSURE

JUXTAGLOMERULAR
APPARATUS SECRETES
RENIN INTO BLOOD

DECREASED
ARTERIAL PRESSURE
IN KIDNEY

RESTORATION OF NORMAL WATER
CONCENTRATION IN BLOOD

INCREASED WATER
CONCENTRATION
IN BLOOD

FACULTATIVE
REABSORPTION OF
WATER BY OSMOSIS
FROM COLLECTING
DUCTS

INHIBITS

INCREASED
PERMEABILITY OF
MEMBRANES OF
COLLECTING
DUCTS

RELEASE OF
HYPOTHALAMIC
ANTI-DIURETIC
HORMONE BY
POSTERIOR
PITUITARY
GLAND

Homeostasis:

endocrine control of kidney action

LOW WATER CONCENTRATION IN
BLOOD

The basic principle of homeostasis (negative feedback) is illustrated by the action of the nephron

DEVIATION FROM
NORM.

CORRECTIVE
MECHANISM
ACTIVATED

NORM.
(optimum solute and
water concentration
in blood)

RETURN TO NORM.

The deviation from the norm. triggers the corrective mechanism which cancels out the deviation.

Kidney failure: if one or both kidneys fail then dialysis is used or a transplant performed to keep urea and solute concentration in the blood constant.

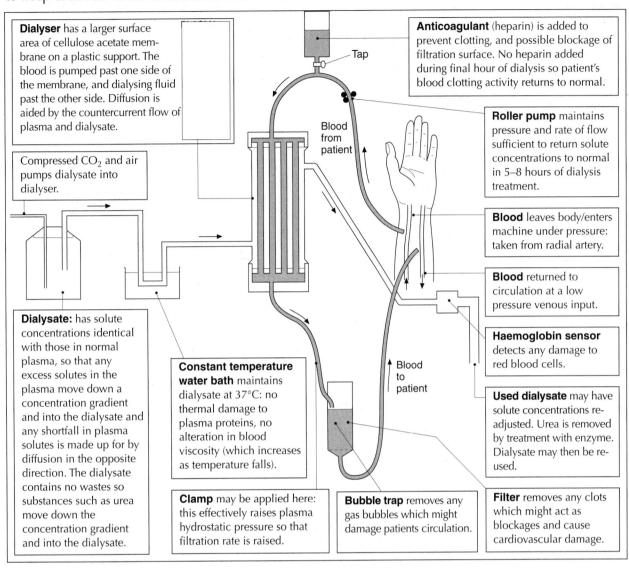

Dialyser has a larger surface area of cellulose acetate membrane on a plastic support. The blood is pumped past one side of the membrane, and dialysing fluid past the other side. Diffusion is aided by the countercurrent flow of plasma and dialysate.

Compressed CO_2 and air pumps dialysate into dialyser.

Dialysate: has solute concentrations identical with those in normal plasma, so that any excess solutes in the plasma move down a concentration gradient and into the dialysate and any shortfall in plasma solutes is made up for by diffusion in the opposite direction. The dialysate contains no wastes so substances such as urea move down the concentration gradient and into the dialysate.

Constant temperature water bath maintains dialysate at 37°C: no thermal damage to plasma proteins, no alteration in blood viscosity (which increases as temperature falls).

Clamp may be applied here: this effectively raises plasma hydrostatic pressure so that filtration rate is raised.

Tap

Blood from patient

Anticoagulant (heparin) is added to prevent clotting, and possible blockage of filtration surface. No heparin added during final hour of dialysis so patient's blood clotting activity returns to normal.

Roller pump maintains pressure and rate of flow sufficient to return solute concentrations to normal in 5–8 hours of dialysis treatment.

Blood leaves body/enters machine under pressure: taken from radial artery.

Blood returned to circulation at a low pressure venous input.

Blood to patient

Haemoglobin sensor detects any damage to red blood cells.

Used dialysate may have solute concentrations re-adjusted. Urea is removed by treatment with enzyme. Dialysate may then be re-used.

Bubble trap removes any gas bubbles which might damage patients circulation.

Filter removes any clots which might act as blockages and cause cardiovascular damage.

KIDNEY TRANSPLANTATION

may be necessary as renal dialysis is inconvenient for the patient and costly.

Kidney transplants have a high success rate because:

1. the vascular connections are simple

2. live donors may be used, so very close blood group matching is possible

3. because of 2 there are fewer immuno-suppression-related problems in which the body's immune system reacts against the new kidney.

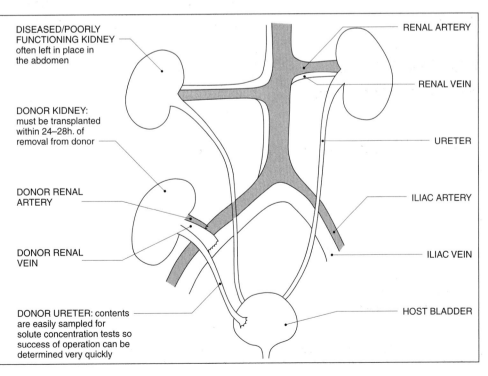

DISEASED/POORLY FUNCTIONING KIDNEY often left in place in the abdomen

DONOR KIDNEY: must be transplanted within 24–28h. of removal from donor

DONOR RENAL ARTERY

DONOR RENAL VEIN

DONOR URETER: contents are easily sampled for solute concentration tests so success of operation can be determined very quickly

RENAL ARTERY

RENAL VEIN

URETER

ILIAC ARTERY

ILIAC VEIN

HOST BLADDER

Control of body temperature in mammals

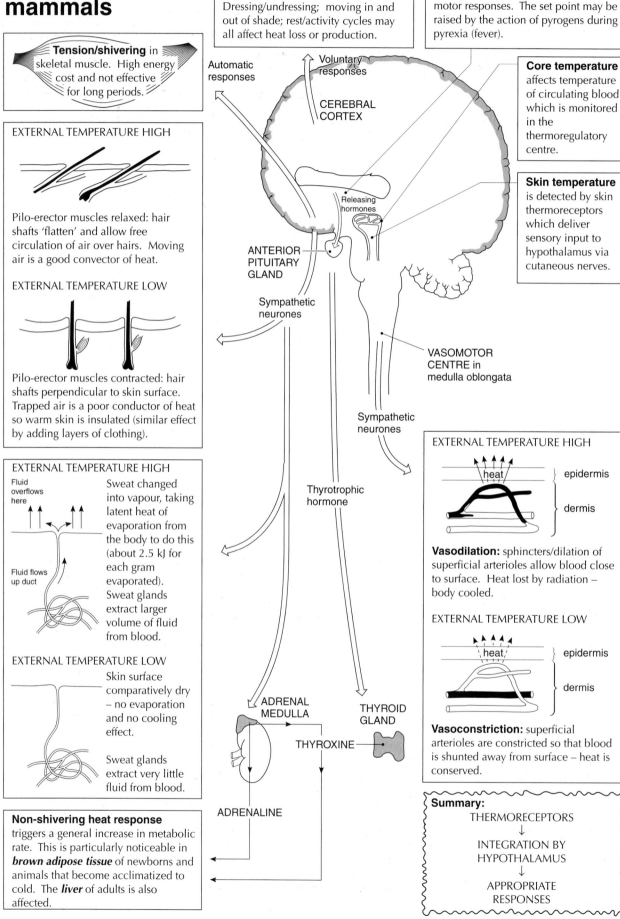

Tension/shivering in skeletal muscle. High energy cost and not effective for long periods.

EXTERNAL TEMPERATURE HIGH

Pilo-erector muscles relaxed: hair shafts 'flatten' and allow free circulation of air over hairs. Moving air is a good convector of heat.

EXTERNAL TEMPERATURE LOW

Pilo-erector muscles contracted: hair shafts perpendicular to skin surface. Trapped air is a poor conductor of heat so warm skin is insulated (similar effect by adding layers of clothing).

EXTERNAL TEMPERATURE HIGH

Fluid overflows here

Fluid flows up duct

Sweat changed into vapour, taking latent heat of evaporation from the body to do this (about 2.5 kJ for each gram evaporated). Sweat glands extract larger volume of fluid from blood.

EXTERNAL TEMPERATURE LOW

Skin surface comparatively dry – no evaporation and no cooling effect.

Sweat glands extract very little fluid from blood.

Non-shivering heat response triggers a general increase in metabolic rate. This is particularly noticeable in *brown adipose tissue* of newborns and animals that become acclimatized to cold. The *liver* of adults is also affected.

Changes in behaviour

Dressing/undressing; moving in and out of shade; rest/activity cycles may all affect heat loss or production.

Hypothalamus contains the thermoregulatory centre which compares sensory input with a set point and initiates the appropriate motor responses. The set point may be raised by the action of pyrogens during pyrexia (fever).

Core temperature affects temperature of circulating blood which is monitored in the thermoregulatory centre.

Skin temperature is detected by skin thermoreceptors which deliver sensory input to hypothalamus via cutaneous nerves.

Automatic responses

Voluntary responses

CEREBRAL CORTEX

Releasing hormones

ANTERIOR PITUITARY GLAND

Sympathetic neurones

VASOMOTOR CENTRE in medulla oblongata

Sympathetic neurones

Thyrotrophic hormone

ADRENAL MEDULLA

THYROID GLAND

THYROXINE

ADRENALINE

EXTERNAL TEMPERATURE HIGH

heat
epidermis
dermis

Vasodilation: sphincters/dilation of superficial arterioles allow blood close to surface. Heat lost by radiation – body cooled.

EXTERNAL TEMPERATURE LOW

heat
epidermis
dermis

Vasoconstriction: superficial arterioles are constricted so that blood is shunted away from surface – heat is conserved.

Summary:
THERMORECEPTORS
↓
INTEGRATION BY HYPOTHALAMUS
↓
APPROPRIATE RESPONSES

Ectotherms attempt to maintain body temperature by *behavioural* rather than *physiological* methods. These methods are less precise and so thermoregulation is more difficult.

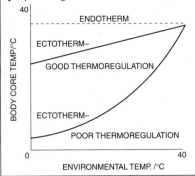

Reorientation of the body with respect to solar *radiation* can vary the surface area exposed to heating. A terrestrial ectotherm may gain heat rapidly by aligning itself at right angles to the Sun's rays but as its body temperature rises it may reduce the exposed surface by reorientating itself parallel to the Sun's rays.

Thermal gaping is used by some larger ectotherms such as alligators and crocodiles. The open mouth allows heat loss by *evaporation* from the moist mucous surfaces. Some tortoises have been observed to use a similar principle by spreading saliva over the neck and front legs which then acts as an evaporative surface.

Colour changes of the skin may alter the ability of the body to absorb *radiated* heat energy. A dark-bodied individual will absorb heat more rapidly than a light-bodied one – thus some ectotherms begin the day with a dark body to facilitate 'warming-up' but then lighten the body as the environmental temperature rises.

Body raising is used by ectotherms to minimize heat gains by *conduction* from hot surfaces such as rocks and sand. The whole body may be lifted and the animal may reduce the area of contact to the absolute minimum by balancing on alternate diagonal pairs of feet.

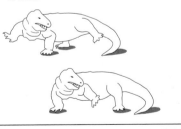

Burrowing is a widely used behavioural device which enables ectotherms to avoid the greater temperature fluctuations on the surface of their habitat. The temperature in even a shallow burrow may only fluctuate by 5°C over a 24 h period whereas the surface temperature may range over 40°C during the same time. Amphibious and semi-aquatic reptiles such as alligators and crocodiles may return to water rather than burrow, since the high heat capacity of water means that its temperature is relatively constant.

The marine iguana and bradycardia
The marine iguana of the Galapagos Islands feeds by browsing on seaweed gathered from the sea around the rocky shores on which it lives. When basking on the rocks it normally maintains a body temperature of 37°C but during the time spent feeding in the sea it is exposed to environmental temperatures of 22–25°C. In order to avoid losing heat rapidly by *conduction and convection* the iguana reduces the flow of blood between its core tissues (at 37°C) and its skin (22°C) by slowing its heart rate (bradycardia).

Heat transfer between the body of an organism and its environment depends on the *magnitude* and *direction* of the *thermal gradient* (i.e. the temperature difference between the organism and its surroundings). Heat may be *lost* or *gained* by
conduction (heat transfer by physical contact)
convection (heat transfer to the air) and
radiation (heat transfer in the form of long-wave, infra-red electromagnetic waves) but can only be *lost* by
evaporation (heat consumption during the conversion of water to water vapour).

Bacteria as pathogens

GENERAL FEATURES OF BACTERIAL DISEASE

To become pathogenic the bacterium must satisfy several conditions:

1. **Attachment to mucous membranes** – necessary to avoid being swept away by mucus, food, or urine. Sometimes involves specific adhesin-receptor interactions such as those between the pili on *E. coli* and surface polysaccharides on gut epithelia of newborn humans.

2. **Penetration of mucous membranes** – necessary if the infection is to spread beyond the point of entry. May involve the production of surface proteins which promote uptake of bacteria by epithelial cells and their transfer to adjacent cells of the host (e.g. *Shigella* sp.).

3. **Growth in host tissue** – to provide increased numbers for colonisation of host. Involves acquisition of nutrient reserves from host tissues – often iron is in low concentrations and may be absorbed by bacterial proteins called siderophores.

4. **Systems for avoidance of host defence** – these include the polysaccharide capsules of pneumococci and meningococci and the protein molecules on the surface of streptococci. These compounds may mask the antigenic molecules on the bacterial cells which are normally targets for complement, antibodies, or phagocytes.

5. **Causing damage to the host** – either by the secretion of toxins (such as the neurotoxin from *C. tetani* or the inhibitor of protein synthesis secreted by diphtheria bacteria) or by hypersensitisation of the immune response (as occurs in tuberculosis where lung damage is caused by accumulation of host phagocytes).

HOST DEFENCES AGAINST BACTERIA

These include the **complement system**, a complex series of proteins which 'cascade' to rapidly increase in concentration at sites of infection. The latter stages of the cascade may destroy bacteria, especially in association with **lysozyme**, which lyses bacterial cell walls. The principal cellular defence is via the phagocytes:

Bacterium is recognised and engulfed by phagocyte.

Lysosome fuses with vacuole enclosing bacterium.

Bacterium destroyed by peroxide ions and lysosomal enzymes.

SOME SITES OF BACTERIAL INVASION

NERVOUS SYSTEM: Botulism is caused by *Clostridium botulinum* which releases powerful neurotoxins – these progressively lead to nausea, blurred vision, and paralysis. **Meningitis** (infection of the meninges) is caused by *Neisseria meningotidis* and is characterised by fever, nausea, photophobia, and neck stiffness. In infants this condition is most commonly caused by *Haemophilus influenzae*.

NASOPHARYNX: may be infected by *Corynebacterium diphtheriae* causing **diphtheria**. This infection may cause local inflammation and hardening of membranes, and a systemic effect due to release of an endotoxin.

HEART: may be infected by a number of streptococci or by *H. influenzae* to cause **pericarditis** or **myocarditis**.

DIRECT ACCESS TO THE BLOOD: may result from wounding. Two possible infections are **tetanus** – caused by release of the potent toxin tetanospasmin by *Clostridium tetani*, leading to widespread muscle spasm – and **gas gangrene**, caused by *Clostridium perfringens*, which leads to oedema and loss of limb function.

LUNGS: particularly prone to infection because of the frequent inhalation of bacteria in air. Some potentially lethal infections are **pneumonia** (most commonly caused by *Streptococcus pneumoniae*), **Legionnaire's disease** (*Legionella pneumophila*), and **tuberculosis** (*Mycobacterium tuberculosis*).

GUT: often infected by bacteria ingested with improperly prepared food. Infections include **food poisoning** (*Staphylococcus aureus*, release of heat-stable enterotoxins – toxins which affect the gut), **cholera** (*Vibrio cholerae*), **diarrhoea** (many causes, including enterotoxigenic *Escherichia coli*), and **dysentery** (*Shigella dysenteriae*). Many gut infections cause severe problems of dehydration.

SKIN: Impetigo may be caused by staphylococci, streptococci, or a combination of the two. **Lupus vulgaris** is a rare infection associated with mycobacterial tuberculosis. **Boils** are infections of the hair follicles, sometimes caused by *S. aureus*. **Leprosy** (Hansen's disease) is caused by *Mycobacterium leprae*.

URETHRA AND GENITALIA: Gonorrhoea is caused by *Neisseria gonorrhoeae* and may spread from urethra to testes, rectum, and vagina. *Chlamydia trachomatis* is the major cause of NGU (Non-Gonococcal urethritis). **Syphillis** is caused by *Trepanoma pallidum*, and may become widespread throughout the body in its later stages.

Growth of a bacterial population

KOCH'S POSTULATES: rules for proving that a particular microbe is responsible for a particular disease.
- it must be shown that the microbe in question is always present in individuals with the disease
- the microbe must be isolated from the diseased host and grown in pure culture containing only that one species of microbe
- microbes obtained from the pure culture, when injected into a healthy susceptible host, must produce the disease in that host
- microbes must be isolated from the experimentally infected host, grown in pure culture, and compared with the microbes in the original culture.

BACTERIAL DISEASE may be
- **Infective**: the growing population directly damages tissues, sometimes by competition for oxygen and nutrients.
- **Toxic**: the bacterial cells release poisons (*toxins*) which irritate or damage cells of the host organism.

Population stable as cell deaths balanced by continued cell division. It is often during this phase that *enterotoxins* (gut poisons) are produced and released.

Death rate > rate of division.
- all nutrients used
- toxins produced kill cells directly
- oxygen concentration reduced for aerobic species.

Complete cycle may be no more than a few hours.

Cells adapt to new environment: e.g. may synthesise enzymes to deal with new food source.

Numbers are so great that a logarithmic scale is used to avoid an enormous *y* axis.

Number of bacterial cells

| LAG PHASE | EXPONENTIAL | STATIONARY | DECLINE |

Time / h

During this phase the population doubles with each unit of time –
1 cell → 2 → 4 → 8 → 16 → 32 → 64 → 128
doubling time may be as little as 20 min – at this rate one bacterial cell → 8000 kg in 24 h!
Food preservation techniques hold population early in this phase by imposing *environmental resistance* i.e. by *limiting availability of factors for growth*.

NUTRIENT SUPPLY
- *Carbon* source for energy, *nitrogen* source to supply amino acids for protein synthesis, *mineral ions* as cofactors.

ATMOSPHERE
- some bacteria require oxygen for aerobic respiration, but a few are totally anaerobic
- ▲ control bacterial growth by modifying atmosphere, e.g. during food preservation a nitrogen atmosphere is totally inert.

WATER
- required for hydrolysis reactions and for cell expansion
- ▲ control bacterial cell division by limiting access to water e.g. by dehydration or vacuum packing of foods.

Bacteria

Growth rate

% water in environment
80   90   100

pH
- bacteria can tolerate a wide range of pH – enzymes are quite resistant.
- ▲ low pH (acid environment) limits bacterial growth, e.g. HCl in stomach, 'pickled' foods.

Growth rate

pH
5   7   9   11

TEMPERATURE
- affects enzyme activity, fluidity of membranes, solubility of molecules, energy release by respiration
- ▲ reduce temperature by refrigeration/freezing (food preservation); increase temperature during fever.

Growth rate

Spore-forming bacteria. e.g. *Clostridium welchii*

Salmonella and *Streptococcus*

Clostridium botulinum

Temperature / °C
20   30   40   50

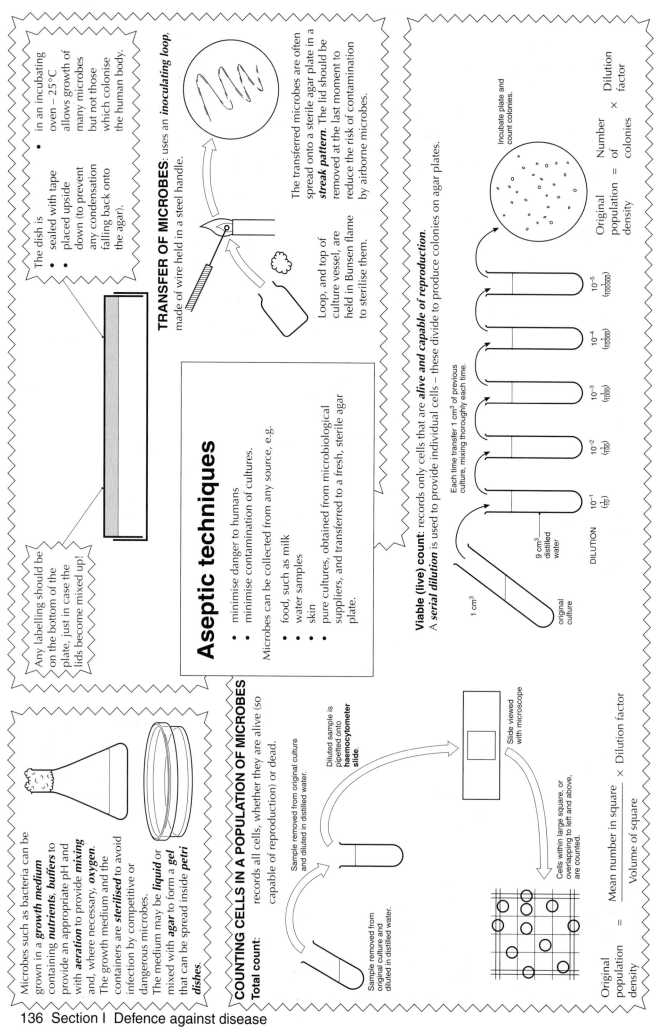

- in an incubating oven – 25°C allows growth of many microbes but not those which colonise the human body.

The dish is
- sealed with tape
- placed upside down (to prevent any condensation falling back onto the agar).

Any labelling should be on the bottom of the plate, just in case the lids become mixed up!

TRANSFER OF MICROBES: uses an *inoculating loop*, made of wire held in a steel handle.

The transferred microbes are often spread onto a sterile agar plate in a *streak pattern*. The lid should be removed at the last moment to reduce the risk of contamination by airborne microbes.

Loop, and top of culture vessel, are held in Bunsen flame to sterilise them.

Aseptic techniques

- minimise danger to humans
- minimise contamination of cultures.

Microbes can be collected from any source, e.g.
- food, such as milk
- water samples
- skin
- pure cultures, obtained from microbiological suppliers, and transferred to a fresh, sterile agar plate.

Viable (live) count: records only cells that are *alive and capable of reproduction*. A *serial dilution* is used to provide individual cells – these divide to produce colonies on agar plates.

Each time transfer 1 cm³ of previous culture, mixing thoroughly each time.

9 cm³ distilled water

DILUTION

10^{-1} ($\frac{1}{10}$) 10^{-2} ($\frac{1}{100}$) 10^{-3} ($\frac{1}{1000}$) 10^{-4} ($\frac{1}{10000}$) 10^{-5} ($\frac{1}{100000}$)

1 cm³
original culture

Incubate plate and count colonies.

Original population density = Number of colonies × Dilution factor

Microbes such as bacteria can be grown in a **growth medium** containing **nutrients**, **buffers** to provide an appropriate pH and with **aeration** to provide **mixing** and, where necessary, **oxygen**. The growth medium and the containers are **sterilised** to avoid infection by competitive or dangerous microbes.
The medium may be **liquid** or mixed with **agar** to form a **gel** that can be spread inside **petri dishes**.

COUNTING CELLS IN A POPULATION OF MICROBES

Total count: records all cells, whether they are alive (so capable of reproduction) or dead.

Sample removed from original culture and diluted in distilled water.

Sample removed from original culture and diluted in distilled water.

Diluted sample is pipetted onto **haemocytometer slide**.

Slide viewed with microscope

Cells within large square, or overlapping to left and above, are counted.

Original population density = $\dfrac{\text{Mean number in square}}{\text{Volume of square}}$ × Dilution factor

Natural defence systems of the body

prevent infection and disease.

SKIN IS THE FIRST BARRIER

many organisms may colonise the body of a human – the tissues are **warm, moist,** and a **good source of nutrients**. Access to these tissues is prevented by the outer layer of the skin, the **epidermis,** which is waxy, acidic, and salty and almost impermeable to pathogens (N.B. some micro-organisms do live on the skin **surface,** and may help to prevent colonisation by more harmful species).

Where there are orifices, essential to permit entry or exit of food, gases, or excretory products, there may be **protective secretions** to limit entry of pathogens, e.g.

ORIFICE	FUNCTION	PROTECTED BY
Mouth, entrance to gut	Entry of food	**lysozyme** in saliva **hydrochloric acid** in stomach
Nose, entrance to respiratory tract	Entry of oxygen in air	**mucus** moved by upwards cilia
Ear	Entry of sound waves	**bacteriocidal** ('bacteria-killing') **wax**

SECOND LINES OF DEFENCE include:

- non-specific immune responses – inflammation and phagocytosis

specific immune responses – cell-mediated and humoral immunity.

antibodies

PATHOGENS TRANSMITTED BY VECTORS

may bypass the skin and be introduced directly to the bloodstream

e.g. *Plasmodium* introduced by mosquito.

To limit defence by clotting the vector may introduce **anticoagulant compounds**.

BLOOD CLOTTING 'PLUGS' WOUNDS

any unnatural gaps in the skin barrier are closed by a set of processes largely due to **plasma proteins** and **platelets** (**thrombocytes**).

EXTRINSIC FACTORS e.g. tissue damage → ACCUMULATION OF PLATELETS → RELEASE OF THROMBOPLASTIN

INTRINSIC FACTORS e.g. contact with collagen in connective tissues

FACTOR VIII

PROTHROMBIN (Inactive) → THROMBIN (Active)

Ca^{2+} ions

FIBRINOGEN (soluble) → FIBRIN (insoluble)

Clotting
- maintains blood volume (**haemostasis**)
- limits entry of pathogens.

Mesh can trap red cells and platelets: when dry forms a **scab**.

Inflammation and phagocytosis

are part of the non-specific defence reactions of the body.

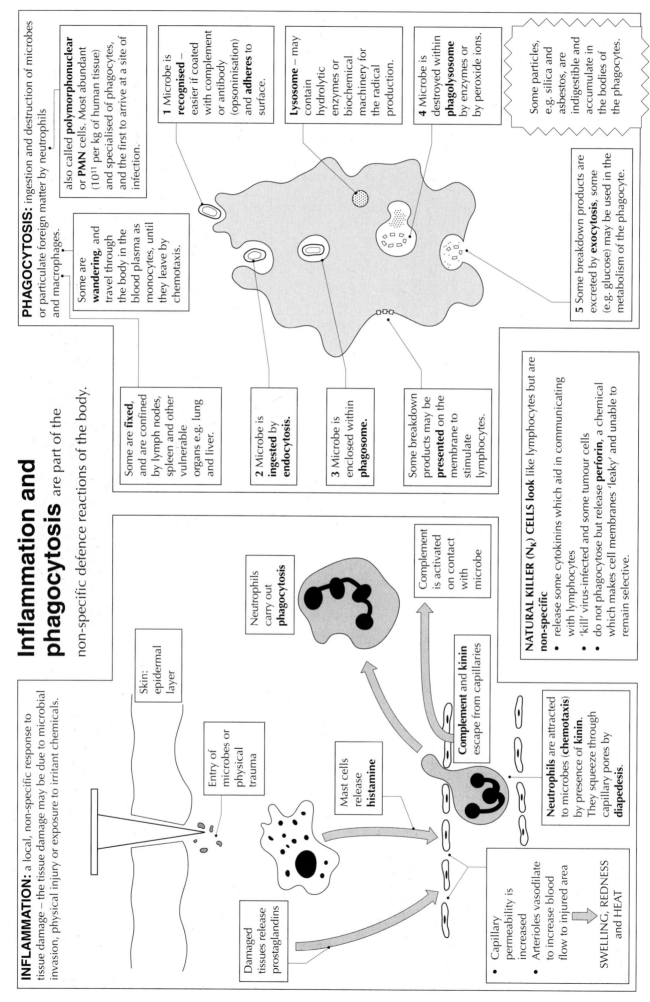

PHAGOCYTOSIS: ingestion and destruction of microbes or particulate foreign matter by neutrophils and macrophages.

also called **polymorphonuclear** or **PMN** cells. Most abundant (10^{11} per kg of human tissue) and specialised of phagocytes, and the first to arrive at a site of infection.

1 Microbe is **recognised** – easier if coated with complement or antibody (opsoninisation) and **adheres** to surface.

Lysosome – may contain hydrolytic enzymes or biochemical machinery for the radical production.

4 Microbe is destroyed within **phagolysosome** by enzymes or by peroxide ions.

Some particles, e.g. silica and asbestos, are indigestible and accumulate in the bodies of the phagocytes.

Some are **wandering**, and travel through the body in the blood plasma as monocytes, until they leave by chemotaxis.

Some are **fixed**, and are confined by lymph nodes, spleen and other vulnerable organs e.g. lung and liver.

2 Microbe is **ingested** by **endocytosis.**

3 Microbe is enclosed within **phagosome.**

Some breakdown products may be **presented** on the membrane to stimulate lymphocytes.

5 Some breakdown products are excreted by **exocytosis,** some (e.g. glucose) may be used in the metabolism of the phagocyte.

INFLAMMATION: a local, non-specific response to tissue damage – the tissue damage may be due to microbial invasion, physical injury or exposure to irritant chemicals.

Skin: epidermal layer

Entry of microbes or physical trauma

Neutrophils carry out **phagocytosis**

Mast cells release **histamine**

Complement is activated on contact with microbe

Complement and **kinin** escape from capillaries

Neutrophils are attracted to microbes (**chemotaxis**) by presence of **kinin.** They squeeze through capillary pores by **diapedesis.**

Damaged tissues release prostaglandins

- Capillary permeability is increased
- Arterioles vasodilate to increase blood flow to injured area

SWELLING, REDNESS and HEAT

NATURAL KILLER (N$_K$) CELLS look like lymphocytes but are **non-specific**
- release some cytokinins which aid in communicating with lymphocytes
- 'kill' virus-infected and some tumour cells
- do not phagocytose but release **perforin,** a chemical which makes cell membranes 'leaky' and unable to remain selective.

Immune response I: cells

This involves a wide range of cells and their products in defence against diseases.

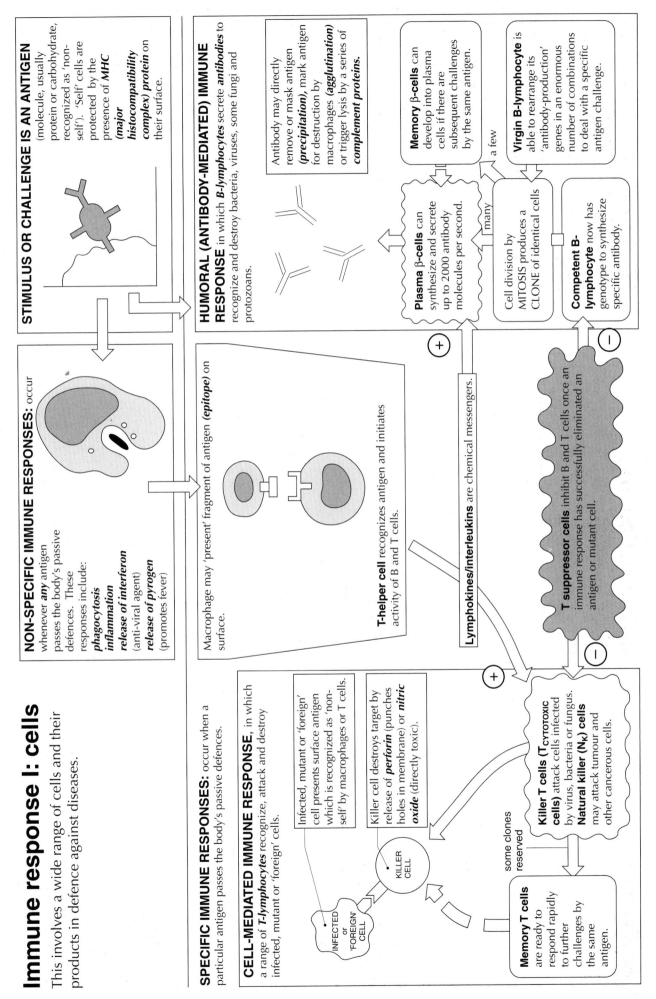

STIMULUS OR CHALLENGE IS AN ANTIGEN

(molecule, usually protein or carbohydrate, recognized as 'non-self'). 'Self' cells are protected by the presence of **MHC (major histocompatibility complex) protein** on their surface.

NON-SPECIFIC IMMUNE RESPONSES: occur

whenever **any** antigen passes the body's passive defences. These responses include:
phagocytosis
inflammation
release of interferon (anti-viral agent)
release of pyrogen (promotes fever)

SPECIFIC IMMUNE RESPONSES: occur when a

particular antigen passes the body's passive defences.

CELL-MEDIATED IMMUNE RESPONSE, in which

a range of **T-lymphocytes** recognize, attack and destroy infected, mutant or 'foreign' cells.

Macrophage may 'present' fragment of antigen (**epitope**) on surface.

Infected, mutant or 'foreign' cell presents surface antigen which is recognized as 'non-self' by macrophages or T cells.

Killer cell destroys target by release of **perforin** (punches holes in membrane) or **nitric oxide** (directly toxic).

KILLER CELL

INFECTED or 'FOREIGN' CELL

Killer T cells (T$_{\text{CYTOTOXIC}}$ cells) attack cells infected by virus, bacteria or fungus.
Natural killer (N$_\text{K}$) cells may attack tumour and other cancerous cells.

some clones reserved

Memory T cells are ready to respond rapidly to further challenges by the same antigen.

T-helper cell recognizes antigen and initiates activity of B and T cells.

Lymphokines/interleukins are chemical messengers.

T suppressor cells inhibit B and T cells once an immune response has successfully eliminated an antigen or mutant cell.

HUMORAL (ANTIBODY-MEDIATED) IMMUNE RESPONSE in which **B-lymphocytes** secrete **antibodies** to

recognize and destroy bacteria, viruses, some fungi and protozoans.

Antibody may directly remove or mask antigen (**precipitation**), mark antigen for destruction by macrophages (**agglutination**) or trigger lysis by a series of **complement proteins.**

Plasma β-cells can synthesize and secrete up to 2000 antibody molecules per second.

Cell division by MITOSIS produces a CLONE of identical cells

Competent B-lymphocyte now has genotype to synthesize specific antibody.

Memory β-cells can develop into plasma cells if there are subsequent challenges by the same antigen.

Virgin B-lymphocyte is able to rearrange its 'antibody-production' genes in an enormous number of combinations to deal with a specific antigen challenge.

a few

many

+

−

+

−

Immune response II: antibodies and immunity

An antibody is a protein molecule synthesized by an animal in response to a specific antigen.

The basic structure of an antibody has the shape of the letter Y. Each molecule is composed of four polypeptide chains, two heavy and two light, all linked by disulphide bridges.

Constant (C) region of light chain

Constant (C) region of heavy chain

The constant regions determine the *general class* of the antibody:

IgG and IgM participate in the *precipitation, agglutination* and *complement* reactions.
IgA in tears, mucous secretions and saliva specifically binds to *surface antigens on bacteria.*
IgD helps to *activate lymphocytes.*
IgE is bound to mast cells and provokes *allergies.*

The **antigen-binding sites ('sticky ends')** have highly variable amino acid sequences which produce a huge number of possible recognition sites to bind to an enormous range of possible antigens.

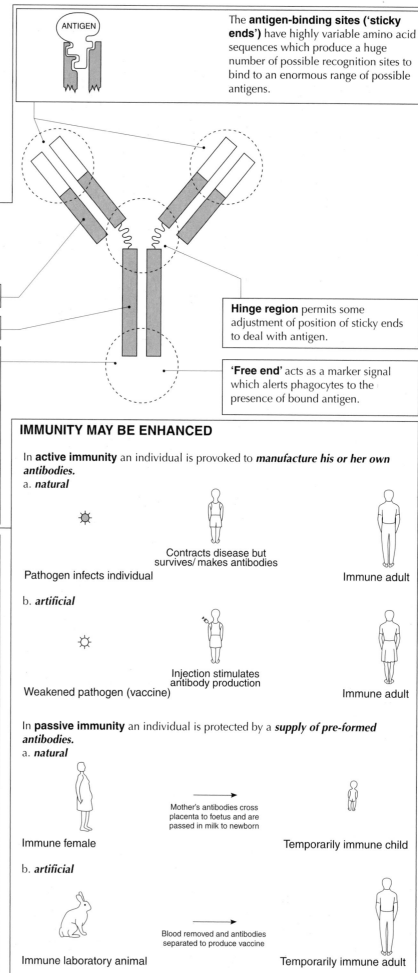

ANTIGEN

Hinge region permits some adjustment of position of sticky ends to deal with antigen.

'Free end' acts as a marker signal which alerts phagocytes to the presence of bound antigen.

MEMORY CELLS SPEED UP IMMUNE RESPONSE

Secondary response: curve is steeper, peak is higher (commonly 10^3 x the primary response), lag period is negligible *due to the presence of B-memory cells.* Dominant antibody is IgG which is more stable and has a greater affinity for the antigen.

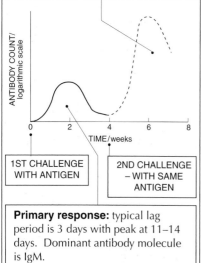

ANTIBODY COUNT/ logarithmic scale

TIME/weeks

1ST CHALLENGE WITH ANTIGEN

2ND CHALLENGE – WITH SAME ANTIGEN

Primary response: typical lag period is 3 days with peak at 11–14 days. Dominant antibody molecule is IgM.

IMMUNITY MAY BE ENHANCED

In **active immunity** an individual is provoked to *manufacture his or her own antibodies.*
a. *natural*

Pathogen infects individual

Contracts disease but survives/ makes antibodies

Immune adult

b. *artificial*

Weakened pathogen (vaccine)

Injection stimulates antibody production

Immune adult

In **passive immunity** an individual is protected by a *supply of pre-formed antibodies.*
a. *natural*

Immune female

Mother's antibodies cross placenta to foetus and are passed in milk to newborn

Temporarily immune child

b. *artificial*

Immune laboratory animal

Blood removed and antibodies separated to produce vaccine

Temporarily immune adult

Human immunodeficiency virus (HIV)

may depress the immune response

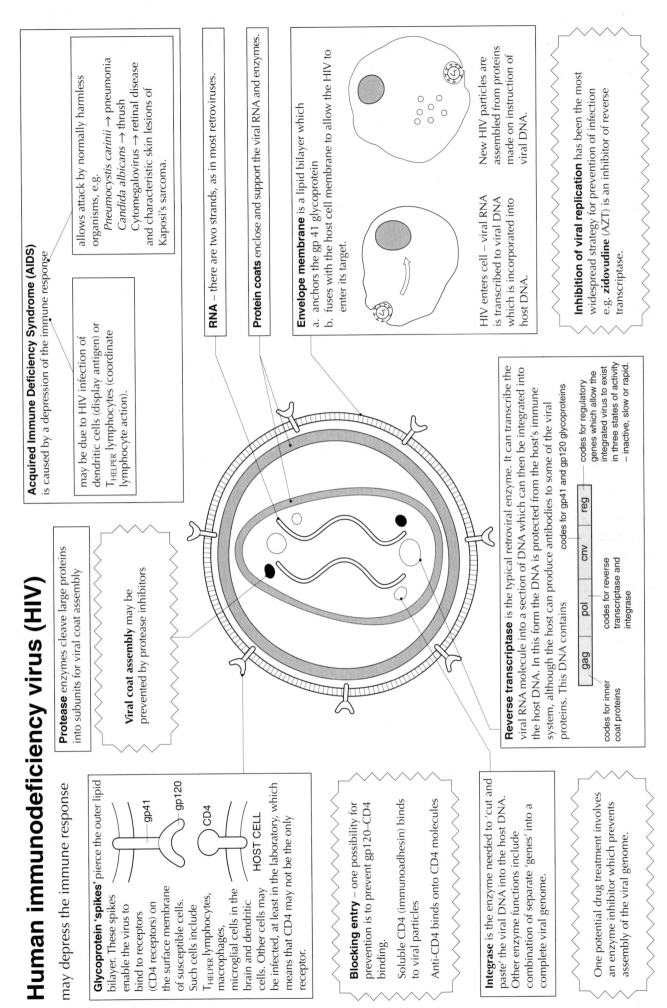

Acquired Immune Deficiency Syndrome (AIDS)
is caused by a depression of the immune response

may be due to HIV infection of dendritic cells (display antigen) or T$_{HELPER}$ lymphocytes (coordinate lymphocyte action).

allows attack by normally harmless organisms, e.g.
Pneumocystis carinii → pneumonia
Candida albicans → thrush
Cytomegalovirus → retinal disease
and characteristic skin lesions of Kaposi's sarcoma.

RNA – there are two strands, as in most retroviruses.

Protein coats enclose and support the viral RNA and enzymes.

Envelope membrane is a lipid bilayer which
a. anchors the gp 41 glycoprotein
b. fuses with the host cell membrane to allow the HIV to enter its target.

New HIV particles are assembled from proteins made on instruction of viral DNA.

HIV enters cell – viral RNA is transcribed to viral DNA which is incorporated into host DNA.

Inhibition of viral replication has been the most widespread strategy for prevention of infection e.g. **zidovudine** (AZT) is an inhibitor of reverse transcriptase.

Glycoprotein 'spikes' pierce the outer lipid bilayer. These spikes enable the virus to bind to receptors (CD4 receptors) on the surface membrane of susceptible cells. Such cells include T$_{HELPER}$ lymphocytes, macrophages, microglial cells in the brain and dendritic cells. Other cells may be infected, at least in the laboratory, which means that CD4 may not be the only receptor.

gp41
gp120
CD4
HOST CELL

Protease enzymes cleave large proteins into subunits for viral coat assembly

Viral coat assembly may be prevented by protease inhibitors

Blocking entry – one possibility for prevention is to prevent gp120–CD4 binding.

Soluble CD4 (immunoadhesin) binds to viral particles

Anti-CD4 binds onto CD4 molecules

Integrase is the enzyme needed to 'cut and paste' the viral DNA into the host DNA. Other enzyme functions include combination of separate 'genes' into a complete viral genome.

Reverse transcriptase is the typical retroviral enzyme. It can transcribe the viral RNA molecule into a section of DNA which can then be integrated into the host DNA. In this form the DNA is protected from the host's immune system, although the host can produce antibodies to some of the viral proteins. This DNA contains

gag	pol	env	reg

codes for inner coat proteins

codes for reverse transcriptase and integrase

codes for gp41 and gp120 glycoproteins

codes for regulatory genes which allow the integrated virus to exist in three states of activity – inactive, slow or rapid.

One potential drug treatment involves an enzyme inhibitor which prevents assembly of the viral genome.

Monoclonal antibodies: production and applications

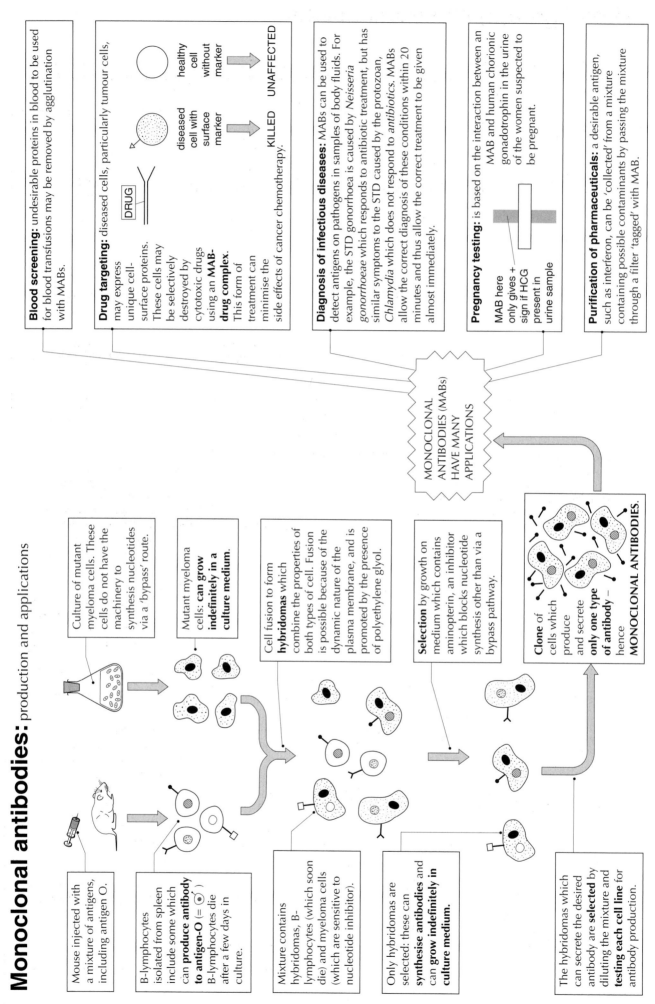

Blood screening: undesirable proteins in blood to be used for blood transfusions may be removed by agglutination with MABs.

Drug targeting: diseased cells, particularly tumour cells, may express unique cell-surface proteins. These cells may be selectively destroyed by cytotoxic drugs using an **MAB-drug complex.** This form of treatment can minimise the side effects of cancer chemotherapy.

healthy cell without marker → UNAFFECTED

diseased cell with surface marker → KILLED

DRUG

Diagnosis of infectious diseases: MABs can be used to detect antigens on pathogens in samples of body fluids. For example, the STD gonorrhoea is caused by *Neisseria gonorrhoeae* which responds to antibiotic treatment, but has similar symptoms to the STD caused by the protozoan, *Chlamydia* which does not respond to *antibiotics*. MABs allow the correct diagnosis of these conditions within 20 minutes and thus allow the correct treatment to be given almost immediately.

Pregnancy testing: is based on the interaction between an MAB and human chorionic gonadotrophin in the urine of the women suspected to be pregnant.

MAB here only gives + sign if HCG present in urine sample

Purification of pharmaceuticals: a desirable antigen, such as interferon, can be 'collected' from a mixture containing possible contaminants by passing the mixture through a filter 'tagged' with MAB.

MONOCLONAL ANTIBODIES (MABs) HAVE MANY APPLICATIONS

Mouse injected with a mixture of antigens, including antigen O.

Culture of mutant myeloma cells. These cells do not have the machinery to synthesis nucleotides via a 'bypass' route.

Mutant myeloma cells: **can grow indefinitely in a culture medium.**

B-lymphocytes isolated from spleen include some which can **produce antibody to antigen-O (= ⊙)**. B-lymphocytes die after a few days in culture.

Cell fusion to form **hybridomas** which combine the properties of both types of cell. Fusion is possible because of the dynamic nature of the plasma membrane, and is promoted by the presence of polyethylene glyol.

Mixture contains hybridomas, B-lymphocytes (which soon die) and myeloma cells (which are sensitive to nucleotide inhibitor).

Selection by growth on medium which contains aminopterin, an inhibitor which blocks nucleotide synthesis other than via a bypass pathway.

Only hybridomas are selected: these can **synthesise antibodies and can grow indefinitely in culture medium.**

Clone of cells which produce and secrete **only one type of antibody** – hence **MONOCLONAL ANTIBODIES.**

The hybridomas which can secrete the desired antibody are **selected** by diluting the mixture and **testing each cell line** for antibody production.

Blood groups and transfusions

The **ABO BLOOD GROUPING** is based on two **agglutinogens**, symbolised as **A** and **B**, which are genetically determined carbohydrate molecules carried on the surface membranes of the **red blood cells**, and two **agglutinins**, **anti-A** and **anti-B**, carried in the blood plasma.

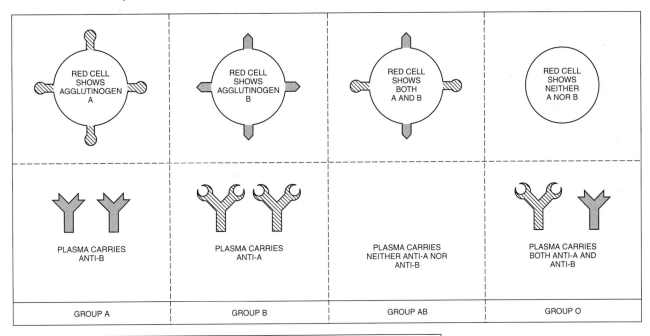

RED CELL SHOWS AGGLUTINOGEN A	RED CELL SHOWS AGGLUTINOGEN B	RED CELL SHOWS BOTH A AND B	RED CELL SHOWS NEITHER A NOR B
PLASMA CARRIES ANTI-B	PLASMA CARRIES ANTI-A	PLASMA CARRIES NEITHER ANTI-A NOR ANTI-B	PLASMA CARRIES BOTH ANTI-A AND ANTI-B
GROUP A	GROUP B	GROUP AB	GROUP O

A **blood transfusion** may be necessary to make up blood volume following haemorrhage or during surgery. Only **compatible** blood should be transfused, or **agglutination** and **haemolysis** may occur. Agglutinated cells may block capillaries and cause kidney or brain damage, or even death.

An **incompatible transfusion**:

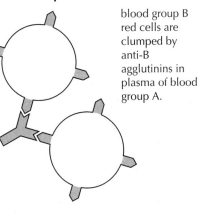

blood group B red cells are clumped by anti-B agglutinins in plasma of blood group A.

About 80% of the population – called **secretors** – release ABO-type antigens in saliva and semen: this has been useful in some criminal investigations.

An enzyme, **α-galactosidase**, isolated from green coffee beans, can be used to convert type B blood to type O, the most useful for transfusions.

RECIPIENT	DONOR			
	A	B	AB	O
A (anti-B)	○ ○ ○ ○ ○ ○ ○ ○ ○ ○	(clumped)	(clumped)	○ ○ ○ ○ ○ ○ ○ ○ ○
B (anti-A)	(clumped)	○ ○ ○ ○ ○ ○ ○ ○ ○ ○	(clumped)	○ ○ ○ ○ ○ ○ ○ ○ ○
AB (neither)	○ ○ ○ ○ ○ ○ ○ ○	○ ○ ○ ○ ○ ○ ○ ○	○ ○ ○ ○ ○ ○ ○ ○	○ ○ ○ ○ ○ ○ ○ ○
O (both)	(clumped)	(clumped)	(clumped)	○ ○ ○ ○ ○ ○ ○ ○ ○

N.B. the agglutinins in the donor blood are ignored – they are in too low concentration to cause major damage.

AB is the **universal recipient** since its plasma contains no agglutinins.

O is the **universal donor** since its red cells carry no agglutinogens.

Penicillin is an antibiotic

The *Penicillium* mould may make products which it secretes into its environment to kill off any disease-causing or competitive micro-organisms. **A compound made by a mould to kill off any other micro-organism is called an ANTIBIOTIC.**

Antibiotics are valuable in the control of bacterial diseases because they affect bacterial cells but rarely affect human or animal cells.

Some important diseases caused by bacteria and thus suitable for antibiotic treatment are:

Dysentery ⎱
Food poisoning ⎰ affect gut

Syphilis ⎱
Gonorrhoea ⎰ affect reproductive organs

Pneumonia - affects lungs

Botulism - affects nervous system

RESISTANT BACTERIAL STRAINS MAY DEVELOP

Bacteria breed rapidly and populations are enormous. Within these populations rare **mutations** may produce cells which are **antibiotic-resistant**. Over-use of antibiotics may destroy 'normal' bacterial cells but allow 'resistant' cells to survive. These may multiply to form a resistant population.

One resistant cell in population → ANTIBIOTIC TREATMENT → Only resistant cell survives → CELL DIVISION → Whole population is now resistant

This is a form of **artificial selection**.

NO TREATMENT - disease symptoms may follow

BACTERIOSTATIC - host defences kill bacteria

BACTERIOCIDAL - antibiotic directly kills bacteria

Treatment begins

Infection of host

Time/h

Number of bacteria

THE EFFECTIVENESS OF ANTIBIOTICS

can be compared using 'multidisks'.

Paper multidisk impregnated with different antibiotics A, B, C, and D.

Bacterial population multiplies on agar growth medium.

Clear area caused by inhibition of bacterial growth.

DON'T BE CONFUSED!
Antibiotics – produced by fungi and kill bacteria *inside* the body

Antiseptics – kill microbes *outside* the body, e.g. on skin.

LARGE SCALE PRODUCTION OF PENICILLIN

is a good example of the commercial use of fermentation reactions.

CULTURE OF PENICILLIUM MOULD

NUTRIENTS

OXYGEN

Probes monitor TEMPERATURE, pH and OXYGEN CONCENTRATION

FILTER - removes mould to be reused

SOLVENT - extracts penicillin from mixture

CRYSTALLIZATION - collects pure penicillin for distribution

There are several *Penicillium* moulds which may be used to produce antibiotics. The most widely used is *Penicillium crysogemum*.

ACTION OF PENICILLIN

Bacteria multiply very rapidly by **binary fission,** i.e. they grow and then divide into two. This can happen very rapidly (every 20 minutes), causing disease as the bacterial metabolism competes with or inhibits the activity of human cells.

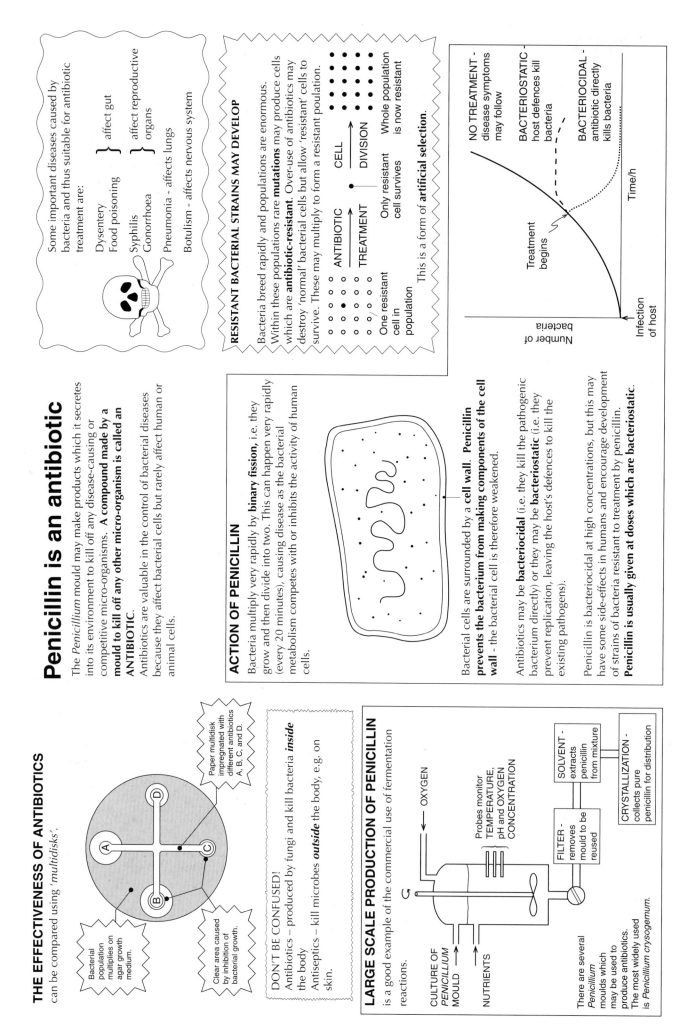

Bacterial cells are surrounded by a **cell wall. Penicillin prevents the bacterium from making components of the cell wall** - the bacterial cell is therefore weakened.

Antibiotics may be **bacteriocidal** (i.e. they kill the pathogenic bacterium directly) or they may be **bacteriostatic** (i.e. they prevent replication, leaving the host's defences to kill the existing pathogens).

Penicillin is bacteriocidal at high concentrations, but this may have some side-effects in humans and encourage development of strains of bacteria resistant to treatment by penicillin. **Penicillin is usually given at doses which are bacteriostatic.**

Malaria

has killed more humans than has any other disease.

Control involves an understanding of the parasite, *Plasmodium*, and the vector, *Anopheles*.

Infected mosquito passes on parasites to uninfected human. *Plasmodium* passes down piercing mouthparts from salivary glands.

Control methods include

- sleeping under mosquito nets to prevent biting
- spraying of insect-repellant chemicals onto the skin
- wearing long-sleeved clothes during the evening (most likely time to be bitten).

Mosquitoes breed in bodies of still water, such as lakes and ponds. Rafts of eggs are laid and hatch into larvae, then pupae, which hang just below the water's surface to obtain oxygen through a short air tube.

Pupa

This is a weak point in the vector's life cycle, and control methods include

- spraying oil on the surface, to block the breathing tubes
- draining marshes and swamps
- introducing small fish, called mosquito fish, which feed on larvae and pupae.

The *Anopheles* mosquito is the **vector** for the pathogen, *Plasmodium*. *Plasmodium* multiplies in the stomach wall of the mosquito, then migrates to the salivary glands.

Adult mosquitoes can be controlled by insecticides, such as DDT, sprayed onto their resting places. This may select **pesticide-resistant** strains of mosquito.

Plasmodium is the protoctist which causes malaria.

It can reproduce sexually inside the mosquito, and the resulting variation makes the production of successful vaccines very difficult.

An uninfected mosquito picks up parasites from the blood of an infected person.

LIVER

VACCINES HAVE LIMITED SUCCESS

- the parasite is 'hidden' from the immune system in liver and red blood cells.
- antigens on the surface of the parasite 'evolve' so that antibodies cannot recognise them.

Parasites multiply rapidly in the liver, and are then released into the blood.

Drugs such as Paludrin and Larium control the disease by limiting entry to the liver and reducing multiplication of the parasite.

Gin contains **quinine**, an effective anti-malarial drug. Early colonists of malarial areas could justify drinking gin and tonic!

Gin

Parasite released from the liver invade red blood cells where they feed on haemoglobin and divide by **multiple fission**. The red cells filled with parasites burst and release the parasites into the blood.

Effects on the red cells cause many of the symptoms of malaria:

- tiredness, since fewer red cells means less oxygen is transported
- fever, as the red cells burst.

People with sickle cell anaemia are protected from malaria. The parasite cannot feed so well inside 'sickle' cells.

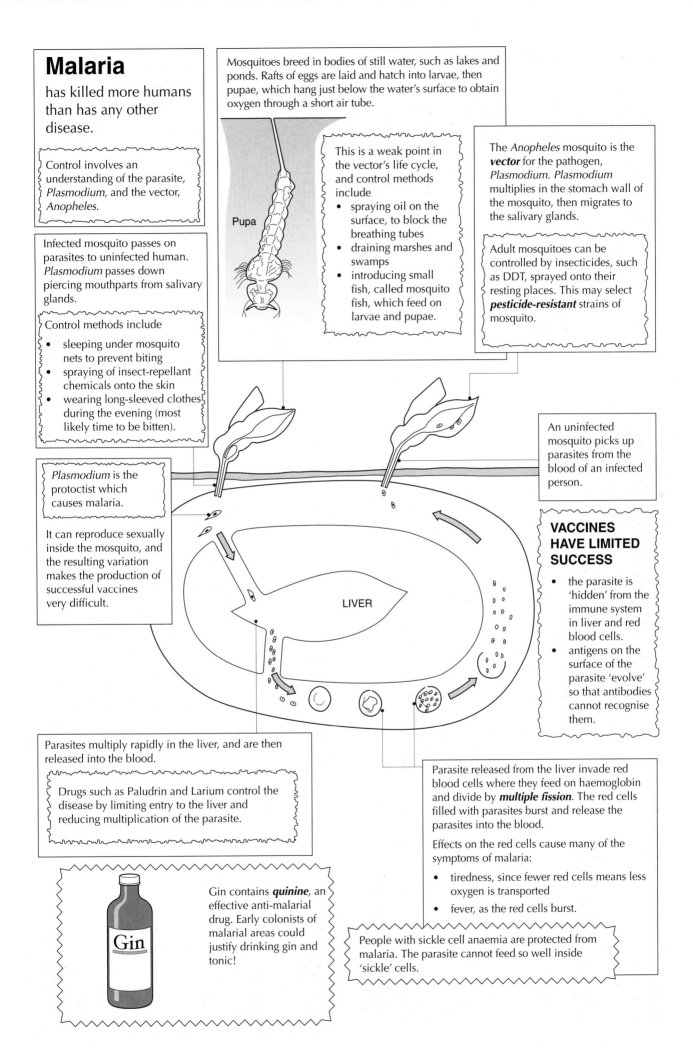

Section J Nerves and muscles

Sensory cells convert stimuli to electrical impulses.

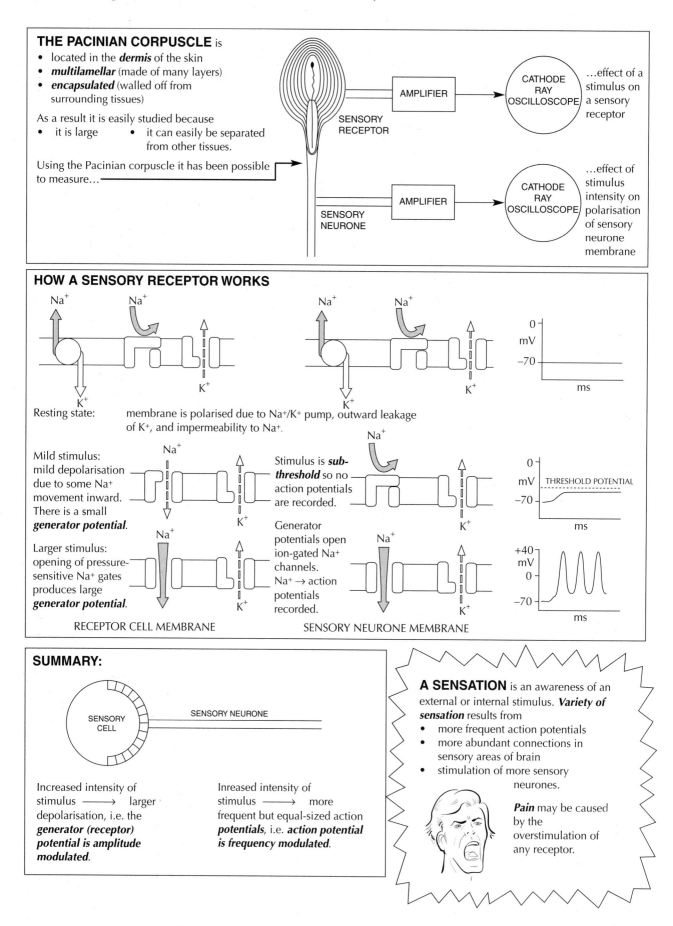

THE PACINIAN CORPUSCLE is
- located in the *dermis* of the skin
- *multilamellar* (made of many layers)
- *encapsulated* (walled off from surrounding tissues)

As a result it is easily studied because
- it is large
- it can easily be separated from other tissues.

Using the Pacinian corpuscle it has been possible to measure…

SENSORY RECEPTOR

AMPLIFIER

CATHODE RAY OSCILLOSCOPE

…effect of a stimulus on a sensory receptor

SENSORY NEURONE

AMPLIFIER

CATHODE RAY OSCILLOSCOPE

…effect of stimulus intensity on polarisation of sensory neurone membrane

HOW A SENSORY RECEPTOR WORKS

Resting state: membrane is polarised due to Na^+/K^+ pump, outward leakage of K^+, and impermeability to Na^+.

Mild stimulus: mild depolarisation due to some Na^+ movement inward. There is a small *generator potential*.

Larger stimulus: opening of pressure-sensitive Na^+ gates produces large *generator potential*.

Stimulus is *sub-threshold* so no action potentials are recorded.

Generator potentials open ion-gated Na^+ channels. $Na^+ \rightarrow$ action potentials recorded.

THRESHOLD POTENTIAL

RECEPTOR CELL MEMBRANE

SENSORY NEURONE MEMBRANE

SUMMARY:

SENSORY CELL

SENSORY NEURONE

Increased intensity of stimulus ⟶ larger depolarisation, i.e. the *generator (receptor) potential is amplitude modulated*.

Inreased intensity of stimulus ⟶ more frequent but equal-sized action *potentials*, i.e. *action potential is frequency modulated*.

A SENSATION is an awareness of an external or internal stimulus. *Variety of sensation* results from
- more frequent action potentials
- more abundant connections in sensory areas of brain
- stimulation of more sensory neurones.

Pain may be caused by the overstimulation of any receptor.

Eye function:
fine focussing in dim and bright light

CORNEA/AQUEOUS HUMOUR: causes **maximum**, but **non-variable** convergence of light due to great difference of refractive index compared with the air.

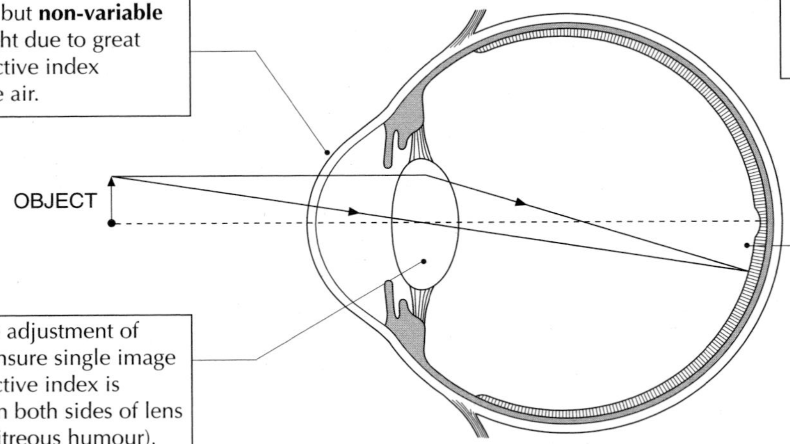

LENS: permits fine adjustment of convergence (to ensure single image on retina) as refractive index is almost identical on both sides of lens (in aqueous and vitreous humour).

RETINA: contains the photosensitive cells. A single INVERTED, DIMINISHED image is formed here.

OBJECT

SUSPENSORY LIGAMENT

MUSCLES OF THE CILIARY BODY { RADIAL CIRCULAR

RELAXED/WAKING EYES - are set for viewing distant objects so that images of close objects (e.g. alarm clock!) are blurred.

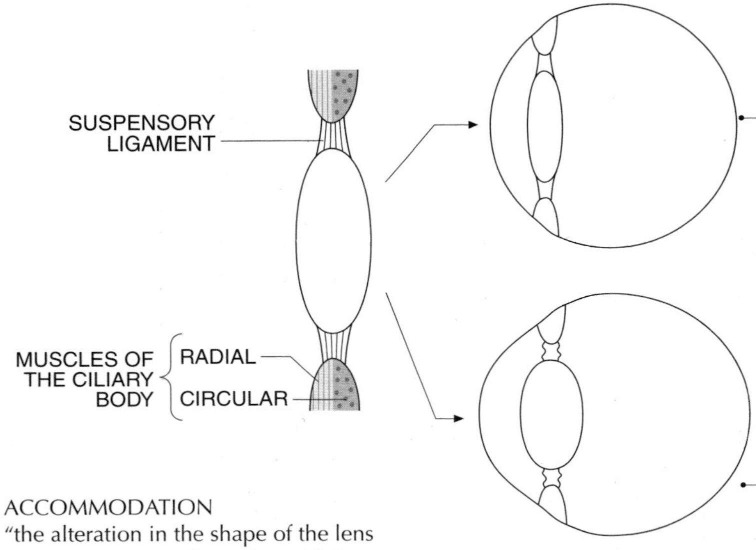

STRUCTURE	CONDITION
Radial muscle	Contracted
Circular muscle	Relaxed
Suspensory ligament	Taut
Lens	Long, thin due to fluid pressure within eyeball

MINIMUM CONVERGENCE: DISTANT OBJECTS

Radial muscle	Relaxed
Circular muscle	Contracted
Suspensory ligaments	Relaxed/looser
Lens	Adopts natural, spherical shape, as eyeball pulled inward

MAXIMUM CONVERGENCE: CLOSE-UP OBJECTS

ACCOMMODATION
"the alteration in the shape of the lens to ensure accurate focussing of light onto the retina"

EYESTRAIN - caused by extended contraction of circular ciliary muscles against the pressure of the humours.

RADIAL MUSCLES OF IRIS DIAPHRAGM: contract - pupil increases in diameter - MORE LIGHT MAY ENTER EYE.

PUPIL: admits light to eye

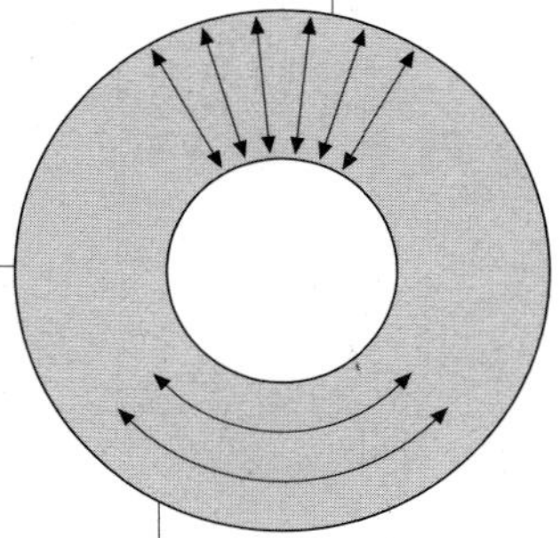

CIRCULAR MUSCLES OF IRIS DIAPHRAGM: contract - pupil decreases in diameter - LESS LIGHT MAY ENTER EYE.

THE PUPIL REFLEX - prevents bleaching of retina by controlling intensity of light which enters the eye.

Light falls on RETINA → Impulse in SENSORY NEURONES (OPTIC NERVE) → Integration in VISUAL CORTEX of the CENTRAL NERVOUS SYSTEM → Impulse in PARASYMPATHETIC MOTOR NEURONES → CIRCULAR MUSCLE CONTRACTS ← RADIAL MUSCLE RELAXES ← PUPIL DECREASES IN DIAMETER

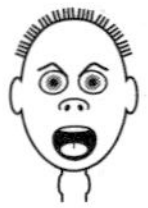

DRUGS AFFECT THE PUPIL REFLEX
• ATROPINE - blocks acetylcholine and thus → PUPIL DILATION
• HEROIN - mimics parasympathetic stimulation → PUPIL CONSTRICTION

ADRENALINE ("FIGHT or FLIGHT") mimics the sympathetic nervous system - PUPIL DILATION

Structure and function of the retina

DIRECTION OF INCIDENT LIGHT

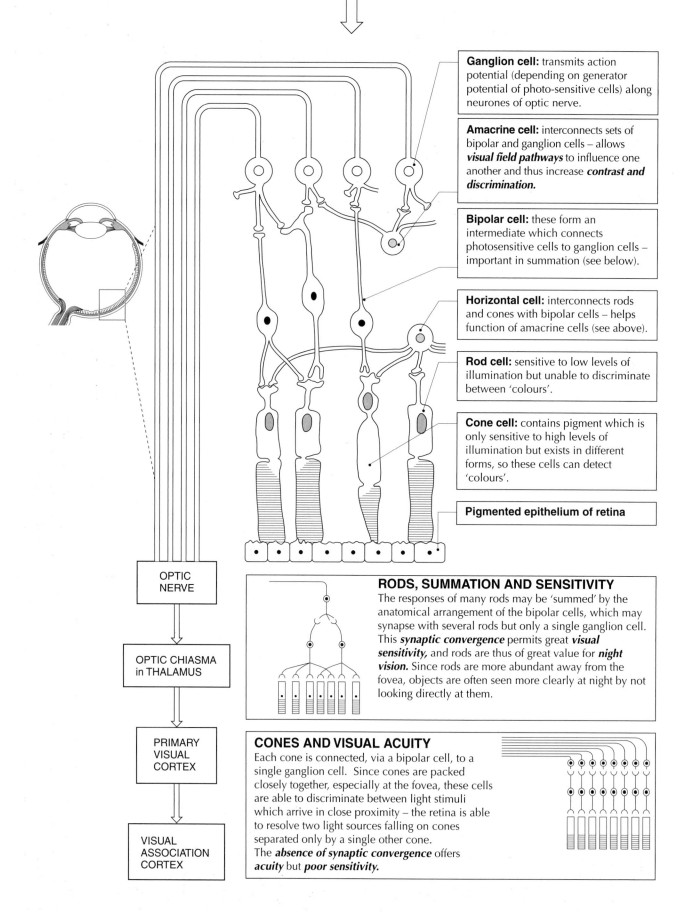

Ganglion cell: transmits action potential (depending on generator potential of photo-sensitive cells) along neurones of optic nerve.

Amacrine cell: interconnects sets of bipolar and ganglion cells – allows *visual field pathways* to influence one another and thus increase *contrast and discrimination.*

Bipolar cell: these form an intermediate which connects photosensitive cells to ganglion cells – important in summation (see below).

Horizontal cell: interconnects rods and cones with bipolar cells – helps function of amacrine cells (see above).

Rod cell: sensitive to low levels of illumination but unable to discriminate between 'colours'.

Cone cell: contains pigment which is only sensitive to high levels of illumination but exists in different forms, so these cells can detect 'colours'.

Pigmented epithelium of retina

OPTIC NERVE

OPTIC CHIASMA in THALAMUS

PRIMARY VISUAL CORTEX

VISUAL ASSOCIATION CORTEX

RODS, SUMMATION AND SENSITIVITY
The responses of many rods may be 'summed' by the anatomical arrangement of the bipolar cells, which may synapse with several rods but only a single ganglion cell. This *synaptic convergence* permits great *visual sensitivity,* and rods are thus of great value for *night vision.* Since rods are more abundant away from the fovea, objects are often seen more clearly at night by not looking directly at them.

CONES AND VISUAL ACUITY
Each cone is connected, via a bipolar cell, to a single ganglion cell. Since cones are packed closely together, especially at the fovea, these cells are able to discriminate between light stimuli which arrive in close proximity – the retina is able to resolve two light sources falling on cones separated only by a single other cone.
The *absence of synaptic convergence* offers *acuity* but *poor sensitivity.*

Photoreceptors

in the retina are the rods and cones.

CONES AND COLOUR VISION

The visual cycle in cones is very similar to that in rods, except that the protein component of the photosensitive compound (**iodopsin** or **photopsin**) is different. In particular the dissociation of the pigment which leads to the generator potential requires a higher energy input – thus cones are not very sensitive in dim light.

PRIMARY CONES STIMULATED			COLOUR PERCEIVED IN BRAIN
BLUE (440 nm)	GREEN (550 nm)	RED (600 nm)	
+	+	+	WHITE
+			BLUE
	+		GREEN
		+	RED
+	+		CYAN
	+	+	ORANGE/YELLOW
+		+	MAGENTA

The trichromatic theory is supported by studies using recombinant DNA techniques which demonstrated separate genes for blue-sensitive, green-sensitive and red-sensitive opsin.

THE VISUAL CYCLE

The light-sensitive compound **rhodopsin** is a conjugate protein of **opsin** and **retinal** (a derivative of a vitamin A).

LIGHT causes conversion of cis- to trans-retinal.

altered retinal shape causes separation from opsin, which undergoes a conformational change.

trans-retinal is reconverted to cis-retinal by the enzyme **retinal isomerase.**

Opsin and cis-retinal recombine to form **rhodopsin**, a slow process called **dark adaption.**

Trichromatic theory of colour vision suggests that there are three variants of iodopsin, each of which is sensitive to a different range of wavelengths, corresponding to the three primary colours **red, blue** and **green.**

Each type of iodopsin is probably located in different cones, and different colours are perceived in the brain from the sensory input from combinations of the three "primary cones".

STRUCTURE AND FUNCTION IN ROD CELLS

OUTER SEGMENT
contains numerous discs or lamellae which are packed with the photosensitive compound **rhodopsin**. Light falling on rhodopsin induces a conformational change in the protein part of the molecule which triggers a cascade via the hydrolysis of cyclic GMP which eventually hyperpolarises the rod cell membrane.

NECK REGION
contains a modified cilium and basal body, and connects the metabolic centre of the inner segment to the photo-sensitive region.

INNER SEGMENT
contains numerous mitochondria (ATP for active transport of Na$^+$ ions), ribosomes (for synthesis of opsin) and an extensive endoplasmic reticulum (Ca^{2+} storage and membrane assembly).

CELL BODY with
nucleus

SYNAPTIC TERMINAL releases an excitatory neurotransmitter. There is maximal release of this compound **in the dark.**

The ear and hearing

PINNA: the fleshy part of the outer ear is supported by cartilage. It is cup-shaped, and collects sound waves which then pass along the ear canal.

EARDRUM (TYMPANUM): taut membrane that vibrates when struck by sound waves. The eardrum transmits the sound waves to the *hammer*, the first of the ossicles, in the middle ear.

OSSICLES: the smallest bones in the body – the hammer, anvil, and stirrup amplify sound waves and transmit them towards the fluid of the inner ear (via the oval window).

THE SEMICIRCULAR CANALS: three fluid-filled canals, at right angles to one another, that give a *sense of balance*.

STIRRUP pressed against *oval window*. The thin membrane of the oval window transmits vibrations to the fluid of the cochlea.

ROUND WINDOW: a membrane that can 'flex' with movements of the fluid in the cochlea. This reduces the effects of these fluid movements on the sensory cells of the organ of Corti so that it is only stimulated for a limited period.

EUSTACHIAN TUBE: leads from the middle ear to the throat. This permits equalisation of pressure on both sides of the ear. Hearing is affected if the tube is blocked, e.g. by mucus when an individual has a cold.

EAR CANAL: the short tube that carries the sound waves from the pinna to the eardrum.

HEARING IS A FUNCTION OF THE BRAIN!

Ears: transducers which convert sound waves to *generator potentials*

Auditory nerve: carries *action potentials* to brain

Auditory centre of brain: analyzes and interprets the incoming action potentials, i.e. only when action potentials reach the brain is the sound 'heard'.

COCHLEA: a spiral tube filled with fluid. Transmits vibrations to the organ of Corti. Sensory cells of this organ convert *mechanical energy* of fluid vibration to *electrical energy* of action potentials in the *auditory nerve*.

fluid movement → movement of tectorial membrane → stimulates hair cells → impulse in auditory neurones

ENDOLYMPH

Section through cochlea

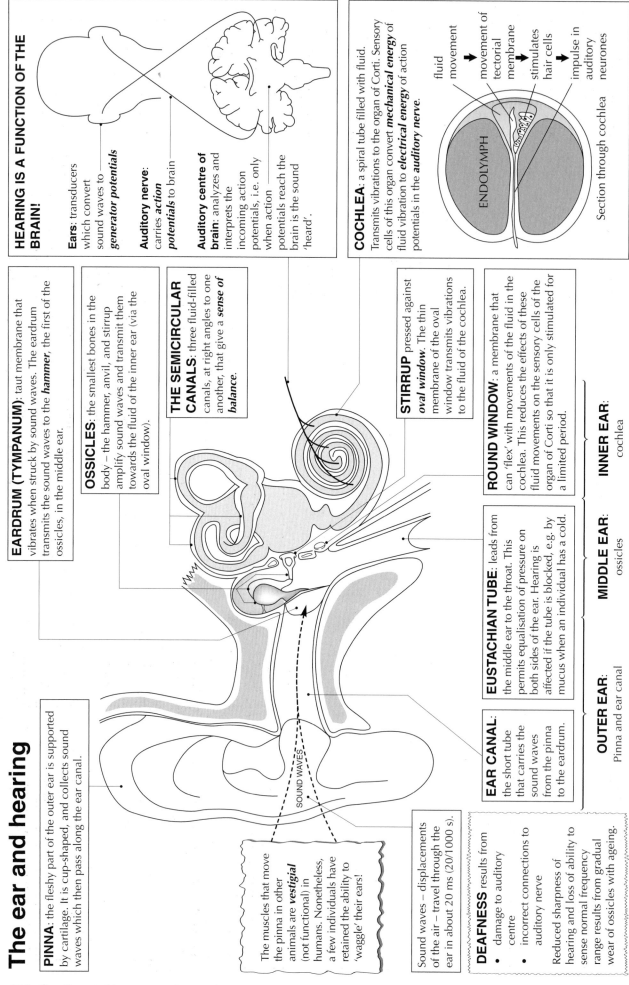

SOUND WAVES

The muscles that move the pinna in other animals are *vestigial* (not functional) in humans. Nonetheless, a few individuals have retained the ability to 'waggle' their ears!

Sound waves – displacements of the air – travel through the ear in about 20 ms (20/1000 s).

DEAFNESS results from
- damage to auditory centre
- incorrect connections to auditory nerve

Reduced sharpness of hearing and loss of ability to sense normal frequency range results from gradual wear of ossicles with ageing.

OUTER EAR: Pinna and ear canal

MIDDLE EAR: ossicles

INNER EAR: cochlea

Endocrine control depends upon chemical messengers secreted from cells and binding to specific hormone receptors

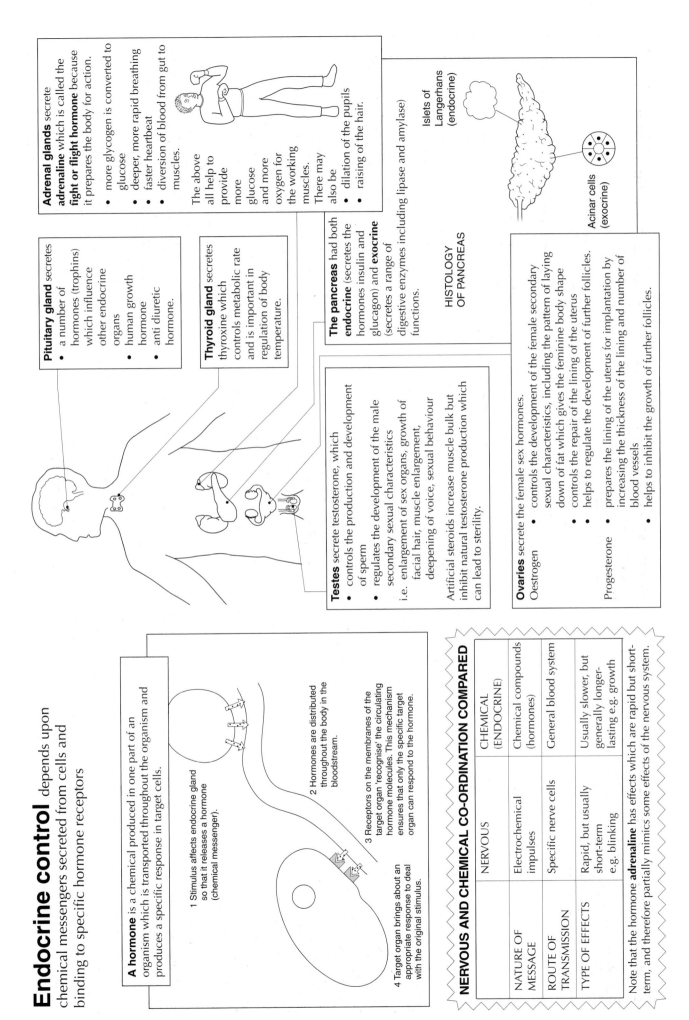

A hormone is a chemical produced in one part of an organism which is transported throughout the organism and produces a specific response in target cells.

1 Stimulus affects endocrine gland so that it releases a hormone (chemical messenger).

2 Hormones are distributed throughout the body in the bloodstream.

3 Receptors on the membranes of the target organ 'recognise' the circulating hormone molecules. This mechanism ensures that only the specific target organ can respond to the hormone.

4 Target organ brings about an appropriate response to deal with the original stimulus.

Adrenal glands secrete **adrenaline** which is called the **fight or flight hormone** because it prepares the body for action.

- more glycogen is converted to glucose
- deeper, more rapid breathing
- faster heartbeat
- diversion of blood from gut to muscles.

The above all help to provide more glucose and more oxygen for the working muscles. There may also be
- dilation of the pupils
- raising of the hair.

Pituitary gland secretes
- a number of hormones (trophins) which influence other endocrine organs
- human growth hormone
- anti diuretic hormone.

Thyroid gland secretes thyroxine which controls metabolic rate and is important in regulation of body temperature.

The pancreas had both **endocrine** (secretes the hormones insulin and glucagon) and **exocrine** (secretes a range of digestive enzymes including lipase and amylase) functions.

Islets of Langerhans (endocrine)

Acinar cells (exocrine)

HISTOLOGY OF PANCREAS

Testes secrete testosterone, which
- controls the production and development of sperm
- regulates the development of the male secondary sexual characteristics

i.e. enlargement of sex organs, growth of facial hair, muscle enlargement, deepening of voice, sexual behaviour

Artificial steroids increase muscle bulk but inhibit natural testosterone production which can lead to sterility.

Ovaries secrete the female sex hormones.

Oestrogen
- controls the development of the female secondary sexual characteristics, including the pattern of laying down of fat which gives the feminine body shape
- controls the repair of the lining of the uterus
- helps to regulate the development of further follicles.

Progesterone
- prepares the lining of the uterus for implantation by increasing the thickness of the lining and number of blood vessels
- helps to inhibit the growth of further follicles.

NERVOUS AND CHEMICAL CO-ORDINATION COMPARED

	NERVOUS	CHEMICAL (ENDOCRINE)
NATURE OF MESSAGE	Electrochemical impulses	Chemical compounds (hormones)
ROUTE OF TRANSMISSION	Specific nerve cells	General blood system
TYPE OF EFFECTS	Rapid, but usually short-term e.g. blinking	Usually slower, but generally longer-lasting e.g. growth

Note that the hormone **adrenaline** has effects which are rapid but short-term, and therefore partially mimics some effects of the nervous system.

Synthetic hormones and control of the oestrous cycle

SYNCHRONISATION OF OESTRUS

is useful in both cows and sheep. For sheep

- lambing season can be advanced for maximum profit
- cycle can be accelerated → 3 lambs in 2 years
- longer milking season from ewes.

Control

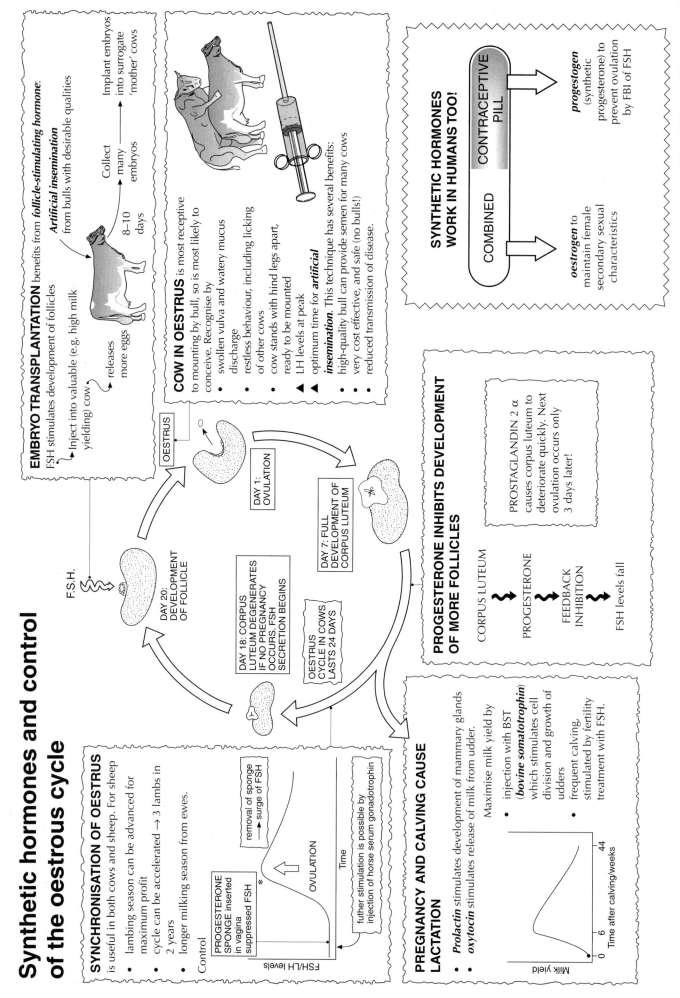

PROGESTERONE SPONGE inserted in vagina suppressed FSH

removal of sponge → surge of FSH

OVULATION

Time

futher stimulation is possible by injection of horse serum gonadotrophin

FSH/LH levels

EMBRYO TRANSPLANTATION benefits from *follicle-stimulating hormone:*

FSH stimulates development of follicles

Inject into valuable (e.g. high milk yielding) cow

releases more eggs

8–10 days

Collect many embryos

Implant embryos into surrogate 'mother' cows

Artificial insemination
from bulls with desirable qualities

COW IN OESTRUS is most receptive

to mounting by bull, so is most likely to conceive. Recognise by

- swollen vulva and watery mucus discharge
- restless behaviour, including licking of other cows
- cow stands with hind legs apart, ready to be mounted
- ▲ LH levels at peak
- ▲ optimum time for *artificial insemination*. This technique has several benefits: high-quality bull can provide semen for many cows very cost effective, and safe (no bulls!) reduced transmission of disease.

OESTRUS

F.S.H.

DAY 1: OVULATION

DAY 20: DEVELOPMENT OF FOLLICLE

DAY 18: CORPUS LUTEUM DEGENERATES IF NO PREGNANCY OCCURS. FSH SECRETION BEGINS

OESTRUS CYCLE IN COWS LASTS 24 DAYS

DAY 7: FULL DEVELOPMENT OF CORPUS LUTEUM

SYNTHETIC HORMONES WORK IN HUMANS TOO!

CONTRACEPTIVE PILL

COMBINED

oestrogen to maintain female secondary sexual characteristics

progestogen (synthetic progesterone) to prevent ovulation by FBI of FSH

PROGESTERONE INHIBITS DEVELOPMENT OF MORE FOLLICLES

CORPUS LUTEUM → PROGESTERONE → FEEDBACK INHIBITION → FSH levels fall

PROSTAGLANDIN 2 α causes corpus luteum to deteriorate quickly. Next ovulation occurs only 3 days later!

PREGNANCY AND CALVING CAUSE LACTATION

- *Prolactin* stimulates development of mammary glands
- *oxytocin* stimulates release of milk from udder.

Maximise milk yield by

- injection with BST (*bovine somatotrophin*) which stimulates cell division and growth of udders
- frequent calving, stimulated by fertility treatment with FSH.

Milk yield

Time after calving/weeks

0   6   44

The human nervous system

can be subdivided into central and peripheral components

CENTRAL NERVOUS SYSTEM (CNS)

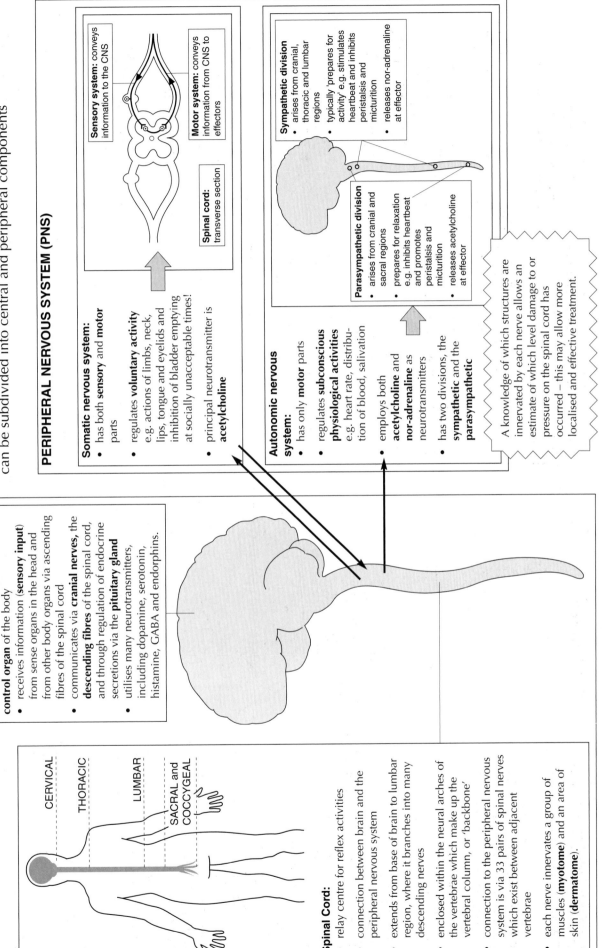

CERVICAL

THORACIC

LUMBAR

SACRAL and COCCYGEAL

Brain: the primary **integration** and **control organ** of the body

- receives information (**sensory input**) from sense organs in the head and from other body organs via ascending fibres of the spinal cord
- communicates via **cranial nerves**, the **descending fibres** of the spinal cord, and through regulation of endocrine secretions via the **pituitary gland**
- utilises many neurotransmitters, including dopamine, serotonin, histamine, GABA and endorphins.

Spinal Cord:

- relay centre for reflex activities
- connection between brain and the peripheral nervous system
- extends from base of brain to lumbar region, where it branches into many descending nerves
- enclosed within the neural arches of the vertebrae which make up the vertebral column, or 'backbone'
- connection to the peripheral nervous system is via 33 pairs of spinal nerves which exist between adjacent vertebrae
- each nerve innervates a group of muscles (**myotome**) and an area of skin (**dermatome**).

PERIPHERAL NERVOUS SYSTEM (PNS)

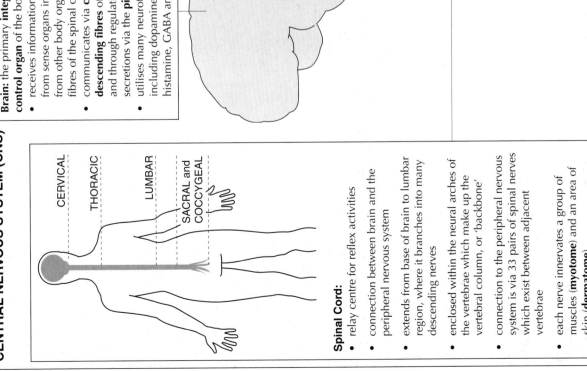

Sensory system: conveys information to the CNS

Motor system: conveys information from CNS to effectors

Spinal cord: transverse section

Somatic nervous system:

- has both **sensory** and **motor** parts
- regulates **voluntary activity** e.g. actions of limbs, neck, lips, tongue and eyelids and inhibition of bladder emptying at socially unacceptable times!
- principal neurotransmitter is **acetylcholine**

Autonomic nervous system:

- has only **motor** parts
- regulates **subconscious physiological activities** e.g. heart rate, distribution of blood, salivation
- employs both **acetylcholine** and **nor-adrenaline** as neurotransmitters
- has two divisions, the **sympathetic** and the **parasympathetic**

Sympathetic division

- arises from cranial, thoracic and lumbar regions
- typically 'prepares for activity' e.g. stimulates heartbeat and inhibits peristalsis and micturition
- releases nor-adrenaline at effector

Parasympathetic division

- arises from cranial and sacral regions
- prepares for relaxation e.g. inhibits heartbeat and promotes peristalsis and micturition
- releases acetylcholine at effector

A knowledge of which structures are innervated by each nerve allows an estimate of which level damage to or pressure on the spinal cord has occurred – this may allow more localised and effective treatment.

Behaviour may be innate (instinctive) or learned

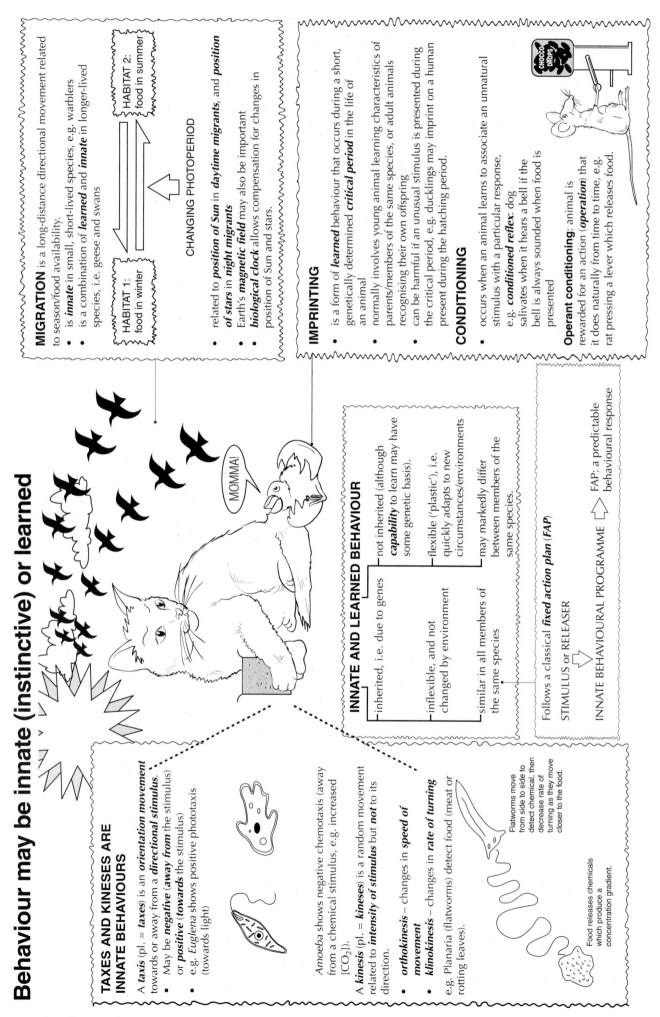

MIGRATION is a long-distance directional movement related to season/food availability.

- is *innate* in small, short-lived species, e.g. warblers
- is a combination of *learned* and *innate* in longer-lived species, i.e. geese and swans

HABITAT 2: food in summer

HABITAT 1: food in winter

CHANGING PHOTOPERIOD

- related to *position of Sun* in *daytime migrants*, and *position of stars* in *night migrants*
- Earth's *magnetic field* may also be important
- *biological clock* allows compensation for changes in position of Sun and stars.

MOMMA!

IMPRINTING

- is a form of *learned* behaviour that occurs during a short, genetically determined *critical period* in the life of an animal
- normally involves young animal learning characteristics of parents/members of the same species, or adult animals recognising their own offspring
- can be harmful if an unusual stimulus is presented during the critical period, e.g. ducklings may imprint on a human present during the hatching period.

CONDITIONING

- occurs when an animal learns to associate an unnatural stimulus with a particular response, e.g. *conditioned reflex*: dog salivates when it hears a bell if the bell is always sounded when food is presented

Operant conditioning: animal is rewarded for an action (*operation*) that it does naturally from time to time, e.g. rat pressing a lever which releases food.

CHOCCO DROPS

TAXES AND KINESES ARE INNATE BEHAVIOURS

A *taxis* (pl. = *taxes*) is an *orientation movement* towards or away from a *directional stimulus*.

- May be *negative* (*away from* the stimulus) or *positive* (*towards* the stimulus) e.g. *Euglena* shows positive phototaxis (towards light)

Amoeba shows negative chemotaxis (away from a chemical stimulus, e.g. increased [CO_2]).

A *kinesis* (pl. = *kineses*) is a random movement related to *intensity of stimulus* but *not* to its direction.

- *orthokinesis* – changes in *speed of movement*
- *klinokinesis* – changes in *rate of turning*

e.g. *Planaria* (flatworms) detect food (meat or rotting leaves).

Flatworms move from side to side to detect chemical, then decrease rate of turning as they move closer to the food.

Food releases chemicals which produce a concentration gradient.

INNATE AND LEARNED BEHAVIOUR

- inherited, i.e. due to genes
- inflexible, and not changed by environment
- similar in all members of the same species

- not inherited (although *capability* to learn may have some genetic basis).
- flexible ('plastic'), i.e. quickly adapts to new circumstances/environments
- may markedly differ between members of the same species.

Follows a classical *fixed action plan* (*FAP*)

STIMULUS or RELEASER ⟹

INNATE BEHAVIOURAL PROGRAMME ⟹ FAP: a predictable behavioural response

Neurones: the dendrites (antennae), axon (cable), synaptic buttons (contacts) are serviced and maintained by the cell body.

Dendrites are extensions of the cell body containing all typical cell body organelles. They provide a large surface area to receive information which they then pass on towards the cell body. The plasma has a high density of **chemically gated ion channels**, important in impulse transmission.

MOTOR NEURONE

Cell body contains a well-developed nucleus and nucleolus and many organelles such as lysosomes and mitochondria. There is **no mitotic apparatus** (centriole/spindle) in neurones more than six months old, which means that damaged neurones can never be replaced (although they may regenerate – see below).

Axon hillock is the point on the neuronal membrane at which a **threshold stimulus** may lead to the initiation of an **action potential.**

Nodes of Ranvier are unmyelinated segments of the neurone. Since these are uninsulated, ion movements may take place which effectively lead to action potentials 'leaping' from one node to another during **saltatory conduction.**

Axon is the communication route between the cell body and the axon terminals.

Schwann cell is a glial cell which encircles the axon. When the two 'ends' of the Schwann cell meet, overlapping occurs which pushes the nucleus and cytoplasm to the outside layer.

Neurilemma is found only around axons of the peripheral nervous system, i.e. typical sensory and motor neurones. The neurilemma plays a part in the regeneration of damaged nerves by forming a tubular sheath around the damaged area within which regeneration may occur (very rare in CNS).

DIRECTION OF IMPULSE TRANSMISSION

DENDRITES

CELL BODY

SENSORY NEURONE

AXON

Axon terminal

Myelin sheath is composed of 20–30 layers of Schwann cell membrane. The high phospholipid content of the sheath offers electrical insulation → **saltatory impulse conduction.** Not complete until late childhood so infants often have slow responses/poor co-ordination. Some axons are not myelinated but are enclosed in Schwann cell cytoplasm.

Synaptic end bulb or **synaptic button** is important in nerve impulse conduction from one neurone to another or from a neurone to an effector. They contain membrane-enclosed sacs (**synaptic vesicles**) which store **neurotransmitters** prior to release and diffusion to the post-synaptic membrane.

Spinal cord and reflex action

Reflex actions are rapid responses to internal or external stimuli which allow the body to maintain homeostasis. They may be *somatic* or *autonomic* but all involve the sequence

RECEPTOR → SENSORY NEURONE → CNS → MOTOR NEURONE → EFFECTOR

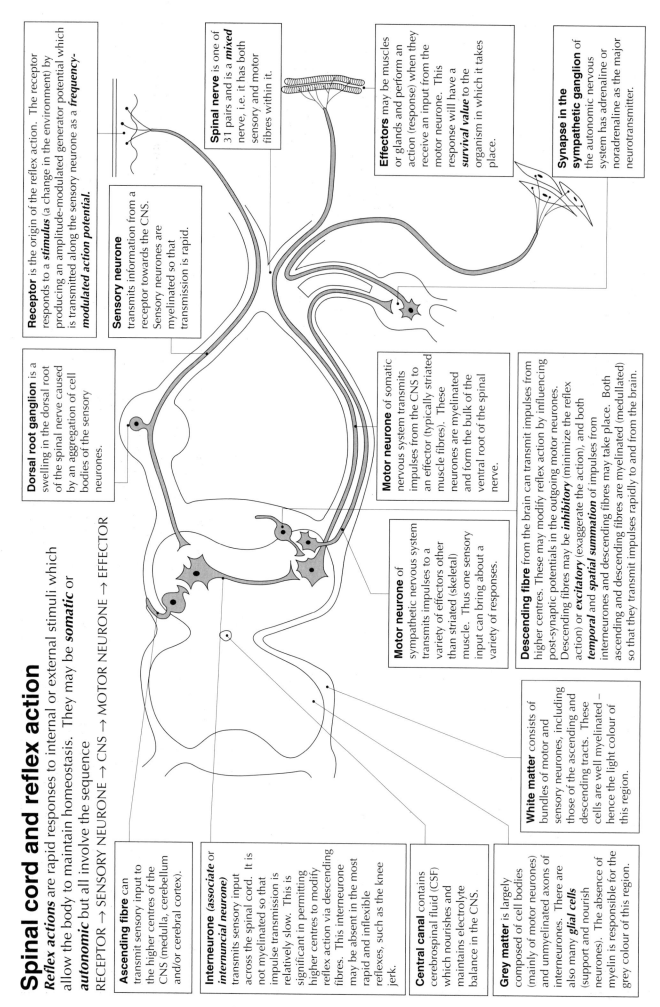

Receptor is the origin of the reflex action. The receptor responds to a *stimulus* (a change in the environment) by producing an amplitude-modulated generator potential which is transmitted along the sensory neurone as a *frequency-modulated action potential.*

Sensory neurone transmits information from a receptor towards the CNS. Sensory neurones are myelinated so that transmission is rapid.

Spinal nerve is one of 31 pairs and is a *mixed* nerve, i.e. it has both sensory and motor fibres within it.

Effectors may be muscles or glands and perform an action (response) when they receive an input from the motor neurone. This response will have a *survival value* to the organism in which it takes place.

Synapse in the sympathetic ganglion of the autonomic nervous system has adrenaline or noradrenaline as the major neurotransmitter.

Dorsal root ganglion is a swelling in the dorsal root of the spinal nerve caused by an aggregation of cell bodies of the sensory neurones.

Motor neurone of somatic nervous system transmits impulses from the CNS to an effector (typically striated muscle fibres). These neurones are myelinated and form the bulk of the ventral root of the spinal nerve.

Motor neurone of sympathetic nervous system transmits impulses to a variety of effectors other than striated (skeletal) muscle. Thus one sensory input can bring about a variety of responses.

Descending fibre from the brain can transmit impulses from higher centres. These may modify reflex action by influencing post-synaptic potentials in the outgoing motor neurones. Descending fibres may be *inhibitory* (minimize the reflex action) or *excitatory* (exaggerate the action), and both *temporal* and *spatial summation* of impulses from interneurones and descending fibres may take place. Both ascending and descending fibres are myelinated (medullated) so that they transmit impulses rapidly to and from the brain.

Ascending fibre can transmit sensory input to the higher centres of the CNS (medulla, cerebellum and/or cerebral cortex).

Interneurone (*associate* or *internuncial neurone*) transmits sensory input across the spinal cord. It is not myelinated so that impulse transmission is relatively slow. This is significant in permitting higher centres to modify reflex action via descending fibres. This interneurone may be absent in the most rapid and inflexible reflexes, such as the knee jerk.

Central canal contains cerebrospinal fluid (CSF) which nourishes and maintains electrolyte balance in the CNS.

Grey matter is largely composed of cell bodies (mainly of motor neurones) and unmyelinated axons of interneurones. There are also many *glial cells* (support and nourish neurones). The absence of myelin is responsible for the grey colour of this region.

White matter consists of bundles of motor and sensory neurones, including those of the ascending and descending tracts. These cells are well myelinated – hence the light colour of this region.

Resting potential and action potential:

The **resting potential** of -70mV is largely the result of a potassium ion (K+) equilibrium.

An **action potential** is the depolarization–repolarization cycle at the neurone membrane following the application of a threshold stimulus. The depolarization is about 110 mV resulting from *an inward flow of Na+ ions*. Since the depolarization– repolarization depend upon ion concentration gradients and upon time of ion channel opening, both of which are effectively fixed, *all action potentials are of the same size*. Thus a nerve cell obeys the *all-or-nothing principle:* if a stimulus is strong enough to generate an action potential, the impulse is conducted along the entire neurone *at a constant and maximum strength* for the existing conditions.

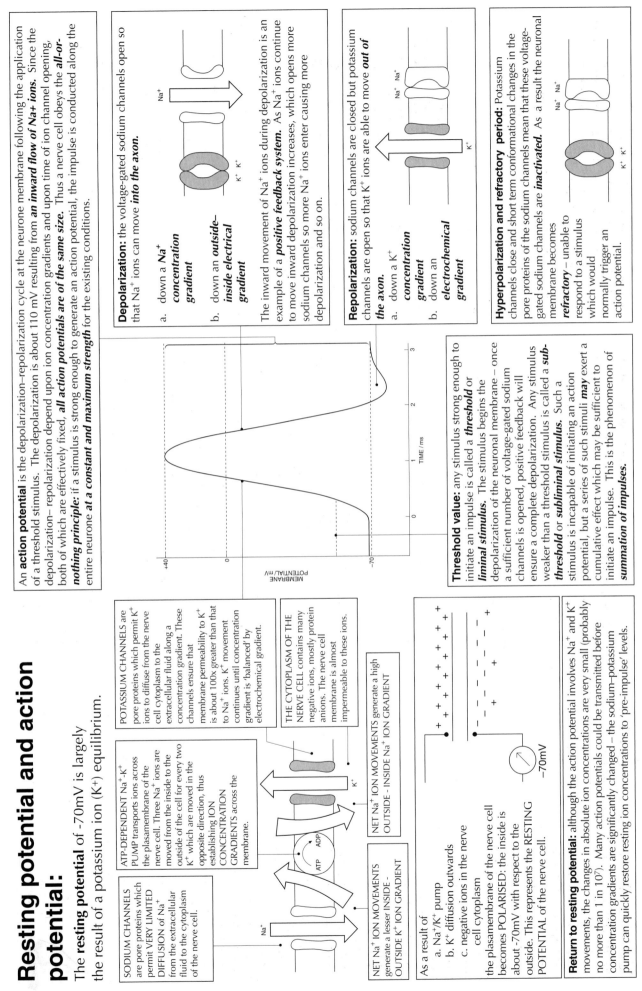

Depolarization: the voltage-gated sodium channels open so that Na+ ions can move *into the axon.*

a. down a *Na+ concentration gradient*

b. down an *outside–inside electrical gradient*

The inward movement of Na+ ions during depolarization is an example of a *positive feedback system.* As Na+ ions continue to move inward depolarization increases, which opens more sodium channels so more Na+ ions enter causing more depolarization and so on.

Repolarization: sodium channels are closed but potassium channels are open so that K+ ions are able to move *out of the axon.*

a. down a K+ *concentration gradient*

b. down an *electrochemical gradient*

Hyperpolarization and refractory period: Potassium channels close and short term conformational changes in the pore proteins of the sodium channels mean that these voltage-gated sodium channels are *inactivated.* As a result the neuronal membrane becomes *refractory* – unable to respond to a stimulus which would normally trigger an action potential.

Threshold value: any stimulus strong enough to initiate an impulse is called a *threshold* or *liminal stimulus.* The stimulus begins the depolarization of the neuronal membrane – once a sufficient number of voltage-gated sodium channels is opened, positive feedback will ensure a complete depolarization. Any stimulus weaker than a threshold stimulus is called a *sub-threshold* or *subliminal stimulus.* Such a stimulus is incapable of initiating an action potential, but a series of such stimuli *may* exert a cumulative effect which may be sufficient to initiate an impulse. This is the phenomenon of *summation of impulses.*

SODIUM CHANNELS are pore proteins which permit VERY LIMITED DIFUSION of Na+ from the extracellular fluid to the cytoplasm of the nerve cell.

ATP-DEPENDENT Na+-K+ PUMP transports ions across the plasamembrane of the nerve cell. Three Na+ ions are moved from the inside to the outside of the cell for every two K+ which are moved in the opposite direction, thus establishing ION CONCENTRATION GRADIENTS across the membrane.

POTASSIUM CHANNELS are pore proteins which permit K+ ions to diffuse from the nerve cell cytoplasm to the extracellular fluid along a concentration gradient. These channels ensure that membrane permeability to K+ is about 100x greater than that to Na+ ions. K+ movement continues until concentration gradient is 'balanced' by electrochemical gradient.

THE CYTOPLASM OF THE NERVE CELL contains many negative ions, mostly protein anions. The nerve cell membrane is almost impermeable to these ions.

NET Na+ ION MOVEMENTS generate a high OUTSIDE - INSIDE Na+ ION GRADIENT

NET K+ ION MOVEMENTS generate a lesser INSIDE - OUTSIDE K+ ION GRADIENT

As a result of

a. Na+/K+ pump

b. K+ diffusion outwards

c. negative ions in the nerve cell cytoplasm

the plasamembrane of the nerve cell becomes POLARISED: the inside is about -70mV with respect to the outside. This represents the RESTING POTENTIAL of the nerve cell.

Return to resting potential: although the action potential involves Na+ and K+ movements, the changes in absolute ion concentrations are very small (probably no more than 1 in 10⁷). Many action potentials could be transmitted before concentration gradients are significantly changed – the sodium-potassium pump can quickly restore resting ion concentrations to 'pre-impulse' levels.

Propagation of action potentials

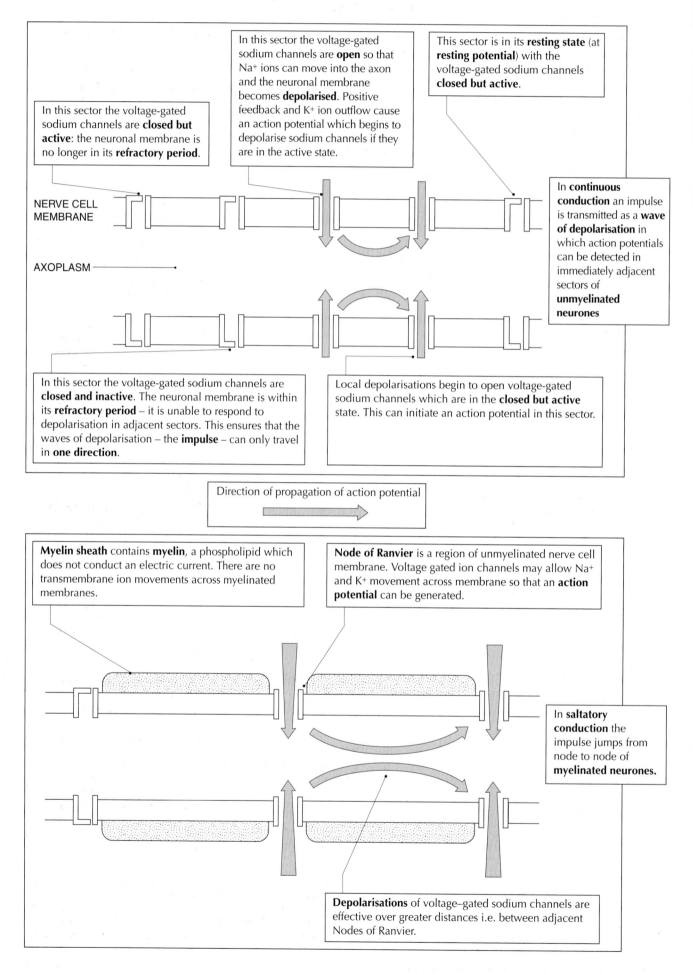

In this sector the voltage-gated sodium channels are **open** so that Na+ ions can move into the axon and the neuronal membrane becomes **depolarised**. Positive feedback and K+ ion outflow cause an action potential which begins to depolarise sodium channels if they are in the active state.

This sector is in its **resting state** (at **resting potential**) with the voltage-gated sodium channels **closed but active**.

In this sector the voltage-gated sodium channels are **closed but active**: the neuronal membrane is no longer in its **refractory period**.

NERVE CELL MEMBRANE

AXOPLASM

In **continuous conduction** an impulse is transmitted as a **wave of depolarisation** in which action potentials can be detected in immediately adjacent sectors of **unmyelinated neurones**

In this sector the voltage-gated sodium channels are **closed and inactive**. The neuronal membrane is within its **refractory period** – it is unable to respond to depolarisation in adjacent sectors. This ensures that the waves of depolarisation – the **impulse** – can only travel in **one direction**.

Local depolarisations begin to open voltage-gated sodium channels which are in the **closed but active** state. This can initiate an action potential in this sector.

Direction of propagation of action potential

Myelin sheath contains **myelin**, a phospholipid which does not conduct an electric current. There are no transmembrane ion movements across myelinated membranes.

Node of Ranvier is a region of unmyelinated nerve cell membrane. Voltage gated ion channels may allow Na+ and K+ movement across membrane so that an **action potential** can be generated.

In **saltatory conduction** the impulse jumps from node to node of **myelinated neurones**.

Depolarisations of voltage–gated sodium channels are effective over greater distances i.e. between adjacent Nodes of Ranvier.

Synapse structure and function

Transmission of an action potential across a chemical synapse involves a uni-directional release of molecules of neurotransmitter from pre-synaptic to post-synaptic membranes

Drugs and poisons may interfere with synaptic transmission by

1. *Mimicry of neurotransmitter,* e.g. *nicotine* mimics both acetylcholine and noradrenaline
2. *Reduced degradation,* e.g. *cocaine* inhibits re-uptake of noradrenaline
3. *Blocking receptors,* e.g. *β-blockers* may help to control rapid heartbeat by blocking receptors on the heart muscle. These β-receptors are normally sensitive to adrenaline.
4. *Reduced release of neurotransmitter,* e.g. *alcohol* alters sleeping patterns by reducing release of serotonin.

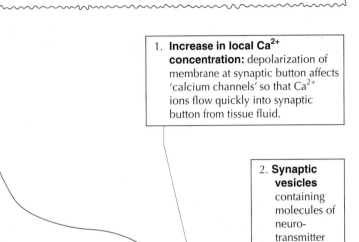

1. **Increase in local Ca^{2+} concentration:** depolarization of membrane at synaptic button affects 'calcium channels' so that Ca^{2+} ions flow quickly into synaptic button from tissue fluid.

Mitochondria are abundant in the synaptic button: release energy for refilling of synaptic vesicles and possibly for pumping of Ca^{2+} to re-establish Ca^{2+} concentration gradient across neurone membrane.

2. **Synaptic vesicles** containing molecules of neuro-transmitter move towards the presynaptic membrane.

6. **Reabsorption of neurotransmitter or products of degradation.** Molecules are resynthesized and reincorporated into synaptic vesicles. *Catecholamines* are often reabsorbed without degradation.

Synaptic cleft represents a barrier to the direct passage of the wave of depolarization from pre-synaptic to post-synaptic membranes.

Na^+

5. **Enzymes degrade neurotransmitter.** These degradative enzymes, which are released from adjacent glial cells or are located on the post-synaptic membrane, remove neurotransmitter molecules so that their effect on the chemically gated ion channels is only short-lived. They include
    monoamine oxidase (degrades *catecholamines)*
    acetylcholine esterase (degrades *acetylcholine)*

4. **Chemically gated ion channels** on post-synaptic membrane – allow influx of Na^+ and efflux of K^+ → depolarization of post-synaptic membrane. Ion channels are 'opened' when triggered by binding of neurotransmitter.

3. **Neurotransmitter molecules** diffuse across synaptic gap when synaptic vesicles fuse with pre-synaptic membrane. Molecules bind to *stereospecific receptors* in the post-synaptic membrane. *Catecholamines* such as adrenaline are released from *adrenergic nerve endings, acetylcholine* from *cholinergic nerve endings,* and *GABA* (*gamma*-aminobutyric acid) and *serotonin* at synapses in the brain.

Excitatory post-synaptic potentials result if the neurotransmitter binding to the receptors on the post-synaptic membrane *opens* chemically gated ion channels, making *depolarization more likely.*

Inhibitory post-synaptic potentials result if the neurotransmitter binding to the receptors on the post-synaptic membrane *keeps* chemically gated ion channels *closed,* promoting *hyperpolarization* and making *depolarization less likely.*

The brain is an integrator — it is able to accept sensory information from a number of sources, compare it with previous experience (learning) and make sure that the appropriate actions are initiated.

DRUGS AND BRAIN FUNCTION

Alcohol has a progressive effect on the brain

- 1–2 units release inhibitions by affecting emotional centres in the forebrain
- 5–6 units affect co-ordinated movements by influencing motor areas of the cerebral cortex
- 7–8 units cause stupor and insensitivity to pain as sensory areas, including the visual centre, are impaired
- more than 10 units may be fatal as the vital centres of the medulla and hypothalamus are severely inhibited.

Heroin can cause euphoria and insensitivity to pain as pain and emotion centres are inhibited.

L.S.D. may cause hallucinations by altering the balance of brain neurotransmitters and causing loss of the medulla's ability to 'filter' information reaching the cerebral cortex.

Skull: the cranium is a bony 'box' which encloses and protects the brain.

Visual centre: this area of the cerebral cortex
a interprets impulses along the optic nerve i.e. responsible for vision
b has the connector neurones for both accommodation and the pupil reflex.

Cerebellum: co-ordinates movement using sensory information from position receptors in various parts of the body.
Helps to maintain posture using sensory information from the inner ear.
Can control learned sequences of activity involved in dancing, athletic pursuits and in the playing of musical instruments.

Cerebral cortex: has motor areas to control voluntary movement, sensory areas which interpret sensations and association areas to link the activity of motor and sensory regions. The centre of intelligence, memory, language and consciousness.

Meninges: membranes which line the skull and cover the brain. They help to protect and nourish the brain tissues, but may be infected either by a virus or a bacterium to cause the potentially fatal condition **meningitis**.

Forebrain: here many emotions are localised, and damage to this area may cause aggression, apathy, extreme sexual behaviour and other emotional disturbances.

Hypothalamus: contains centres which control thirst, hunger and thermoregulation.

Pituitary gland: is a link between the central nervous system and the endocrine system. Secretes a number of hormones, including follicle stimulating hormone which regulates development of female gametes, and anti-diuretic hormone which controls water retention by the kidney.

Medulla: the link between the spinal cord and the brain, and relays information between these two structure. Has a number of reflex centres which control
a vital reflexes which regulate heartbeat, breathing and blood vessel diameter
b non-vital reflexes which co-ordinate swallowing, salivation, coughing and sneezing.

Spine (vertebral column): composed of 33 separate vertebra which surround and protect the spinal cord; between each pair of vertebrae two **spinal nerves** carry sensory information into the spinal cord and motor information out of it. Dislocation of the vertebrae may compress the spinal nerves, causing great pain, or even crush the spinal cord, leading to paralysis.

Synovial joints have a space between the articulating bones and are freely movable.

THE MOVEMENTS POSSIBLE AT SYNOVIAL JOINTS ARE:

1. **Gliding:** one part slides upon another without any angular or rotary motion, e.g. joints between carpals and between tarsals.

ANGULAR MOVEMENTS: increase or decrease the angle between bones.

2. **Abduction:** moving the part away from the midline of the body.
3. **Adduction:** bringing the part towards the midline.
4. **Flexion:** decreasing the angle between two bones, includes bending the head forward (joint between the occipital bone and the atlas).
5. **Extension:** increasing the angle between two bones, includes returning the head to the anatomical position.

6. and 7. **Rotation:** turning upon an axis.
8. **Circumduction:** moving the extremity of the part around a circle so that the whole part describes a cone.

ROTARY MOVEMENTS

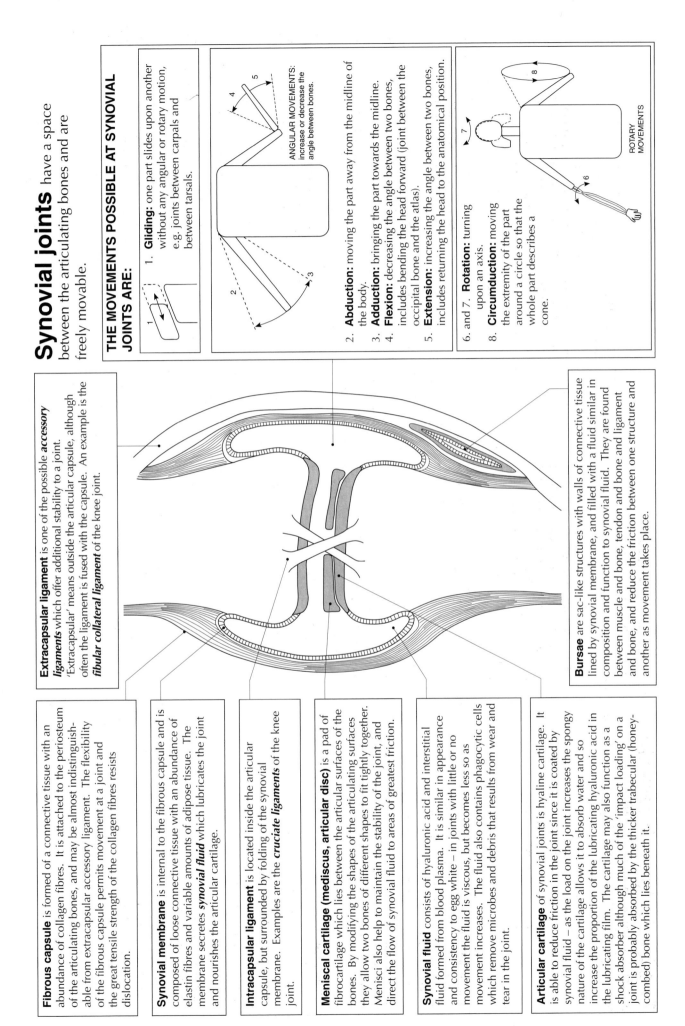

Extracapsular ligament is one of the possible *accessory ligaments* which offer additional stability to a joint. 'Extracapsular' means outside the articular capsule, although often the ligament is fused with the capsule. An example is the *fibular collateral ligament* of the knee joint.

Bursae are sac-like structures with walls of connective tissue lined by synovial membrane, and filled with a fluid similar in composition and function to synovial fluid. They are found between muscle and bone, tendon and bone and ligament and bone, and reduce the friction between one structure and another as movement takes place.

Fibrous capsule is formed of a connective tissue with an abundance of collagen fibres. It is attached to the periosteum of the articulating bones, and may be almost indistinguishable from extracapsular accessory ligament. The flexibility of the fibrous capsule permits movement at a joint and the great tensile strength of the collagen fibres resists dislocation.

Synovial membrane is internal to the fibrous capsule and is composed of loose connective tissue with an abundance of elastin fibres and variable amounts of adipose tissue. The membrane secretes *synovial fluid* which lubricates the joint and nourishes the articular cartilage.

Intracapsular ligament is located inside the articular capsule, but surrounded by folding of the synovial membrane. Examples are the *cruciate ligaments* of the knee joint.

Meniscal cartilage (mediscus, articular disc) is a pad of fibrocartilage which lies between the articular surfaces of the bones. By modifying the shapes of the articulating surfaces they allow two bones of different shapes to fit tightly together. Menisci also help to maintain the stability of the joint, and direct the flow of synovial fluid to areas of greatest friction.

Synovial fluid consists of hyaluronic acid and interstitial fluid formed from blood plasma. It is similar in appearance and consistency to egg white – in joints with little or no movement the fluid is viscous, but becomes less so as movement increases. The fluid also contains phagocytic cells which remove microbes and debris that results from wear and tear in the joint.

Articular cartilage of synovial joints is hyaline cartilage. It is able to reduce friction in the joint since it is coated by synovial fluid – as the load on the joint increases the spongy nature of the cartilage allows it to absorb water and so increase the proportion of the lubricating hyaluronic acid in the lubricating film. The cartilage may also function as a shock absorber although much of the 'impact loading' on a joint is probably absorbed by the thicker trabecular (honey-combed) bone which lies beneath it.

Movement of the forelimb

illustrates the action of *muscle groups*. Skeletal muscles produce movement by exerting forces of contraction on tendons, which in turn pull on bones.

Length of forelimb in humans permits accurate use of arms in feeding.

HUMAN FORELIMB

Separate ulna and radius allow considerable rotation of forearm. This allows wide range of movement at wrist – vital for efficient use of tools.

Humerus

Triceps is the main *extensor* of the elbow joint. When the biceps contracts and the elbow flexes, the triceps is the *antagonist* of the biceps.

CONTRACTION OF TRICEPS CAUSES EXTENSION

CONTRACTION OF BICEPS CAUSES FLEXION

Three origins of the *triceps*.

Origin: the attachment of a muscle tendon to the stationary bone. Note that the *biceps* has two origins.

Tendon: inelastic but with some limited flexibility due to parallel arrangements of densely packed collagen fibres. Inelasticity is essential so that contraction of muscle can be transmitted to the moving bone. *Ligaments* are *elastic* to allow movement of bones at joints when muscles contract.

Radius

Ulna

Biceps brachii (commonly 'biceps') is the principal *flexor* of the elbow joint. During flexation of the elbow the biceps is the *prime mover* or *agonist* and the triceps brachii is the *antagonist*. The biceps also supinates the palm and forearm (turns them upward or forward).

Insertion: the attachment of a muscle tendon to the movable bone. This tendon runs across the joint.

The fleshy part of a muscle (the *belly* or *gaster*) does not generally cover the moving part – instead the inserting tendon extends across the joint which permits the movement.

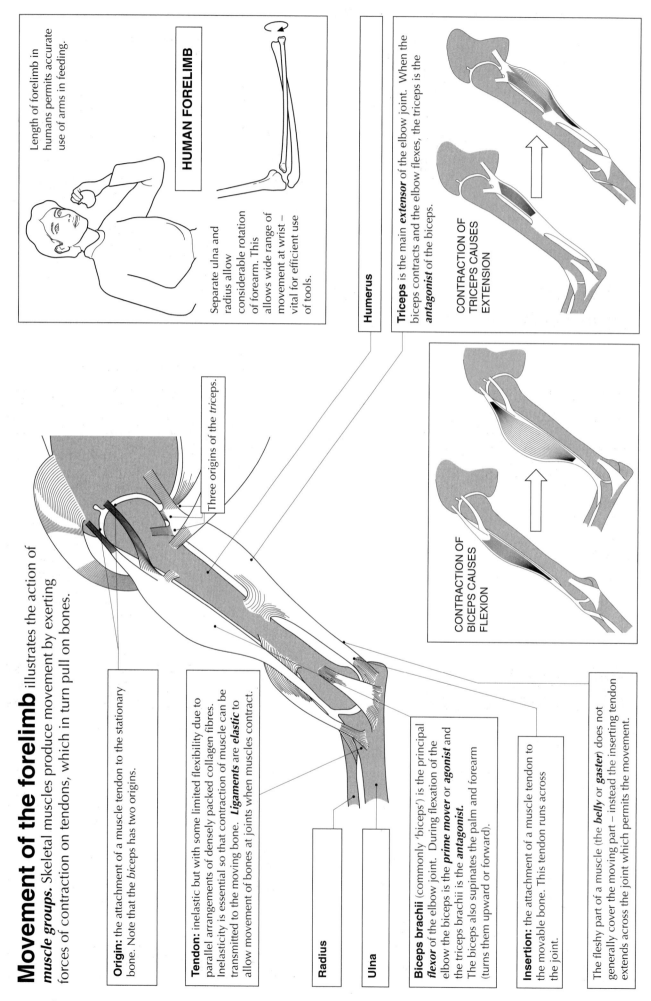

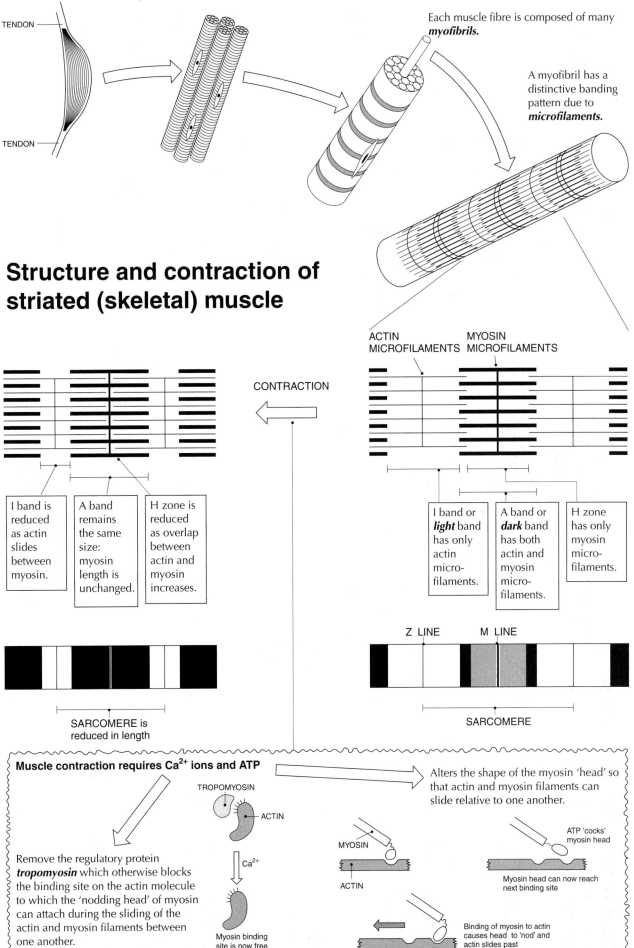

An individual muscle is composed of hundreds of **muscle fibres.**

TENDON

TENDON

Each muscle fibre is composed of many **myofibrils.**

A myofibril has a distinctive banding pattern due to **microfilaments.**

Structure and contraction of striated (skeletal) muscle

ACTIN MICROFILAMENTS

MYOSIN MICROFILAMENTS

CONTRACTION

I band is reduced as actin slides between myosin.

A band remains the same size: myosin length is unchanged.

H zone is reduced as overlap between actin and myosin increases.

I band or **light** band has only actin microfilaments.

A band or **dark** band has both actin and myosin microfilaments.

H zone has only myosin microfilaments.

Z LINE

M LINE

SARCOMERE is reduced in length

SARCOMERE

Muscle contraction requires Ca^{2+} ions and ATP

Alters the shape of the myosin 'head' so that actin and myosin filaments can slide relative to one another.

TROPOMYOSIN

ACTIN

Remove the regulatory protein **tropomyosin** which otherwise blocks the binding site on the actin molecule to which the 'nodding head' of myosin can attach during the sliding of the actin and myosin filaments between one another.

Ca^{2+}

Myosin binding site is now free

MYOSIN

ACTIN

ATP 'cocks' myosin head

Myosin head can now reach next binding site

Binding of myosin to actin causes head to 'nod' and actin slides past

Muscles: effects of exercise

SHORT TERM

Blood flow to muscles may increase by up to 25 times during exercise.

Respiration and oxygen debt: once the creatine phosphate supply has been exhausted ATP may be generated by anaerobic respiration: exercise may generate an oxygen debt of 10–12 dm^3 (up to 18–20 dm^3 in trained athletes).

Fatigue and exhaustion: "Fatigue" is the inability to repeat muscular contraction with the same force and is at least partially explained by toxic effects of accumulated lactate.

Eventually fatigue gives way to exhaustion, which is associated with depleted K$^+$ concentration in muscle cells.

Damage: for example tearing or straining through over-stretching without warming-up.

Heavy exercise leads to shorter, tighter muscles which are more prone to such damage.

Muscle soreness following exercise is due to minor inflammation and associated tissue swelling during the recovery period

Cramp is a powerful, sustained and uncontrolled contraction resulting from over-exercise, especially of muscles with inadequate circulation.

It is made worse by lactate, chilling and low concentrations of oxygen or salt.

Glycogen and potassium depletion: these are associated with fatigue and exhaustion. Several days of recovery may be necessary to return to optimum concentrations of these solutes.

Glycogen can be an energy 'store'. This compound can be 'loaded' into muscles by following a period of carbohydrate starvation (48 hours) with a high intake of simple sugars (24 hours) – useful marathon preparation!

LONG TERM

Muscle size is genetically determined; high testosterone concentration (anabolic steriod) increases muscle growth.

Exercise may increase muscle size by up to 60%, mainly by increased diameter of individual fibres.

Co-ordination: improved response, especially between antagonistic pairs, since as prime mover increases its speed of contraction the antagonist must be allowed to relax more quickly.

Biochemical changes associated with an increase in the number and size of the mitochondria is an increased activity of enzymes of the TCA cycle, electron transport and fatty acid oxidation (leading to a doubling of mitochondrial efficiency).

Stores of creatine phosphate, glycogen, fat and myoglobin are doubled and there is a more rapid release of fatty acids from the stores.

Blood supply: number of vessels and extent of capillary beds increases, both to improve *delivery of substrates and clearance of toxic products*

SPRINT OR MARATHON? FAST AND SLOW MUSCLE FIBRES

	FAST (WHITE)	SLOW (RED)
STRUCTURE	Few mitochondria	Many mitochondria
	Little myoglobin	Much myoglobin
	Much glycogen	Little glycogen
LOCATION	Relatively superficial	Deep-seated within limbs
GENERAL PROPERTIES	Relatively excitable – 'all-or-nothing' response	Low excitability – 'graded' response
	Fast contraction	Slow contraction
	Fatigues quickly	Fatigues slowly
	Rapid formation of oxygen debt–anaerobic	Respiration mainly aerobic – **can** build up oxygen debt
FUNCTION	Immediate, fast contraction: sprint	Sustained contraction: marathon, maintenance of posture

Section K Reproduction

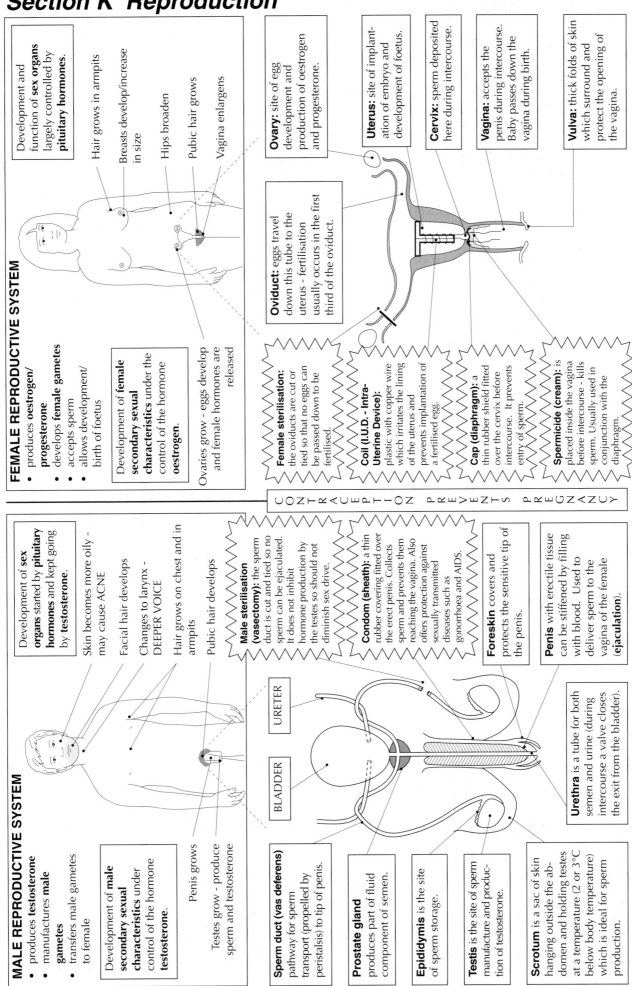

FEMALE REPRODUCTIVE SYSTEM

Development and function of **sex organs** largely controlled by **pituitary hormones.**

- produces **oestrogen/ progesterone**
- develops **female gametes**
- accepts sperm
- allows development/ birth of foetus

Development of **female secondary sexual characteristics** under the control of the hormone **oestrogen.**

Ovaries grow - eggs develop and female hormones are released

Hair grows in armpits

Breasts develop/increase in size

Hips broaden

Pubic hair grows

Vagina enlargens

Ovary: site of egg development and production of oestrogen and progesterone.

Oviduct: eggs travel down this tube to the uterus - fertilisation usually occurs in the first third of the oviduct.

Uterus: site of implant-ation of embryo and development of foetus.

Cervix: sperm deposited here during intercourse.

Vagina: accepts the penis during intercourse. Baby passes down the vagina during birth.

Vulva: thick folds of skin which surround and protect the opening of the vagina.

CONTRACEPTION PREVENTS PREGNANCY

Female sterilisation: the oviducts are cut or tied so that no eggs can be passed down to be fertilised.

Coil (I.U.D. - Intra-Uterine Device): plastic with copper wire which irritates the lining of the uterus and prevents implantation of a fertilised egg.

Cap (diaphragm): a thin rubber shield fitted over the cervix before intercourse. It prevents entry of sperm.

Spermicide (cream): is placed inside the vagina before intercourse - kills sperm. Usually used in conjunction with the diaphragm.

MALE REPRODUCTIVE SYSTEM

Development of **sex organs** started by **pituitary hormones** and kept going by **testosterone.**

- produces **testosterone**
- manufactures **male gametes**
- transfers male gametes to female

Development of **male secondary sexual characteristics** under control of the hormone **testosterone.**

Penis grows

Testes grow - produce sperm and testosterone

Skin becomes more oily - may cause ACNE

Facial hair develops

Changes to larynx - DEEPER VOICE

Hair grows on chest and in armpits

Pubic hair develops

Male sterilisation (vasectomy): the sperm duct is cut and tied so no sperm can be ejaculated. It does not inhibit hormone production by the testes so should not diminish sex drive.

Condom (sheath): a thin rubber covering fitted over the erect penis. Collects sperm and prevents them reaching the vagina. Also offers protection against sexually transmitted diseases such as gonorrhoea and AIDS.

Foreskin covers and protects the sensitive tip of the penis.

Penis with erectile tissue can be stiffened by filling with blood. Used to deliver sperm to the vagina of the female (**ejaculation**).

URETER

BLADDER

Urethra is a tube for both semen and urine (during intercourse a valve closes the exit from the bladder).

Sperm duct (vas deferens) pathway for sperm transport (propelled by peristalsis) to tip of penis.

Prostate gland produces part of fluid component of semen.

Epididmis is the site of sperm storage.

Testis is the site of sperm manufacture and produc-tion of testosterone.

Scrotum is a sac of skin hanging outside the ab-domen and holding testes at a temperature (2 or 3°C below body temperature) which is ideal for sperm production.

Seminiferous tubule and sperm

SERTOLI CELL is joined to other Sertoli cells by junctions which form a blood-testis barrier which allows the spermatozoa to evade the host's immune response.

Sertoli cells also

1. nourish all spermatogenic cells
2. remove degenerating spermatogenic cells by phagocytosis
3. control release of spermatozoa into the lumen of the tubule
4. secrete the hormone **inhibin** which is a feedback inhibitor of gonadotrophin release
5. secrete **androgen-binding protein** which concentrates testosterone in the seminiferous tubules where it promotes spermatogenesis
6. remove excess cytoplasm as spermatids develop into spermatozoa.

Acrosome is effectively an enclosed lysosome. It develops from the Golgi complex and contains hydrolytic enzymes – a hyaluronidase and several proteinases – which aid in the penetration of the granular layer and plasma membrane of the oocyte immediately prior to fertilization.

BASEMENT MEMBRANE

SPERMATOGONIUM (Sperm mother cell) divides mitotically ($2n \nearrow^{2n}_{\searrow 2n}$): the two products have different fates

– one remains to prevent depletion of the reservoir of these stem cells, the other loses contact with the basement membrane and becomes a primary spermatocyte.

PRIMARY SPERMATOCYTE has been formed from sperm mother cell and now undergoes a meiotic (reduction) division: the first meiotic division, which involves synapsis and offers the possibility of crossover, produces two secondary spermatocytes, and the second meiotic division, which is effectively a modified mitosis, produces four spermatids i.e.

Primary Spermatocyte (diploid) $\xrightarrow{\text{Meiosis I}}$ 2 Secondary Spermatocytes (haploid)

$\xrightarrow{\text{Meiosis II}}$ 4 Spermatids (haploid)

SPERMATIDS become embedded in Sertoli cells and mature into spermatozoa. This involves the development of an acrosome and a flagellum, and since no division is involved.

1 Spermatid (haploid) $\longrightarrow$ 1 Spermatozoan (haploid)

SECONDARY SPERMATOCYTES remain attached by cytoplasmic bridges which persist until sperm development is complete. This may be necessary since half the sperm carry an X chromosome and half a Y chromosome. The genes carried on the X chromosome and absent on the Y-bearing sperm may be essential for survival of Y-bearing sperm, and hence for the production of male offspring.

Nucleus contains the haploid number of chromosomes derived by meiosis from the male germinal cells – thus this genetic complement will be either an X or a Y heterosome plus 22 autosomes. Since the head delivers no cytoplasm, the male contributes no extranuclear genes or organelles.

Centriole: one of a pair, which lie at right angles to one another. One of the centrioles produces microtubules which elongate and run the entire length of the spermatozoon, forming the axial filament of the flagellum.

Mitochondria are arranged in a spiral surrounding the flagellum. They complete the aerobic stages of respiration to release the ATP required for contraction of the filaments, leading to 'beating' of the flagellum and movement of the spermatozoon.

PERIPHERAL MICROTUBULE

DYNEIN 'ARM' (acts as ATPase)

CENTRAL PAIR OF MICROTUBULES

Flagellum has the 9 + 2 arrangement of microtubules typical of such structures. The principal role of the flagella is to allow sperm to move close to the oocyte and to orientate themselves correctly prior to digestion of the oocytemembranes. The sperm are moved close to the oocyte by muscular contractions of the walls of the uterus and the oviduct.

HUMAN SPERMATO-ZOON (v.s.)

HEAD 5 μm   MID PIECE 7 μm   TAIL PIECE 45 μm

Structure of the ovary

reflects events of the menstrual cycle

CAPILLARY

TUNICA ALBUGINEA: a tough, white connective tissue coat which surrounds the ovary

SUPPORT LIGAMENT: part of mesovarium which contains blood and lymph vessels which service the ovarian tissues

CORPUS LUTEUM develops from the remains of the Graafian Follicle under the influence of LH. The corpus luteum secretes progesterone which maintains the endometrium during the second half of the menstrual cycle. If there is no fertilisation and consequently no implantation the corpus luteum is allowed to degenerate (it is degraded by an enzyme called luteolysin), plasma progesterone concentration falls and the endometrium is shed at menstruation.

STROMA is composed of connective tissue, blood vessels and cells secreting the enzymes which convert testosterone to oestrogen, the hormone which promotes development of female secondary sexual characteristics.

GERMINAL EPITHELIUM provides **primordial germ cells.** During foetal development these enlarge and undergo mitosis to form a large number of **oogonia.** Each of these is still diploid (2n = 46) by repeated mitosis and cell growth approximately 2 000 000 **primary oocytes** are formed. Each is enclosed in a single layer of cells to become a **primary follicle.**

OVULATION, the release of the secondary oocyte in a burst of follicular fluid, occurs once every 56 days from each ovary. There are complex endocrine signals involving peak blood plasma levels of LH (luteinising hormone) FSH and prostaglandins. Hydrolytic enzymes rupture the ovary wall, and ovulation is often accompanied by a small (0.5–1.0°C) rise in body temperature.

MATURE GRAAFIAN FOLLICLE contains the haploid (n = 23) secondary oocyte surrounded by nutritive and secretory cells. The thecae are two membranes – some of the thecal cells secrete testosterone (a precursor of oestrogen). Development of the follicle until the oocyte is released at ovulation is stimulated by the follicle stimulating hormone (FSH) released from the anterior pituitary gland.

By puberty there are perhaps 300 000 remaining primary follicles of which only 400–450 will develop further.

Prior to ovulation the primary oocyte enters meiosis but only completes the first stage – the reduction division

PRIMARY OOCYTE (2n) → MEIOSIS I → SECONDARY OOCYTE (n)

Meiosis II does not occur until the secondary oocyte is fertilised

SECONDARY OOCYTE (n) → MEIOSIS II → OVUM (n)

At each of the two meiotic divisions one haploid set of chromosomes is packaged in a polar body. The two polar bodies play no part in oogenesis and eventually degenerate.

HUMAN OOCYTE (v.s.)

140 μm

First polar body contains 23 chromosomes from the first meiotic division of the germ wall.

Cumulus cells which once synthesized proteins and nucleic acids into the oocyte cytoplasm.

Zona pellucida will undergo structural changes at fertilization and form a barrier to the entry of more than one sperm.

23 chromosomes will complete second meiotic division on fertilization to provide **female haploid nucleus** and a second polar body (the ovum).

Cytoplasm may contribute extranuclear genes and organelles to the zygote.

Cortical granules contain enzymes which are released at fertilization and alter the structure of the zona pellucida, preventing further sperm penetration, which would upset the just restored diploid number.

Events of the menstrual cycle

Follicle-stimulating hormone (FSH) initiates the development of several primary follicles (each containing a primary oocyte): one follicle continues to develop but the others degenerate by the process of follicular atresia. FSH also increases the activity of the enzymes responsible for formation of oestrogen.

Oestrogen is produced by enzyme modification (in the stroma) of testosterone produced by the thecae of the developing follicle. Oestrogen has several effects:
1. It stimulates further growth of the follicle.
2. It promotes repair of the endometrium.
3. It acts as a feedback inhibitor of the secretion of FSH from the anterior pituitary gland.
4. From about day 11 it has a positive feedback action on the secretion of both LH and FSH.

Development of the follicle within the ovary is initiated by FSH but continued by LH. The Graafian follicle (A) is mature by day 10–11 and ovulation occurs at day 14 (B) following a surge of LH. The remains of the follicle become the corpus luteum (C), which secretes steroid hormones. These steroid hormones inhibit LH secretion so that the corpus luteum degenerates and becomes the corpus albicans (D).

The **endometrium** begins to thicken and become more vascular under the influence of the ovarian hormone oestrogen. Because of this thickening, to 4–6 mm, the time between menstruation and ovulation is sometimes called the *proliferative phase.*

Luteinizing hormone (LH) triggers the secretion of testosterone by the thecae of the follicle, and when its concentration 'surges' it causes release of enzymes which rupture the wall of the ovary, allowing the secondary oocyte to be released at ovulation. After ovulation LH promotes development of the corpus luteum from the remains of the Graafian follicle.

Progesterone is secreted by the corpus luteum. It has several effects:
1. It prepares the endometrium for implantation of a fertilized egg by increasing vascularization, thickening and the storage of glycogen.
2. It begins to promote growth of the mammary glands.
3. It acts as a feedback inhibitor of FSH secretion, thus arresting development of any further follicles.

This is the basis of the contraceptive 'pill'. The pill contains progesterone at a concentration which feedback-inhibits ovulation.

Body temperature rises by about 1°C at the time of ovulation. This 'heat' is used to determine the 'safe period' for the rhythm method of contraception.

During the post-ovulatory or luteal phase the **endometrium** becomes thicker with more tortuously coiled glands and greater vascularization of the surface layer, and retains more tissue fluid.

Menstruation is initiated by falling concentrations of oestrogen and progesterone as the corpus luteum degenerates.

At **menstruation** the *stratum functionalis* of the endometrium is shed, leaving the *stratum basilis* to begin proliferation of a new *functionalis*.

TIME / days

0   14   28

OVARIAN (FOLLICULAR) PHASE   LUTEAL PHASE

A   B   C   D

Implantation and fertilisation

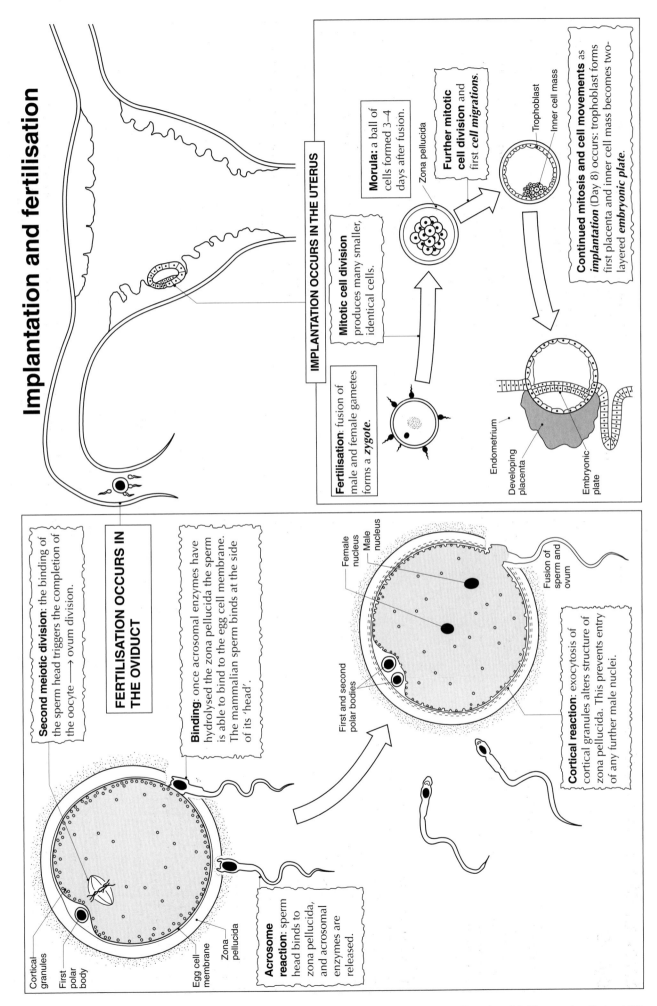

IMPLANTATION OCCURS IN THE UTERUS

Fertilisation: fusion of male and female gametes forms a *zygote*.

Mitotic cell division produces many smaller, identical cells.

Morula: a ball of cells formed 3–4 days after fusion.

Zona pellucida

Further mitotic cell division and first *cell migrations*.

Trophoblast

Inner cell mass

Continued mitosis and cell movements as *implantation* (Day 8) occurs: trophoblast forms first placenta and inner cell mass becomes two-layered *embryonic plate*.

Endometrium

Developing placenta

Embryonic plate

FERTILISATION OCCURS IN THE OVIDUCT

Second meiotic division: the binding of the sperm head triggers the completion of the oocyte → ovum division.

Binding: once acrosomal enzymes have hydrolysed the zona pellucida the sperm is able to bind to the egg cell membrane. The mammalian sperm binds at the side of its 'head'.

Female nucleus

Male nucleus

First and second polar bodies

Fusion of sperm and ovum

Cortical reaction: exocytosis of cortical granules alters structure of zona pellucida. This prevents entry of any further male nuclei.

Cortical granules

First polar body

Egg cell membrane

Zona pellucida

Acrosome reaction: sperm head binds to zona pellucida, and acrosomal enzymes are released.

Methods of contraception

include barrier, surgical, and hormonal devices. Failure rate is expressed per HWY, i.e. the number of pregnancies if 100 women were to use the same method for one year.

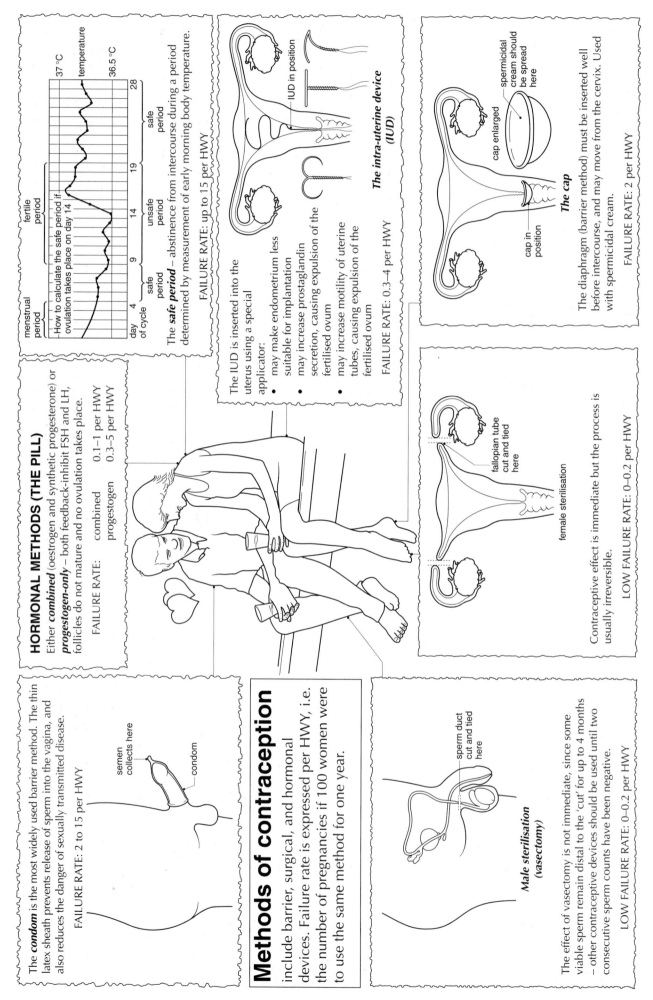

The **condom** is the most widely used barrier method. The thin latex sheath prevents release of sperm into the vagina, and also reduces the danger of sexually transmitted disease.

semen collects here

condom

FAILURE RATE: 2 to 15 per HWY

HORMONAL METHODS (THE PILL)

Either **combined** (oestrogen and synthetic progesterone) or **progestogen-only** – both feedback-inhibit FSH and LH, follicles do not mature and no ovulation takes place.

FAILURE RATE: combined   0.1–1 per HWY
              progestogen 0.3–5 per HWY

temperature

37 °C

36.5 °C

menstrual period

fertile period

How to calculate the safe period if ovulation takes place on day 14

safe period

unsafe period

safe period

day of cycle   4   9   14   19   28

The **safe period** – abstinence from intercourse during a period determined by measurement of early morning body temperature.

FAILURE RATE: up to 15 per HWY

The IUD is inserted into the uterus using a special applicator:

- may make endometrium less suitable for implantation
- may increase prostaglandin secretion, causing expulsion of the fertilised ovum
- may increase motility of uterine tubes, causing expulsion of the fertilised ovum

IUD in position

The intra-uterine device (IUD)

FAILURE RATE: 0.3–4 per HWY

cap enlarged

spermicidal cream should be spread here

cap in position

The cap

The diaphragm (barrier method) must be inserted well before intercourse, and may move from the cervix. Used with spermicidal cream.

FAILURE RATE: 2 per HWY

fallopian tube cut and tied here

female sterilisation

Contraceptive effect is immediate but the process is usually irreversible.

LOW FAILURE RATE: 0–0.2 per HWY

sperm duct cut and tied here

Male sterilisation (vasectomy)

The effect of vasectomy is not immediate, since some viable sperm remain distal to the 'cut' for up to 4 months – other contraceptive devices should be used until two consecutive sperm counts have been negative.

LOW FAILURE RATE: 0–0.2 per HWY

Functions of the placenta

Immune protection: protective molecules (possibly including HCG) cover the surface of the early placenta and 'camouflage' the embryo which, with its complement of paternal genes, might be identified by the maternal immune system and rejected as tissue of 'non-self' origin.

Branch of uterine artery delivers maternal blood to the lacunae. Blood transports oxygen, soluble nutrients, hormones and antibodies, but also drugs and viruses.

Chorionic villi are the sites of exchange of many solutes between maternal and fetal circulations. Oxygen transfer to the fetus is aided by fetal haemoglobin with its high oxygen affinity, and soluble nutrients such as glucose and amino acids are selectively transported by membrane-bound carrier proteins. Carbon dioxide and urea diffuse from fetus to mother along diffusion gradients maintained over larger areas of the placenta by countercurrent flow of fetal and maternal blood. In the later stages of pregnancy antibodies pass from mother to fetus – these confer immunity in the young infant, particularly to some gastro-intestinal infections. **There is no direct contact between maternal and foetal blood.**

Branch of uterine vein removes blood from lacunae. Blood contains increased concentrations of carbon dioxide, urea and placental hormones.

After expulsion of the fetus, further contractions of the uterus cause detachment of the placenta (spontaneous constriction of uterine artery and vein limit blood loss) – the placenta, once delivered, is referred to as the *afterbirth*. The placenta may be used as a source of hormones (e.g. in 'fertility pills'), as tissue for burn grafts and to supply veins for blood vessel grafts.

Umbilical arteries

Umbilical vein

Barrier: the placenta limits the transfer of solutes and blood components from maternal to fetal circulation. Cells of the maternal immune system do not cross – this minimizes the possibility of immune rejection (although antibodies may cross which may cause haemolysis of fetal blood cells if **Rhesus** antibodies are present in the maternal circulation). The placenta is **not** a barrier to heavy metals such as lead, to nicotine, to HIV and other viruses, to heroin and to toxins such as DDT. Thus the **Rubella** (German measles) virus may cross and cause severe damage to eyes, ears, heart and brain of the fetus, the sedative Thalidomide caused severely abnormal limb development, and some children are born already addicted to heroin or HIV positive.

Umbilical cord provides the connection between the abdomen of the foetus and the placenta. It is composed of the two spiralling umbilical arteries (carrying deoxygenated blood from the foetal circulation) and the umbilical vein (returning oxygenated, nutrient-enriched blood to the foetus) embedded in Wharton's jelly.

Endocrine function: cells of the **chorion** secrete a number of hormones:

1. **HCG (human chorionic gonadotrophin)** maintains the corpus luteum so that this body may continue to secrete the progesterone necessary to continue the development of the endometrium. HCG is principally effective during the first 3 months of pregnancy and the overflow of this hormone into the urine is used in diagnosis of the pregnant condition.

2. **Oestrogen and progesterone** are secreted as the production of these hormones from the degenerating corpus luteum diminishes.

3. **Human placental lactogen** promotes milk production in the mammary glands as birth approaches.

4. **Prostaglandins** are released under the influence of fetal adrenal steroid hormones. These prostaglandins are powerful stimulators of contractions of the smooth muscle of the uterus, the contractions which constitute labour and eventually expel the fetus from the uterus.

Amnion is an extra-embryonic membrane which surrounds the umbilical cord and encloses the **amniotic fluid** to form the **amniotic sac**. This fluid supports the embryo and provides samples of embryonic cells via the technique of **amniocentesis**. Rupture of the amnion, and release of the 'waters' is often the first sign that parturition is imminent.

Hormones and parturition

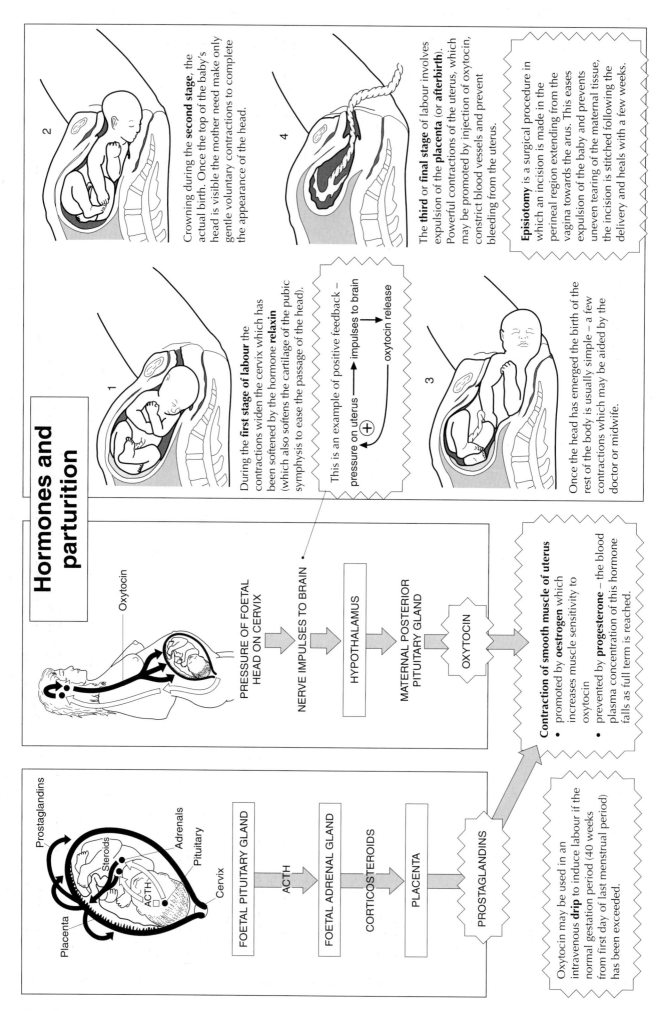

Crowning during the **second stage**, the actual birth. Once the top of the baby's head is visible the mother need make only gentle voluntary contractions to complete the appearance of the head.

The **third** or **final stage** of labour involves expulsion of the **placenta** (or **afterbirth**). Powerful contractions of the uterus, which may be promoted by injection of oxytocin, constrict blood vessels and prevent bleeding from the uterus.

Episiotomy is a surgical procedure in which an incision is made in the perineal region extending from the vagina towards the arus. This eases expulsion of the baby and prevents uneven tearing of the maternal tissue, the incision is stitched following the delivery and heals with a few weeks.

During the **first stage of labour** the contractions widen the cervix which has been softened by the hormone **relaxin** (which also softens the cartilage of the pubic symphysis to ease the passage of the head).

This is an example of positive feedback – pressure on uterus → impulses to brain → oxytocin release (+)

Once the head has emerged the birth of the rest of the body is usually simple – a few contractions which may be aided by the doctor or midwife.

Oxytocin

PRESSURE OF FOETAL HEAD ON CERVIX

NERVE IMPULSES TO BRAIN

HYPOTHALAMUS

MATERNAL POSTERIOR PITUITARY GLAND

OXYTOCIN

Contraction of smooth muscle of uterus
- promoted by **oestrogen** which increases muscle sensitivity to oxytocin
- prevented by **progesterone** – the blood plasma concentration of this hormone falls as full term is reached.

Oxytocin may be used in an intravenous **drip** to induce labour if the normal gestation period (40 weeks from first day of last menstrual period) has been exceeded.

Prostaglandins
Adrenals
Pituitary
Steroids
ACTH
Cervix
Placenta

FOETAL PITUITARY GLAND

ACTH

FOETAL ADRENAL GLAND

CORTICOSTEROIDS

PLACENTA

PROSTAGLANDINS

Measurement of growth

METHODS OF GROWTH MEASUREMENT – all involve measuring some parameter of the individual and then plotting the results against time

- take **samples** from populations wherever possible, and calculate **means** to minimise the effect of any individual result

- choose the correct parameter

Supine length – eliminates the problem of spinal compression or poor muscular tone. Most useful directly comparative linear measurement

Standing height – the most commonly used linear measurement, but does not take any account of growth in other directions (e.g. girth) nor of growth in different parts of the body as a whole.

Body mass – takes volume into account but does not distinguish between the tissue growth (e.g. addition of protein) and temporary changes (e.g. fat disposition). The measured value may also be affected by variable water intake (**dry** mass measurement is not acceptable in humans!)

WHAT IS 'NORMAL' GROWTH?

Polygenes — hormones — 'size' — nutrition

so many variables mean that there is no fixed **growth norm**, but attempts **can** be made to determine boundaries of normal growth patterns

Longitudinal studies measure height and mass of individuals from birth to early adulthood: *very time consuming*

Cross-sectional studies measure height and mass of a large number of individuals of different ages: COMPLETED IN A SHORT TIME **but**
a. *samples may be unrepresentative and*
b. *nutritional factors vary from one generation to another.*

Actual growth – this is the cumulative increase in the measured parameter over a period of time. The general pattern is typically an S-shaped curve – growth stops altogether when adulthood is reached typically in the early twenties.

Time/years	Mass/kg	Growth increment/kg
9	32.0	
10	33.0	1.0
11	34.5	1.5
12	36.5	2.0
13	40.0	3.5
14	45.0	5.0
15	48.5	3.5
16	50.5	2.0
17	52.0	1.5
18	53.0	1.0

TIME/YEARS

RATE OF GROWTH this is a re-expression of the basic growth curve. It describes the **change** in the measured parameter which takes place during successive time intervals i.e. it plots GROWTH INCREMENTS against TIME

Growth patterns in humans

Changes in FORM (the **qualitative** changes alongside growth) are caused by two factors

Timing: when do organs commence growth

Rate: organs may grow at different rates from one another and from the body as a whole = ALLOMETRIC GROWTH

LYMPHOID TISSUE: this tissue is found throughout the body in lymph nodes, and aggregated in tonsils, adenoids and thymus gland. A major function is the production of antibodies and the rapid growth up to the early teens corresponds to the child's rapid acquisition of immunity to a considerable number of infectious conditions which may be encountered for the first time.

BRAIN: at birth the brain is already 25% of the eventual adult size, by one year it is 75% of the adult size and by 7–8 years of age brain growth is almost complete. At birth the brain contains its full complement of neurones so that brain growth after birth is due to increase in

- size (length) of individual cells
- number of subsidiary fibres (e.g. dendrons) as cross-connections are formed between cells
- development of the insulating myelin sheath around the neurones.

REPRODUCTIVE ORGANS: both **primary** (ovary, testis) and **secondary** (vulva and breasts, penis) sexual organs show little change until puberty. At this time PITUITARY GONADOTROPHINS followed by SEX STEROIDS (oestrogen, testosterone) initiate rapid growth which continues through adolescence.

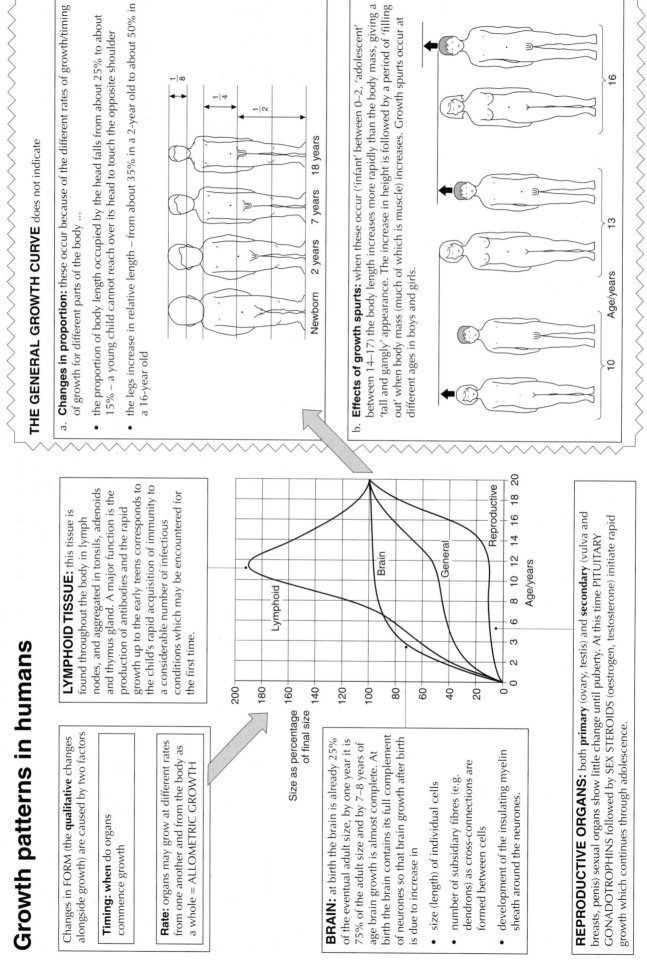

Size as percentage of final size

THE GENERAL GROWTH CURVE does not indicate

a. **Changes in proportion:** these occur because of the different rates of growth/timing of growth for different parts of the body …

- the proportion of body length occupied by the head falls from about 25% to about 15% – a young child cannot reach over its head to touch the opposite shoulder
- the legs increase in relative length – from about 35% in a 2-year old to about 50% in a 16-year old

Newborn   2 years   7 years   18 years

b. **Effects of growth spurts:** when these occur ('infant' between 0–2, 'adolescent' between 14–17) the body length increases more rapidly than the body mass, giving a 'tall and gangly' appearance. The increase in height is followed by a period of 'filling out' when body mass (much of which is muscle) increases. Growth spurts occur at different ages in boys and girls.

Ageing

involves **age-related** (i.e. programmed from birth) and **age-associated** (accumulated throughout life) changes

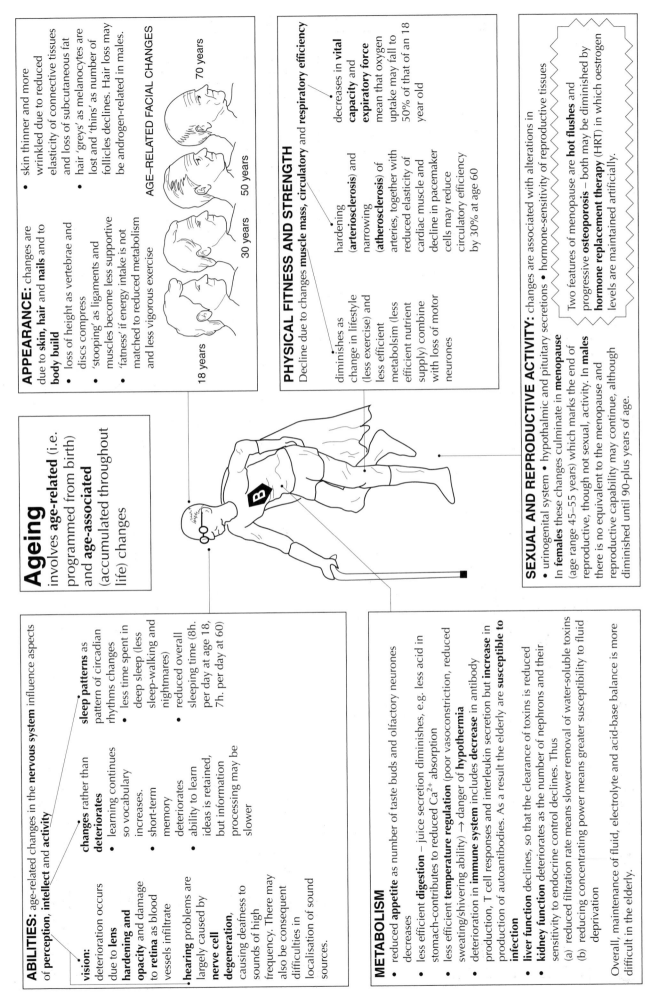

APPEARANCE: changes are due to **skin, hair** and **nails** and to **body build**

- loss of height as vertebrae and discs compress
- 'stooping' as ligaments and muscles become less supportive
- 'fatness' if energy intake is not matched to reduced metabolism and less vigorous exercise

- skin thinner and more wrinkled due to reduced elasticity of connective tissues and loss of subcutaneous fat
- hair 'greys' as melanocytes are lost and 'thins' as number of follicles declines. Hair loss may be androgen-related in males.

AGE-RELATED FACIAL CHANGES

18 years   30 years   50 years   70 years

ABILITIES: age-related changes in the **nervous system** influence aspects of **perception, intellect** and **activity**

vision: deterioration occurs due to **lens hardening and opacity** and damage to **retina** as blood vessels infiltrate

hearing problems are largely caused by **nerve cell degeneration,** causing deafness to sounds of high frequency. There may also be consequent difficulties in localisation of sound sources.

changes rather than **deteriorates**
- learning continues so vocabulary increases.
- short-term memory deteriorates
- ability to learn ideas is retained, but information processing may be slower

sleep patterns as pattern of circadian rhythms changes
- less time spent in deep sleep (less sleep-walking and nightmares)
- reduced overall sleeping time (8h. per day at age 18, 7h. per day at 60)

METABOLISM

- reduced **appetite** as number of taste buds and olfactory neurones decreases
- less efficient **digestion** – juice secretion diminishes, e.g. less acid in stomach–contributes to reduced Ca^{2+} absorption
- less efficient **temperature regulation** (poor vasoconstriction, reduced sweating/shivering ability) → danger of **hypothermia**
- deterioration in **immune system** includes **decrease** in antibody production, T cell responses and interleukin secretion but **increase** in production of autoantibodies. As a result the elderly are **susceptible to infection**
- **liver function** declines, so that the clearance of toxins is reduced
- **kidney function** deteriorates as the number of nephrons and their sensitivity to endocrine control declines. Thus
  (a) reduced filtration rate means slower removal of water-soluble toxins
  (b) reducing concentrating power means greater susceptibility to fluid deprivation

Overall, maintenance of fluid, electrolyte and acid-base balance is more difficult in the elderly.

PHYSICAL FITNESS AND STRENGTH
Decline due to changes **muscle mass, circulatory** and **respiratory efficiency**

diminishes as change in lifestyle (less exercise) and less efficient metabolism (less efficient nutrient supply) combine with loss of motor neurones

hardening (**arteriosclerosis**) and narrowing (**atherosclerosis**) of arteries, together with reduced elasticity of cardiac muscle and decline in pacemaker cells may reduce circulatory efficiency by 30% at age 60

decreases in **vital capacity** and **expiratory force** mean that oxygen uptake may fall to 50% of that of an 18 year old

SEXUAL AND REPRODUCTIVE ACTIVITY: changes are associated with alterations in
- urinogenital system • hypothalmic and pituitary secretions • hormone-sensitivity of reproductive tissues
In **females** these changes culminate in **menopause** (age range 45–55 years) which marks the end of reproductive, though not sexual, activity. In **males** there is no equivalent to the menopause and reproductive capability may continue, although diminished until 90-plus years of age.

Two features of menopause are **hot flushes** and progressive **osteoporosis** – both may be diminished by **hormone replacement therapy** (HRT) in which oestrogen levels are maintained artificially.

Experiments on DNA function

GRIFFITH'S EXPERIMENT: DNA IS THE GENETIC MATERIAL

Griffith used two strains of *Pneumococcus*:

Smooth strain has capsule and causes pneumonia

Rough strain has no capsule and is harmless

BOTH STRAINS CAN BE KILLED BY HEAT TREATMENT.

Experiment 1

Living rough pneumococcus bacteria injected into mouse

Mouse remains healthy

Interpretation
Rough pneumococcus bacteria are not infective.

Experiment 2

Living smooth pneumococcus bacteria injected into mouse

Mouse gets pneumonia – smooth pneumococcus bacteria isolated from dead mouse

Interpretation
Smooth pneumococcus bacteria are infective.

Experiment 3

Heat-killed smooth pneumococcus bacteria injected into mouse

Mouse remains healthy

Interpretation
Smooth pneumococcus bacteria that are killed by heat are not infective.

Experiment 4

Living rough and heat-killed smooth pneumococcus bacteria injected into mouse

Mouse gets pneumonia – smooth pneumococcus bacteria isolated from dead mouse

Interpretation
Non-infective rough bacteria have been transformed into smooth bacteria as a result of being mixed with heat-killed smooth bacteria.

The *transforming principle* discovered by Griffith

a. is *not* affected by *protease*
b. *is* destroyed by *DNAase*
i.e. *the transforming principle is DNA.*

Genetic engineering shows that transfer of DNA alters characteristics of organisms.

This is the most compelling evidence that DNA is the genetic material.

MESELSOHN AND STAHL DEMONSTRATE THAT DNA REPLICATION IS SEMI-CONSERVATIVE

- Used *Escherischia coli*, a harmless gut bacterium
- Grew colonies of *E.coli* with NH_4Cl as the nitrogen source for DNA synthesis
- Were able to use *density gradient centrifugation* to identify DNA
- A 'heavy' (not radioactive) isotope of nitrogen, ^{15}N, is available.

DNA extracted from bacteria grown on $^{15}NH_4Cl$. All DNA is 'heavy'

DNA molecules have two 'heavy' strands

Bacteria grown on $^{15}NH_4Cl$ for many generations are transferred to $^{14}NH_4Cl$ for several generations.

After one generation all DNA is 'intermediate' in mass

After two generations ½ DNA is 'intermediate' and ½ is 'light

'light' strand

'heavy' strand

- The presence of 'intermediate' DNA after growth in $^{14}NH_4Cl$ supports the *semi-conservative principle*.
- *Conservative replication* would produce two bands, one 'heavy' and one 'light', after one generation in $^{14}NH_4Cl$.

DNA replication and chromosomes

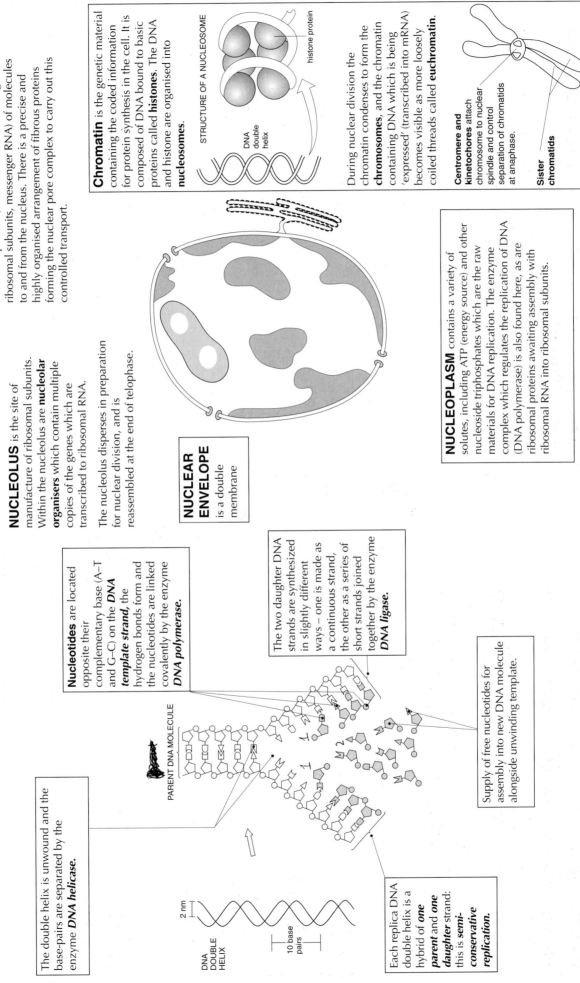

NUCLEAR PORE can regulate the entry (e.g. ribosomal proteins, nucleotides) and exit (e.g., ribosomal subunits, messenger RNA) of molecules to and from the nucleus. There is a precise and highly organised arrangement of fibrous proteins forming the nuclear pore complex to carry out this controlled transport.

Chromatin is the genetic material containing the coded information for protein synthesis in the cell. It is composed of DNA bound to basic proteins called **histones**. The DNA and histone are organised into **nucleosomes.**

STRUCTURE OF A NUCLEOSOME

histone protein

DNA double helix

During nuclear division the chromatin condenses to form the **chromosomes**, and the chromatin containing DNA which is being 'expressed' (transcribed into mRNA) becomes visible as more loosely coiled threads called **euchromatin.**

Centromere and kinetochores attach chromosome to nuclear spindle and control separation of chromatids at anaphase.

Sister chromatids

NUCLEOLUS is the site of manufacture of ribosomal subunits. Within the nucleolus are **nucleolar organisers** which contain multiple copies of the genes which are transcribed to ribosomal RNA.

The nucleolus disperses in preparation for nuclear division, and is reassembled at the end of telophase.

NUCLEAR ENVELOPE is a double membrane

NUCLEOPLASM contains a variety of solutes, including ATP (energy source) and other nucleoside triphosphates which are the raw materials for DNA replication. The enzyme complex which regulates the replication of DNA (DNA polymerase) is also found here, as are ribosomal proteins awaiting assembly with ribosomal RNA into ribosomal subunits.

Nucleotides are located opposite their complementary base (A–T and G–C) on the *DNA template strand*, the hydrogen bonds form and the nucleotides are linked covalently by the enzyme *DNA polymerase.*

The two daughter DNA strands are synthesized in slightly different ways – one is made as a continuous strand, the other as a series of short strands joined together by the enzyme *DNA ligase.*

The double helix is unwound and the base-pairs are separated by the enzyme *DNA helicase.*

PARENT DNA MOLECULE

Supply of free nucleotides for assembly into new DNA molecule alongside unwinding template.

2 nm

DNA DOUBLE HELIX

10 base pairs

Each replica DNA double helix is a hybrid of *one parent* and *one daughter* strand: this is *semi-conservative replication.*

Genes control cell characteristics

DNA in nucleus contains the *genes* (specific sequences of bases which code for particular proteins), *regulators* (which switch genes 'on' or 'off') and many stretches with no apparent function (*junk* or *redundant DNA*).

DNA $\xrightarrow{\text{TRANSCRIPTION}}$ RNA $\xrightarrow{\text{TRANSLATION}}$ PROTEIN $\downarrow$ CHARACTERISTICS

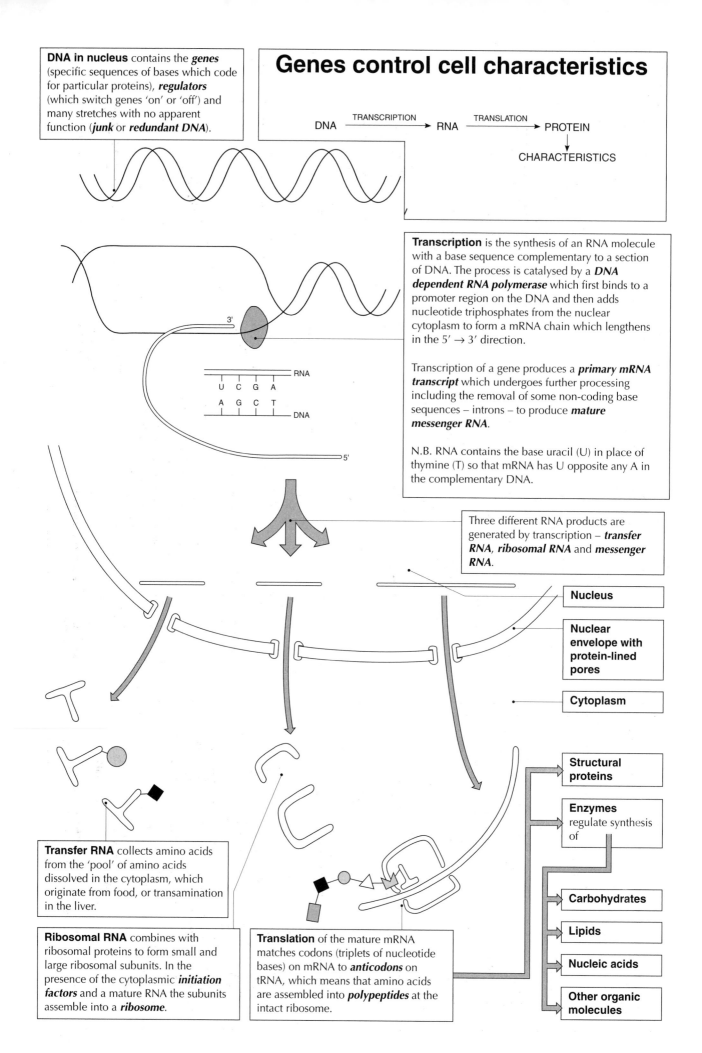

Transcription is the synthesis of an RNA molecule with a base sequence complementary to a section of DNA. The process is catalysed by a **DNA dependent RNA polymerase** which first binds to a promoter region on the DNA and then adds nucleotide triphosphates from the nuclear cytoplasm to form a mRNA chain which lengthens in the $5' \rightarrow 3'$ direction.

Transcription of a gene produces a **primary mRNA transcript** which undergoes further processing including the removal of some non-coding base sequences – introns – to produce **mature messenger RNA**.

N.B. RNA contains the base uracil (U) in place of thymine (T) so that mRNA has U opposite any A in the complementary DNA.

Three different RNA products are generated by transcription – **transfer RNA**, **ribosomal RNA** and **messenger RNA**.

Nucleus

Nuclear envelope with protein-lined pores

Cytoplasm

Structural proteins

Enzymes regulate synthesis of

Carbohydrates

Lipids

Nucleic acids

Other organic molecules

Transfer RNA collects amino acids from the 'pool' of amino acids dissolved in the cytoplasm, which originate from food, or transamination in the liver.

Ribosomal RNA combines with ribosomal proteins to form small and large ribosomal subunits. In the presence of the cytoplasmic **initiation factors** and a mature RNA the subunits assemble into a **ribosome**.

Translation of the mature mRNA matches codons (triplets of nucleotide bases) on mRNA to **anticodons** on tRNA, which means that amino acids are assembled into **polypeptides** at the intact ribosome.

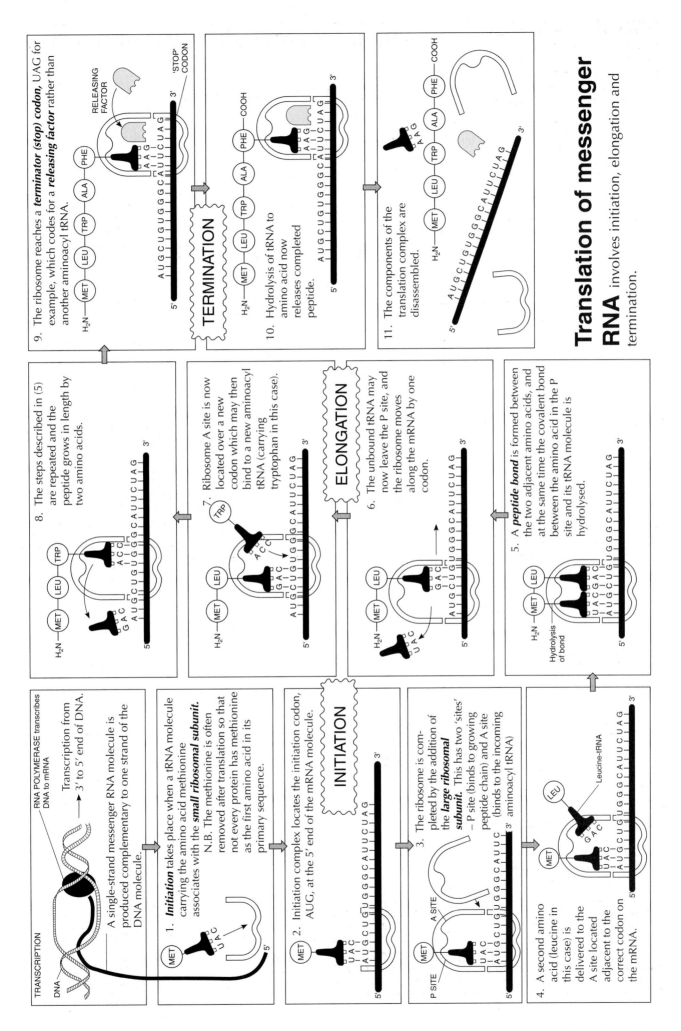

Translation of messenger RNA

involves initiation, elongation and termination.

TRANSCRIPTION

RNA POLYMERASE transcribes DNA to mRNA

Transcription from → 3' to 5' end of DNA.

A single-strand messenger RNA molecule is produced complementary to one strand of the DNA molecule.

INITIATION

1. **Initiation** takes place when a tRNA molecule carrying the amino acid methionine associates with the **small ribosomal subunit.** N.B. The methionine is often removed after translation so that not every protein has methionine as the first amino acid in its primary sequence.

2. Initiation complex locates the initiation codon, AUG, at the 5' end of the mRNA molecule.

3. The ribosome is completed by the addition of the **large ribosomal subunit.** This has two 'sites' – P site (binds to growing peptide chain) and A site (binds to the incoming aminoacyl tRNA)

4. A second amino acid (leucine in this case) is delivered to the A site located adjacent to the correct codon on the mRNA.

Leucine-tRNA

ELONGATION

5. A **peptide bond** is formed between the two adjacent amino acids, and at the same time the covalent bond between the amino acid in the P site and its tRNA molecule is hydrolysed.

Hydrolysis of bond

6. The unbound tRNA may now leave the P site, and the ribosome moves along the mRNA by one codon.

7. Ribosome A site is now located over a new codon which may then bind to a new aminoacyl tRNA (carrying tryptophan in this case).

8. The steps described in (5) are repeated and the peptide grows in length by two amino acids.

9. The ribosome reaches a **terminator (stop) codon,** UAG for example, which codes for a **releasing factor** rather than another aminoacyl tRNA.

RELEASING FACTOR

'STOP' CODON

TERMINATION

10. Hydrolysis of tRNA to amino acid now releases completed peptide.

11. The components of the translation complex are disassembled.

Regulation of gene activity:
control of transcription of DNA to mRNA.

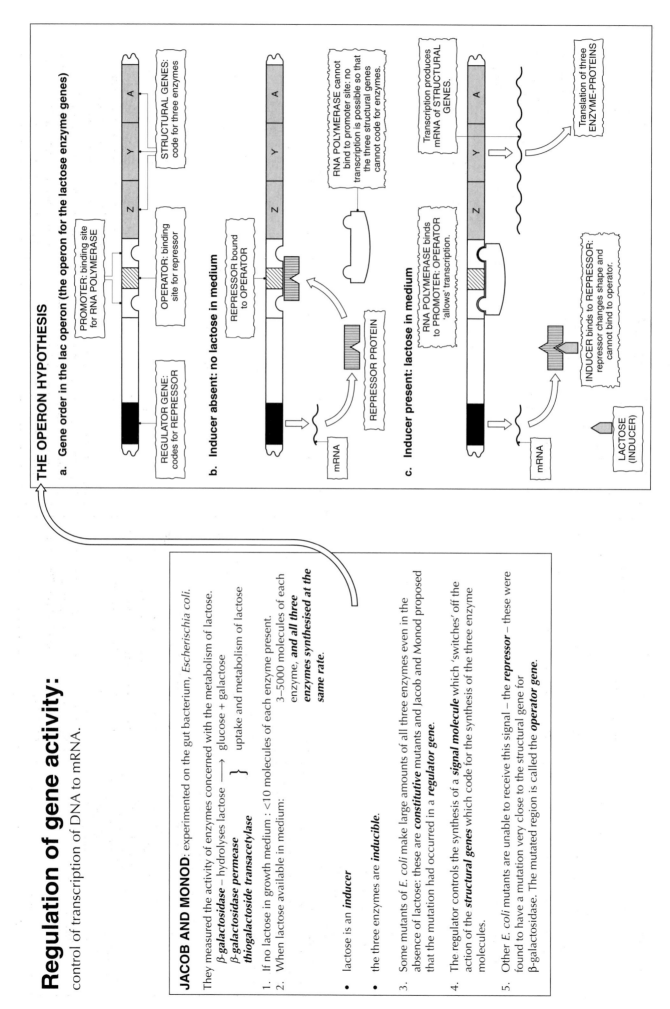

THE OPERON HYPOTHESIS

a. Gene order in the lac operon (the operon for the lactose enzyme genes)

PROMOTER: binding site for RNA POLYMERASE

OPERATOR: binding site for repressor

STRUCTURAL GENES: code for three enzymes

REGULATOR GENE: codes for REPRESSOR

b. Inducer absent: no lactose in medium

RNA POLYMERASE cannot bind to promoter site: no transcription is possible so that the three structural genes cannot code for enzymes.

REPRESSOR bound to OPERATOR

mRNA

REPRESSOR PROTEIN

c. Inducer present: lactose in medium

Transcription produces mRNA of STRUCTURAL GENES.

Translation of three ENZYME-PROTEINS

RNA POLYMERASE binds to PROMOTER: OPERATOR 'allows' transcription.

INDUCER binds to REPRESSOR: repressor changes shape and cannot bind to operator.

mRNA

LACTOSE (INDUCER)

JACOB AND MONOD: experimented on the gut bacterium, *Escherischia coli*.

They measured the activity of enzymes concerned with the metabolism of lactose.

β-galactosidase – hydrolyses lactose ⟶ glucose + galactose
β-galactosidase permease
thiogalactoside transacetylase } uptake and metabolism of lactose

1. If no lactose in growth medium : <10 molecules of each enzyme present.
2. When lactose available in medium: 3–5000 molecules of each enzyme, ***and all three enzymes synthesised at the same rate.***

*   lactose is an ***inducer***

*   the three enzymes are ***inducible***.

3. Some mutants of *E. coli* make large amounts of all three enzymes even in the absence of lactose: these are ***constitutive*** mutants and Jacob and Monod proposed that the mutation had occurred in a ***regulator gene***.

4. The regulator controls the synthesis of a ***signal molecule*** which 'switches' off the action of the ***structural genes*** which code for the synthesis of the three enzyme molecules.

5. Other *E. coli* mutants are unable to receive this signal – the ***repressor*** – these were found to have a mutation very close to the structural gene for β-galactosidase. The mutated region is called the ***operator gene***.

Mitosis and growth

The significance of mitosis is that it involves duplication of the genetic material and its equal distribution to each of two 'daughter' cells: *variation is minimal.*

Importance of mitosis

1. It is the process which provides the cells required for the *growth* of zygote into a functioning multicellular organism – this requires an increase in number of cells from one to 6 x 10^{13} in humans.

2. It supplies the cells to *repair* worn out or damaged tissues. In the human the replacement of skin, gut and lung linings and blood cells requires about 1 x 10^{11} cells per day.

3. It maintains the chromosome number. Daughter cells have identical sets of chromosomes and so function harmoniously as part of the tissue, organ or organism.

4. *Asexual reproduction* provides offspring which are genetically identical to the parent – ideal when rapidly establishing a population. Mitosis provides the cells which make up the fragments of the parent body dispersed during this form of reproduction.

The cell cycle typically lasts from 8 to 24 h in humans – the nuclear division (mitosis) occupies about 10% of this time.

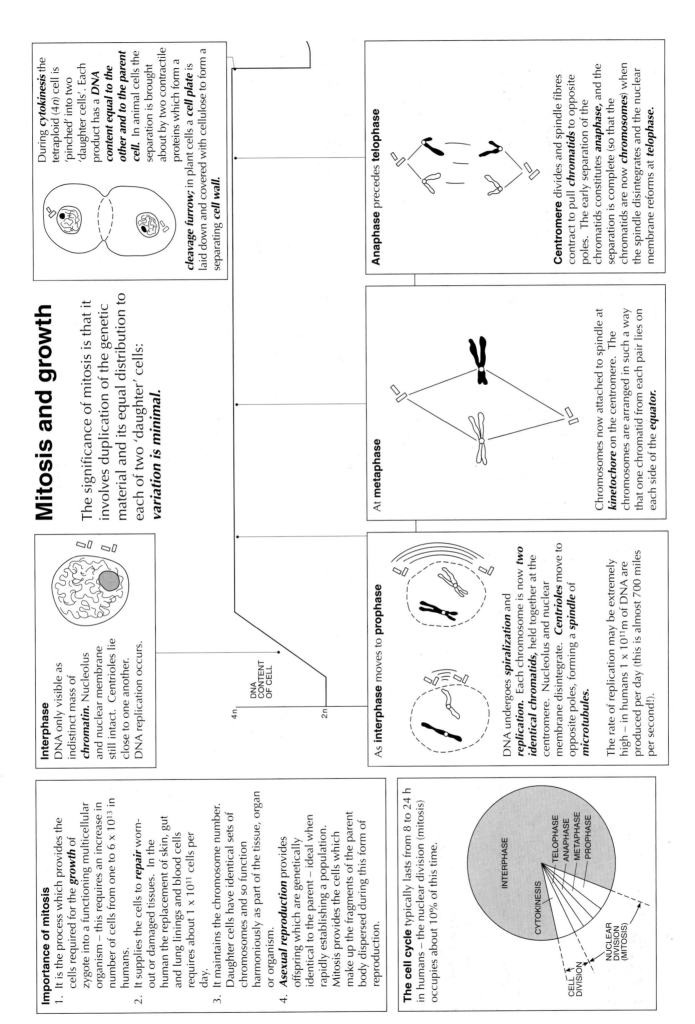

Interphase
DNA only visible as indistinct mass of *chromatin.* Nucleolus and nuclear membrane still intact. Centrioles lie close to one another. DNA replication occurs.

As **interphase** moves to **prophase**

DNA undergoes *spiralization* and *replication.* Each chromosome is now *two identical chromatids,* held together at the centromere. Nucleolus and nuclear membrane disintegrate. *Centrioles* move to opposite poles, forming a *spindle* of *microtubules.*

The rate of replication may be extremely high – in humans 1 x 10^{11}m of DNA are produced per day (this is almost 700 miles per second!).

At metaphase

Chromosomes now attached to spindle at *kinetochore* on the centromere. The chromosomes are arranged in such a way that one chromatid from each pair lies on each side of the *equator.*

Anaphase precedes **telophase**

Centromere divides and spindle fibres contract to pull *chromatids* to opposite poles. The early separation of the chromatids constitutes *anaphase,* and the separation is complete (so that the chromatids are now *chromosomes*) when the spindle disintegrates and the nuclear membrane reforms at *telophase.*

During *cytokinesis* the tetraploid (4n) cell is 'pinched' into two 'daughter cells'. Each product has a *DNA content equal to the other and to the parent cell.* In animal cells the separation is brought about by two contractile proteins which form a *cell plate* is

cleavage furrow; in plant cells a *cell plate* is laid down and covered with cellulose to form a separating *cell wall.*

DNA
CONTENT
OF CELL

4n

2n

Meiosis and variation

Meiosis separates chromosomes, halving the diploid number, and introduces variation to the haploid products.

During **prophase I** each replicated chromosome (comprising two chromatids) pairs with its **homologous partner**, i.e. the diploid number of chromosomes produces the haploid number of homologous pairs.

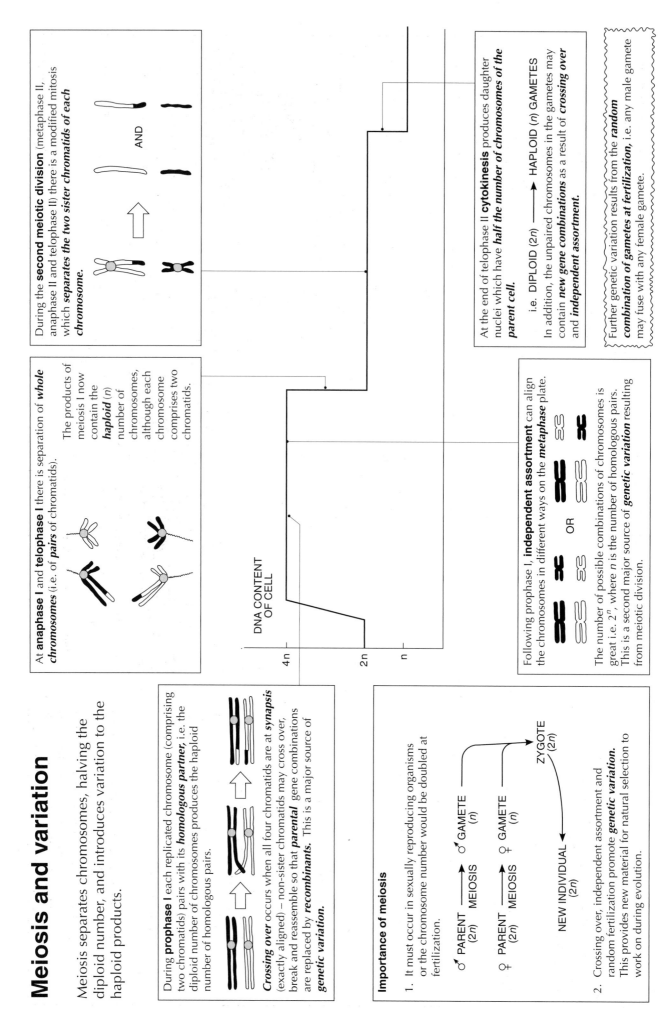

Crossing over occurs when all four chromatids are at *synapsis* (exactly aligned) – non-sister chromatids may cross over, break and reassemble so that *parental* gene combinations are replaced by *recombinants*. This is a major source of *genetic variation*.

At **anaphase I** and **telophase I** there is separation of *whole chromosomes* (i.e. of *pairs* of chromatids).

The products of meiosis I now contain the *haploid (n)* number of chromosomes, although each chromosome comprises two chromatids.

During the **second meiotic division** (metaphase II, anaphase II and telophase II) there is a modified mitosis which *separates the two sister chromatids of each chromosome.*

AND

DNA CONTENT OF CELL

4n

2n

n

Following prophase I, **independent assortment** can align the chromosomes in different ways on the **metaphase** plate.

OR

The number of possible combinations of chromosomes is great i.e. 2^n, where n is the number of homologous pairs. This is a second major source of *genetic variation* resulting from meiotic division.

At the end of telophase II **cytokinesis** produces daughter nuclei which have *half the number of chromosomes of the parent cell.*

i.e. DIPLOID (2n) $\longrightarrow$ HAPLOID (n) GAMETES

In addition, the unpaired chromosomes in the gametes may contain *new gene combinations* as a result of *crossing over* and *independent assortment.*

Further genetic variation results from the *random combination of gametes at fertilization*, i.e. any male gamete may fuse with any female gamete.

Importance of meiosis

1. It must occur in sexually reproducing organisms or the chromosome number would be doubled at fertilization.

♂ PARENT $\longrightarrow$ ♂ GAMETE
(2n)   MEIOSIS   (n)

♀ PARENT $\longrightarrow$ ♀ GAMETE
(2n)   MEIOSIS   (n)

ZYGOTE (2n)

NEW INDIVIDUAL (2n)

2. Crossing over, independent assortment and random fertilization promote *genetic variation.* This provides new material for natural selection to work on during evolution.

Variation is the basis of evolution.

Origins of variation

May be **non-heritable** (e.g. sunburn in a light-skinned individual) or **heritable** (e.g. skin colour in different races). The second type, which result from genetic changes, are the most significant in evolution.

Mutation is any change in the structure or the amount of DNA in an organism.

A **gene** or **point mutation** occurs at a single locus on a chromosome – most commonly by **deletion**, **addition** or **substitution** of a nucleotide base. Examples are sickle cell anaemia, phenylketonuria and cystic fibrosis.

A **change in chromosome structure** occurs when a substantial portion of a chromosome is altered. For example, Cri-du-chat syndrome results from **deletion** of a part of human chromosome 5, and a form of white blood cell cancer follows **translocation** of a portion of chromosome 8 to chromosome 14.

Aneuploidy (typically the **loss or gain of a single chromosome**) results from **non-disjunction** in which chromosomes fail to separate at anaphase of meiosis. The best known examples are Down's syndrome (extra chromosome 21), Klinefelter's syndrome (male with extra X chromosome) and Turner's syndrome (female with one fewer X chromosome).

Polyploidy (the presence of additional **whole sets of chromosomes**) most commonly occurs when one or both gametes is diploid, forming a polyploid on fertilization. Polyploidy is rare in animals, but there are many important examples in plants, e.g. bananas are triploid, and tetraploid tomatoes are larger and richer in vitamin C.

Discontinuous variation occurs when a characteristic is either present or absent (the two extremes) and there are no intermediate forms.

Such variations do not give normal distribution curves but bar charts are often used to illustrate the distribution of a particular characteristic in a population.

Examples are human blood groups in the ABO system (O, A, B or AB), basic fingerprint forms (loop, whorl or arch) and tongue-rolling (can or cannot).

A characteristic which shows discontinuous variation is normally controlled by a single gene – there may be two or more alleles of this gene.

Continuous variation occurs when there is a gradation between one extreme and the other of some given characteristic – all individuals exhibit the characteristic but to differing extents.

If a frequency distribution is plotted for such a characteristic a **normal** or **Gaussian distribution** is obtained.

The **mean** is the average number of such a group (i.e. the total number of individuals divided by the number of groups), the **mode** is the most common of the groups and the **median** is the central value of a set of values.

Typical examples are height, mass, handspan, or number of leaves on a plant.

Characteristics which show continuous variation are controlled by the combined effect of a number of genes, called **polygenes**, and are therefore **polygenic characteristics.**

Sexual or genetic recombination is a most potent force in evolution, since it reshuffles genes into new combinations. It may involve

free assortment in gamete formation
crossing over during meiosis
random fusion during zygote formation.

Gene mutation and sickle cell anaemia

Sickle cell anaemia is the result of a ***single gene (point) mutation*** and the resulting ***error in protein synthesis.***

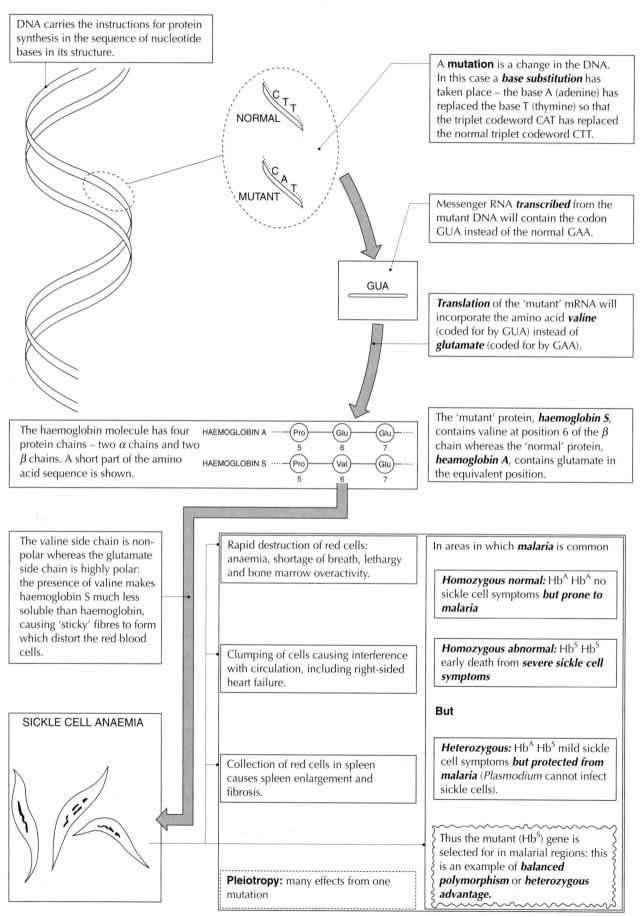

DNA carries the instructions for protein synthesis in the sequence of nucleotide bases in its structure.

A **mutation** is a change in the DNA. In this case a ***base substitution*** has taken place – the base A (adenine) has replaced the base T (thymine) so that the triplet codeword CAT has replaced the normal triplet codeword CTT.

NORMAL

MUTANT

Messenger RNA ***transcribed*** from the mutant DNA will contain the codon GUA instead of the normal GAA.

GUA

Translation of the 'mutant' mRNA will incorporate the amino acid ***valine*** (coded for by GUA) instead of ***glutamate*** (coded for by GAA).

The haemoglobin molecule has four protein chains – two α chains and two β chains. A short part of the amino acid sequence is shown.

HAEMOGLOBIN A ···· Pro — Glu — Glu ····
                    5   6   7

HAEMOGLOBIN S ···· Pro — Val — Glu ····
                    5   6   7

The 'mutant' protein, ***haemoglobin S***, contains valine at position 6 of the β chain whereas the 'normal' protein, ***heamoglobin A***, contains glutamate in the equivalent position.

The valine side chain is non-polar whereas the glutamate side chain is highly polar: the presence of valine makes haemoglobin S much less soluble than haemoglobin, causing 'sticky' fibres to form which distort the red blood cells.

Rapid destruction of red cells: anaemia, shortage of breath, lethargy and bone marrow overactivity.

In areas in which ***malaria*** is common

Homozygous normal: HbA HbA no sickle cell symptoms ***but prone to malaria***

Clumping of cells causing interference with circulation, including right-sided heart failure.

Homozygous abnormal: HbS HbS early death from ***severe sickle cell symptoms***

But

SICKLE CELL ANAEMIA

Heterozygous: HbA HbS mild sickle cell symptoms ***but protected from malaria*** (*Plasmodium* cannot infect sickle cells).

Collection of red cells in spleen causes spleen enlargement and fibrosis.

Thus the mutant (HbS) gene is selected for in malarial regions: this is an example of ***balanced polymorphism*** or ***heterozygous advantage.***

Pleiotropy: many effects from one mutation

Chromosome mutation and Down's syndrome

Down's syndrome (trisomy-21) is a chromosome mutation caused by **non-disjunction**.

In a normal meiotic division chromosomes are distributed equally between the gametes.

CELL IN GONADS

Diploid number of chromosomes: 46 in humans

MEIOSIS

Haploid number $n = 23$ SEX CELL   Haploid number $n = 23$

During **non-disjunction** there is an uneven distribution of the parental chromosomes at meiosis.

CELL IN GONADS

46

MEIOSIS

24   SEX CELL   22

has both parental chromosomes 21

IF FERTILIZATION OCCURS

Trisomic cell 47 three chromosomes 21

Down's syndrome (trisomy-21) is characterized by a number of distinctive physical features.

BROAD FOREHEAD

FOLD IN EYELID

SPOTS IN IRIS

DOWNWARD–SLOPING EYES

SHORT NOSE

PROTRUDING TONGUE

SHORT NECK

In addition there are congenital heart defects (30% of sufferers die before the age of 10) and mental retardation.

A **karyotype** is obtained by cutting out and rearranging photographic images of chromosomes stained during mitotic metaphase.

KARYOTYPE OF DOWN'S SYNDROME FEMALE

1 2 3   4 5

6 7 8 9 10 11 12

13 14 15   16 17 18   19 20

21 22   X Y

NOTE THE EXTRA CHROMOSOME 21

The non-disjunction is most usually the result of the failure of chromosomes to separate at anaphase I – the probability of this happening increases with the length of time the cell remains in prophase I. In the human female all meioses are initiated before puberty so there is an age-related incidence of Down's syndrome – the longer the oocyte takes to complete development the less accurate are the chromosome separations which follow.

Paternal non-disjunction accounts for only about 15% of cases of Down's syndrome.

INCIDENCE OF DOWN'S SYNDROME

0 20 30 40 50
MOTHER'S AGE / years

Two other significant examples of non-disjunction

Klinefelter's syndrome (XXY) caused by an **extra X chromosome** and resulting in a **sterile male** with **some breast development**.

Turner's syndrome (XO) caused by a **deleted X chromosome** and resulting in a **female** with **underdeveloped sexual characteristics**.

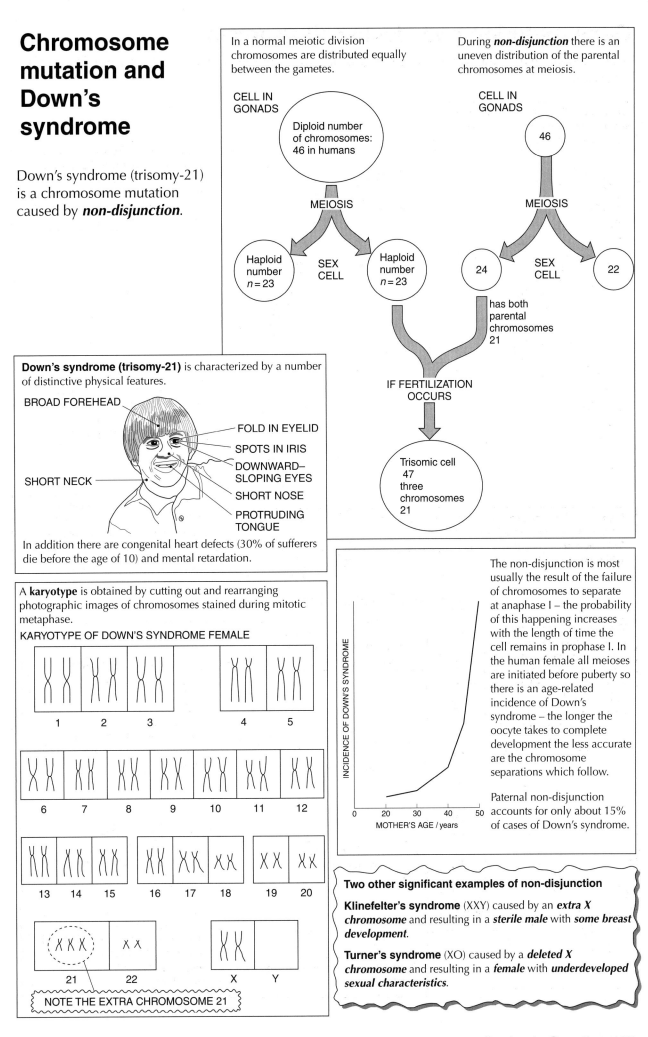

REPLICATION is DNA $\longrightarrow$ DNA and occurs during *mitosis* and *meiosis*.

DON'T CONFUSE THESE PROCESSES!

TRANSCRIPTION is DNA $\longrightarrow$ mRNA

TRANSLATION is mRNA $\longrightarrow$ protein

and occur during *protein synthesis*.

The **GENOTYPE** of a cell or organism is the total DNA in its genes.

The **PHENOTYPE** of an organism is the total of its measurable characteristics.

PHENOTYPE results from *genotype + effects of the environment*.

Definitions in genetics

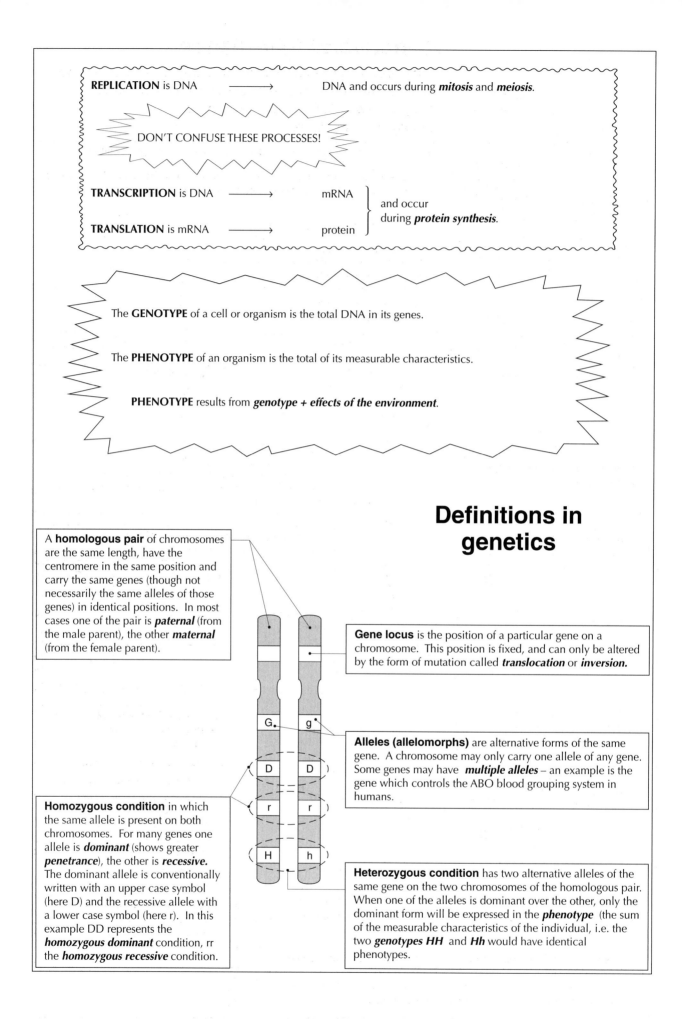

A **homologous pair** of chromosomes are the same length, have the centromere in the same position and carry the same genes (though not necessarily the same alleles of those genes) in identical positions. In most cases one of the pair is *paternal* (from the male parent), the other *maternal* (from the female parent).

Gene locus is the position of a particular gene on a chromosome. This position is fixed, and can only be altered by the form of mutation called *translocation* or *inversion.*

Alleles (allelomorphs) are alternative forms of the same gene. A chromosome may only carry one allele of any gene. Some genes may have **multiple alleles** – an example is the gene which controls the ABO blood grouping system in humans.

Homozygous condition in which the same allele is present on both chromosomes. For many genes one allele is *dominant* (shows greater *penetrance*), the other is *recessive.* The dominant allele is conventionally written with an upper case symbol (here D) and the recessive allele with a lower case symbol (here r). In this example DD represents the *homozygous dominant* condition, rr the *homozygous recessive* condition.

Heterozygous condition has two alternative alleles of the same gene on the two chromosomes of the homologous pair. When one of the alleles is dominant over the other, only the dominant form will be expressed in the *phenotype* (the sum of the measurable characteristics of the individual, i.e. the two *genotypes HH* and *Hh* would have identical phenotypes.

Monohybrid inheritance

Cystic fibrosis is caused by faulty ion transport, and is the most common inherited fatal disease in Caucasian populations. ***Phenylketonuria in humans*** is an example of ***monohybrid inheritance.***

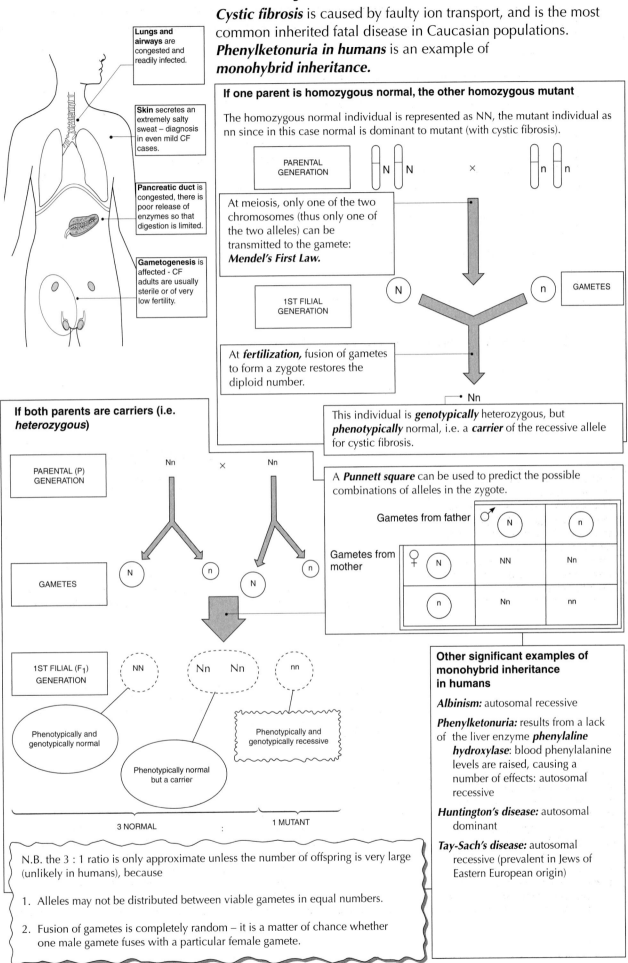

Lungs and airways are congested and readily infected.

Skin secretes an extremely salty sweat – diagnosis in even mild CF cases.

Pancreatic duct is congested, there is poor release of enzymes so that digestion is limited.

Gametogenesis is affected - CF adults are usually sterile or of very low fertility.

If one parent is homozygous normal, the other homozygous mutant

The homozygous normal individual is represented as NN, the mutant individual as nn since in this case normal is dominant to mutant (with cystic fibrosis).

PARENTAL GENERATION   N N   ×   n n

At meiosis, only one of the two chromosomes (thus only one of the two alleles) can be transmitted to the gamete: ***Mendel's First Law.***

1ST FILIAL GENERATION

N   n   GAMETES

At ***fertilization,*** fusion of gametes to form a zygote restores the diploid number.

• Nn

This individual is ***genotypically*** heterozygous, but ***phenotypically*** normal, i.e. a ***carrier*** of the recessive allele for cystic fibrosis.

If both parents are carriers (i.e. heterozygous)

PARENTAL (P) GENERATION   Nn   ×   Nn

GAMETES   N   n   N   n

1ST FILIAL (F$_1$) GENERATION   NN   Nn   Nn   nn

A ***Punnett square*** can be used to predict the possible combinations of alleles in the zygote.

Gametes from father ♂	N	n
Gametes from mother ♀ N	NN	Nn
n	Nn	nn

Phenotypically and genotypically normal

Phenotypically normal but a carrier

Phenotypically and genotypically recessive

3 NORMAL   :   1 MUTANT

Other significant examples of monohybrid inheritance in humans

Albinism: autosomal recessive

Phenylketonuria: results from a lack of the liver enzyme ***phenylaline hydroxylase***: blood phenylalanine levels are raised, causing a number of effects: autosomal recessive

Huntington's disease: autosomal dominant

Tay-Sach's disease: autosomal recessive (prevalent in Jews of Eastern European origin)

N.B. the 3 : 1 ratio is only approximate unless the number of offspring is very large (unlikely in humans), because

1. Alleles may not be distributed between viable gametes in equal numbers.

2. Fusion of gametes is completely random – it is a matter of chance whether one male gamete fuses with a particular female gamete.

The Hardy–Weinberg equation

Calculation of allele and genotype frequencies in a population

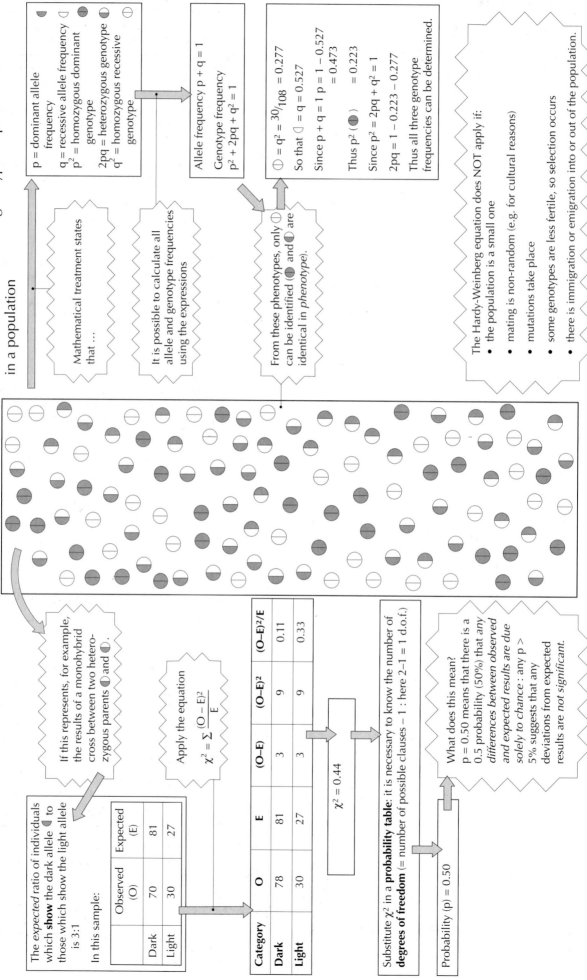

p = dominant allele frequency
q = recessive allele frequency
p^2 = homozygous dominant genotype
$2pq$ = heterozygous genotype
q^2 = homozygous recessive genotype

Mathematical treatment states that ….

It is possible to calculate all allele and genotype frequencies using the expressions

Allele frequency $p + q = 1$
Genotype frequency
$p^2 + 2pq + q^2 = 1$

From these phenotypes, only ⊖ can be identified (⬤ and ◑ are identical in *phenotype*).

⊖ $= q^2 = {}^{30}/_{108} = 0.277$

So that ⊖ $= q = 0.527$

Since $p + q = 1$ $p = 1 - 0.527$
$= 0.473$

Thus p^2 (⬤) $= 0.223$

Since $p^2 = 2pq + q^2 = 1$

$2pq = 1 - 0.223 - 0.277$

Thus all three genotype frequencies can be determined.

The Hardy-Weinberg equation does NOT apply if:
- the population is a small one
- mating is non-random (e.g. for cultural reasons)
- mutations take place
- some genotypes are less fertile, so selection occurs
- there is immigration or emigration into or out of the population.

Calculation of allele and genotype frequencies in a population

The Chi-squared (χ^2) test

Analysis of data consisting of discrete (discontinuous) variables

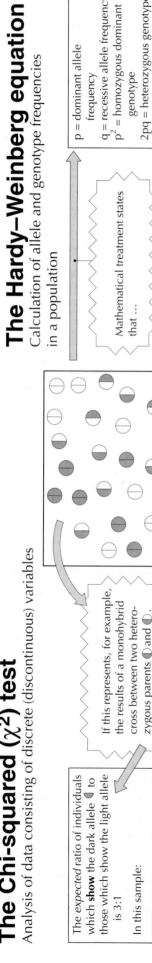

The *expected* ratio of individuals which **show** the dark allele ◑ to those which show the light allele is 3:1

In this sample:

	Observed (O)	Expected (E)
Dark	70	81
Light	30	27

If this represents, for example, the results of a monohybrid cross between two heterozygous parents ◑ and ◑.

Apply the equation
$$\chi^2 = \sum \frac{(O - E)^2}{E}$$

Category	O	E	(O–E)	(O–E)²	(O–E)²/E
Dark	78	81	3	9	0.11
Light	30	27	3	9	0.33

$\chi^2 = 0.44$

Substitute χ^2 in a **probability table**: it is necessary to know the number of **degrees of freedom** (= number of possible clauses – 1 : here 2–1 = 1 d.o.f.)

Probability (p) = 0.50

What does this mean?
p = 0.50 means that there is a 0.5 probability (50%) that *any* differences between observed and expected results are due *solely to chance* : any p > 5% suggests that any deviations from expected results are *not significant.*

Dihybrid inheritance involves

the transmission of *two pairs of alleles* at the same time but independently of one another.

Gregor Mendel was an Austrian monk who studied patterns of inheritance in the garden pea. He made several proposals concerning these paterns: these proposals have been found to hold true for many other organisms. One organism which has been widely used for experiments on inheritance is the fruit fly, *Drosophilia melanogaster*, which has a number of advantages for such studies:

1. It has a rapid generation time (10 days) and produces many offspring from each mating so that stasistical analysis can be applied to results.

2. It is easily cultured in small, convenient containers (milk bottles!) on a simple growth medium.

3. Males and females are easily distinguished so that controlled matings are possible.

4. The flies have a number of obvious external characteristics which are easily mutated. These include wing length, body colour and eye shape.

Flies which differ in *two pairs of characteristics* (e.g. *body colour* and *wing shape*) may be mated.

PARENTAL (P) GENERATION	Grey body, long wings	×	Black body, vestigial wings	
1ST FILIAL (F₁) GENERATION	All have grey body, long wings			
	Self-fertilization between F₁ individuals			
2ND FILIAL (F₂) GENERATION	Grey body, long wing			9
	Grey body, vestigal wing			3
	Black body, long wing			9
	Black body, vestigal wing			1

A phenotypic ratio of 9:3:3:1 would seem to be complex, but Mendel explained this as *two seperate monohybrid crosses* (i.e. *grey v. black* and *long v. vestigial*) occuring at the same time.

i.e. GREY v. BLACK $= (9+3) : (3+1)$
$= 12 : 4$
$= 3 : 1$

LONG v. VESTIGAL $= (9+3) : (3+1)$
$= 12 : 4$
$= 3 : 1$

i.e. $9 : 3 : 3 : 1$ is the same as $3 : 1 \times 3 : 1$

Thus the inheritance of *body colour* had not influenced the inheritance of *wing shape*.

Mendel's Second Law (the law of independent assortment)

'Each member of a pair of alleles may combine randomly with either of another pair'

In this example, the allele for grey body may combine equally often with the allele for *long* wing or with the allele for *vestigal* wing.

Using genetic symbols

Let G = grey, g = black, L = long, l = vestigal

P G G L L × g g l l

Gametes (GL) (gl)

F₁ G g L l

Gametes (GL) (Gl) (gL) (gl)

These will be produced in equal numbers, according to Mendel's Second law.

F₂ The possible combinations of gamets are most easily derived using a Punnett square.

♂ GAMETES / ♀ GAMETES	GL	Gl	gL	gl
GL	GGLL	GGLl	GgLL	GgLl
Gl	GGLl	GGll	GgLl	Ggll
gL	GgLl	GgLl	ggLL	ggLl
gl	GgLl	Ggll	ggLl	ggll

or, phenotypically
9 GREY BODY, LONG WING (both G and L in zygote)
3 GREY BODY, VESTIGAL WING (G and ll in zygote)
3 BLACK BODY, LONG WING (gg and L in zygote)
3 BLACK BODY, VESTIGAL WING (ggll in zygote)

Linkage between genes prevents free

recombination of alleles.

Consider this dihybrid cross between two pure-breeding fruit flies

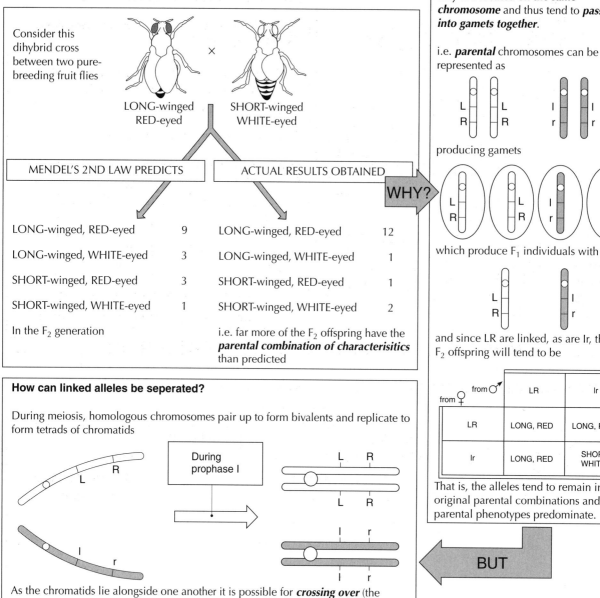

LONG-winged
RED-eyed
×
SHORT-winged
WHITE-eyed

MENDEL'S 2ND LAW PREDICTS		ACTUAL RESULTS OBTAINED	
LONG-winged, RED-eyed	9	LONG-winged, RED-eyed	12
LONG-winged, WHITE-eyed	3	LONG-winged, WHITE-eyed	1
SHORT-winged, RED-eyed	3	SHORT-winged, RED-eyed	1
SHORT-winged, WHITE-eyed	1	SHORT-winged, WHITE-eyed	2

In the F₂ generation

i.e. far more of the F₂ offspring have the **parental combination of characterisitics** than predicted

The genes for **wing length** and **eye colour** are **linked**. This means that t**hey are located on the same chromosome** and thus tend to **pass into gamets together**.

i.e. **parental** chromosomes can be represented as

L L | l l
R R | r r

producing gamets

L R | L R | l r | l r

which produce F₁ individuals with

L l
R r

and since LR are linked, as are lr, the F₂ offspring will tend to be

from ♀ \ from ♂	LR	lr
LR	LONG, RED	LONG, RED
lr	LONG, RED	SHORT, WHITE

That is, the alleles tend to remain in the original parental combinations and so parental phenotypes predominate.

BUT

How can linked alleles be seperated?

During meiosis, homologous chromosomes pair up to form bivalents and replicate to form tetrads of chromatids

During
prophase I

L R
L R

l r
l r

As the chromatids lie alongside one another it is possible for **crossing over** (the exchange of genetic material between adjacent members of a homologous pair) to occur.

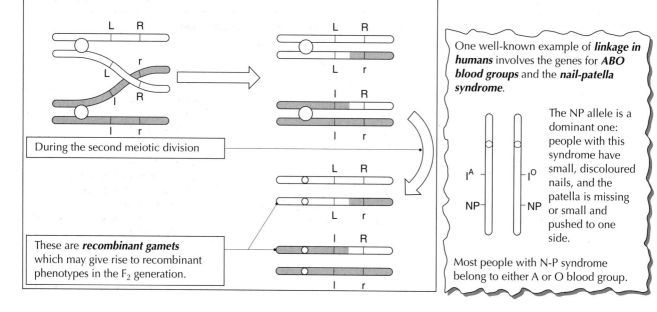

During the second meiotic division

These are **recombinant gamets** which may give rise to recombinant phenotypes in the F₂ generation.

One well-known example of **linkage in humans** involves the genes for **ABO blood groups** and the **nail-patella syndrome**.

I^A — | — I^O

NP — | — NP

The NP allele is a dominant one: people with this syndrome have small, discoloured nails, and the patella is missing or small and pushed to one side.

Most people with N-P syndrome belong to either A or O blood group.

CODOMINANCE occurs when *both alleles of a gene express themselves equally in the phenotype*, e.g. flower colour in snapdragons.

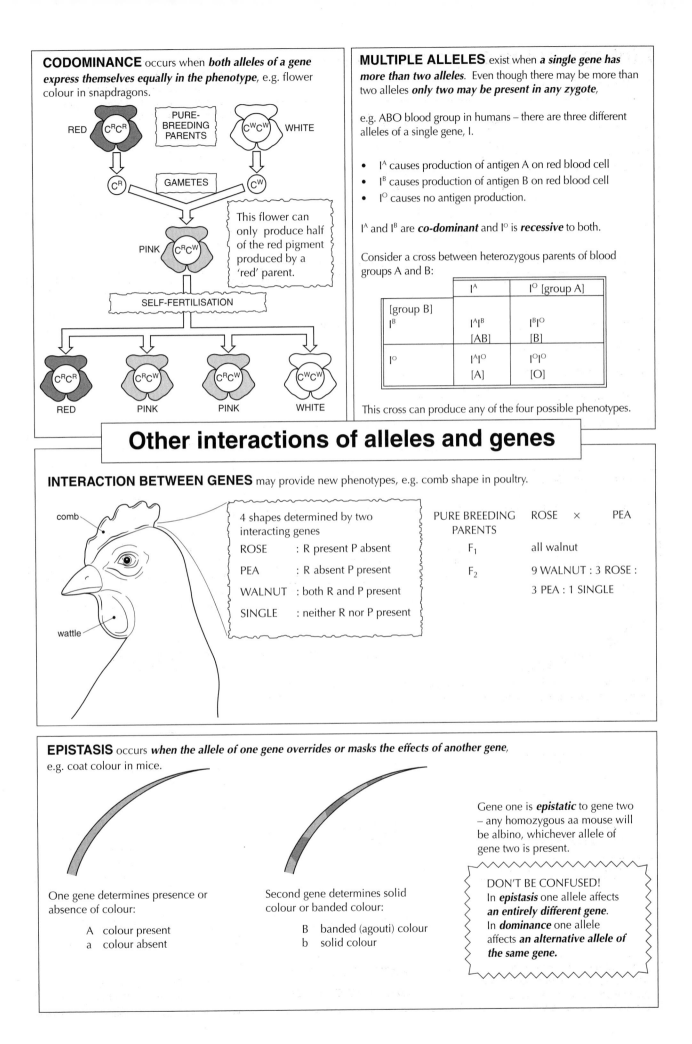

RED C^RC^R PURE-BREEDING PARENTS C^WC^W WHITE

GAMETES C^R C^W

PINK C^RC^W

This flower can only produce half of the red pigment produced by a 'red' parent.

SELF-FERTILISATION

C^RC^R RED C^RC^W PINK C^RC^W PINK C^WC^W WHITE

MULTIPLE ALLELES exist when *a single gene has more than two alleles*. Even though there may be more than two alleles *only two may be present in any zygote*,

e.g. ABO blood group in humans – there are three different alleles of a single gene, I.

- I^A causes production of antigen A on red blood cell
- I^B causes production of antigen B on red blood cell
- I^O causes no antigen production.

I^A and I^B are *co-dominant* and I^O is *recessive* to both.

Consider a cross between heterozygous parents of blood groups A and B:

	I^A	I^O [group A]
[group B] I^B	I^AI^B [AB]	I^BI^O [B]
I^O	I^AI^O [A]	I^OI^O [O]

This cross can produce any of the four possible phenotypes.

Other interactions of alleles and genes

INTERACTION BETWEEN GENES may provide new phenotypes, e.g. comb shape in poultry.

comb

wattle

4 shapes determined by two interacting genes

ROSE   : R present P absent

PEA   : R absent P present

WALNUT : both R and P present

SINGLE   : neither R nor P present

PURE BREEDING PARENTS	ROSE × PEA
F₁	all walnut
F₂	9 WALNUT : 3 ROSE : 3 PEA : 1 SINGLE

EPISTASIS occurs *when the allele of one gene overrides or masks the effects of another gene*, e.g. coat colour in mice.

Gene one is *epistatic* to gene two – any homozygous aa mouse will be albino, whichever allele of gene two is present.

DON'T BE CONFUSED!
In *epistasis* one allele affects *an entirely different gene*. In *dominance* one allele affects *an alternative allele of the same gene.*

One gene determines presence or absence of colour:

A   colour present
a   colour absent

Second gene determines solid colour or banded colour:

B   banded (agouti) colour
b   solid colour

Sex linkage and the inheritance of sex

Any genes on the X chromosomes will be inherited by both sexes, but whereas the male can only receive **one** of the alleles (he will be XY, and therefore must be **homozygous** for the X-linked allele) the female will be XX and thus may be either **homozygous** or **heterozygous**. This gives females a tremendous genetic advantage since any recessive lethal allele will not be expressed in the heterozygote.

For example, **haemophilia** is an X-linked condition.

Normal gene = H, mutant gene = h

PARENTS $X^H X^h$ × $X^H Y$
            female, carrier   male, normal

i.e. **both parents have normal phenotype** …

GAMETES   (X^H) (X^h)   (X^H) (Y)

F₁ OFFSPRING

	♂ GAMETES	
♀ GAMETES	X^H	Y
X^H	$X^H X^H$	$X^H Y$
X^h	$X^H X^h$	$X^h Y$

ie.

$X^H X^H$	$X^H X^h$	$X^H Y$	$X^h Y$
normal, female	carrier, female	normal, male	haemophiliac, male

… **but may have a haemophiliac son.**

Other significant X-linked conditions include *Duchenne muscular dystrophy*, *red-green colour blindness* and *coat colour in cats* (where the alleles for ginger (G) and black (g) produce tortoiseshell in the heterozygote $X^G X^g$: thus there should, in theory, be *no male tortoiseshell cats!*

In mammals sex is determined by two chromosomes which are very different to one another. These are the *heterosomes* – the male is *heterogametic* (XY: can produce both X and Y gametes) and the female is *homogametic* (can only produce X gametes).

♀ (XX)   ♂ (XY)

These sections are *homologous* and carry no genes of sex determination.

These *non-homologous* sections carry the genes concerned with sex determination but are of sufficient size to carry other genes. Such genes are *sex-linked*.

Inheritance of sex is a special form of Mendelian segregation.

♂ XY   ×   ♀ XX

GAMETES   (X) (Y)   (X) (X)

F₁ generation: sex of offspring can be determined from a Punnett square.

	♂ GAMETES	
♀ GAMETES	X	Y
X	XX (female)	XY (male)
X	XX (female)	XY (male)

Theoretically there should be a 1 : 1 ratio of male : female offspring. In humans various factors can upset the ratio – the Y sperm tend to have greater mobility; the XY zygote and embryo is more delicate than the XX embryo. The balance is just about maintained.

Any genes carried on the Y chromosome will be received by *all* the male offspring – there is little space for other than sex genes, but one well-known example concerns *webbed toes*.

$X Y^W$ — only Y chromosome carries W gene   × $X X$

GAMETES   (X) (Y^W)   (X) (X)

OFFSPRING $X Y^W$ $X Y^W$ X X   X X

            males with webbed toes   normal females

Natural selection may be a potent force in *evolution.*

Much variation is of the ***continuous type*** , i.e. a range of phenotypes exists between two extremes. The range of phenotypes within the environment will typically show a ***normal distribution.***

PHENOTYPIC CLASSES

The modern ***neo-Darwinian*** theory accepts that:

1. Some harmful alleles may survive, but the reproductive potential of the individual possessing such an allele will be reduced.

2. Selective advantages and disadvantages of an allele relate to one environment at one particular time, i.e. an allele does not always contribute to 'fitness' but only under certain conditions.

Plants and animals in Nature produce more offspring than can possibly survive, yet the population remains relatively constant. There must be many deaths in Nature.

Overproduction of this type leads to ***competition*** – for food, shelter and breeding sites, for example. There is thus a ***struggle for existence.*** Those factors in the environment for which competition occurs represent ***selection pressures.***

Within a population of individuals there may be considerable ***variation*** in genotype and thus in phenotype.

Variation means that some individuals possess characteristics which would be advantageous in the struggle for existence (and some would be the opposite, of course).

Those possessing the best combination of characteristics would be more competitive in the struggle for existence: they would be more 'fit' to cope with the selection pressures imposed by the environment. This is ***natural selection*** and promotes ***survival of the fittest.***

If variation is ***heritable*** (i.e. caused by an alteration in genotype) new generations will tend to contain a higher proportion of individuals suited to survival.

Stabilizing selection favours intermediate phenotypic classes and operates against extreme forms – there is thus a ***decrease*** in the frequency of alleles representing the extreme forms.

Stabilizing selection operates when the phenotype corresponds with optimal environmental conditions, and competition is not severe. It is probable that this form of selection has favoured heterozygotes for ***sickle cell anaemia*** in an environment in which ***malaria*** is common, and also works against ***extremes of birth weight*** in humans.

Directional selection favours one phenotype at one extreme of the range of variation. It moves the phenotype towards a new optimum environment; then stabilizing selection takes over. There is a change in the allele frequencies corresponding to the new phenotype.

Directional selection has occurred in the case of the peppered moth, *Biston betularia*, where the dark form was favoured in the sooty suburban environments of Britain during the industrial revolution: ***industrial melanism.*** Another significant example is the development of ***antibiotic resistance*** in populations of bacteria – mutant genes confer an advantage in the presence of an antibiotic.

Disruptive selection is the rarest form of selection and is associated with a variety of selection pressures operating within one environment.

This form of selection promotes the co-existence of more than one phenotype, the condition of ***polymorphism (balanced)*** polymorphism when no one selective agent is more important than any other). Important examples are:
1. ***Colour*** (yellow/brown) and ***banding pattern*** (from 0–5) in *Cepaea nemoralis.*
2. ***Three phenotypes*** (corresponding to HbHb/HbHb^s/Hb^{s}Hb^s) show an uneven distribution of the sickle-cell allele in different areas of the world.

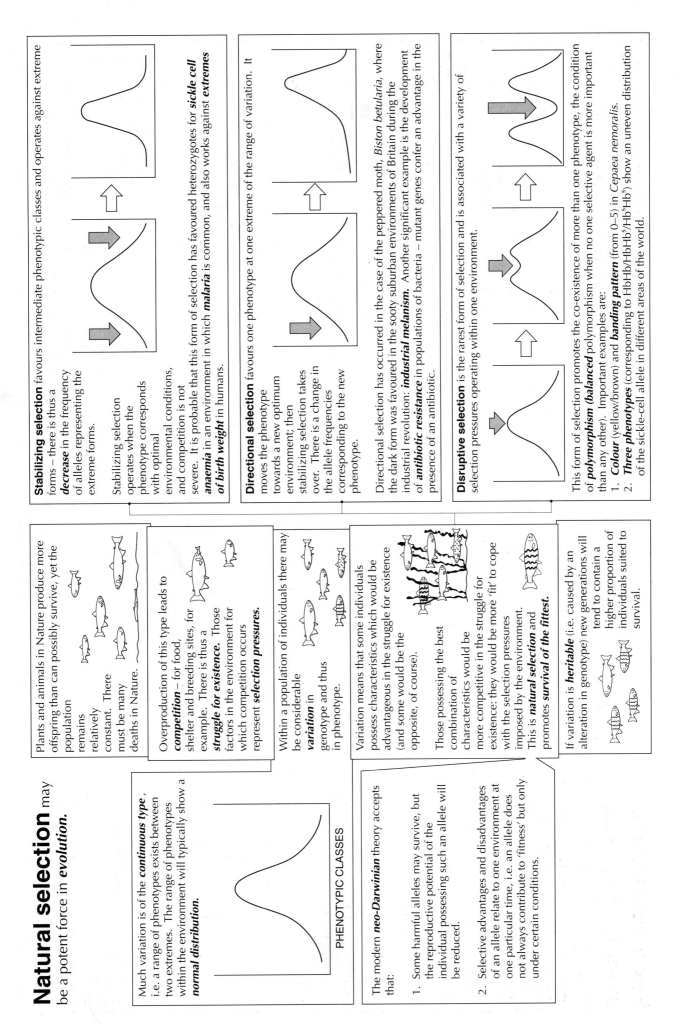

Artificial selection

Artificial selection occurs when humans, rather than environmental factors, determine which genotypes will pass to successive generations.

INBREEDING AND OUTBREEDING

Outbreeding occurs when there is selective controlled reproduction between members of genetically distant populations (different strains or even, for plants, closely related but different species).

Inbreeding occurs when there is selective reproduction between closely related individuals, e.g. between offspring of same litter or between parent and child.

Tends to **introduce new and superior phenotypes:** the progeny are known as **hybrids** and the development of improved characteristics is called **heterosis** or **hybrid vigour.**

e.g. introduction of disease resistance from wild sheep to domestic strains;

combination of shorter-stemmed 'wild' wheat and heavy yielding 'domestic' wheat.

This may result from increased numbers of dominant alleles or from new opportunities for gene interaction.

Tends to **maintain desirable characteristics**

e.g. uniform height in maize (easier mechanical harvesting);

maximum oil content of linseed (more economical extraction);

milk production by Jersey cows (high cream content).

But it may cause **reduced fertility** and **lowered disease resistance** as genetic variation is reduced.

Thus inbreeding is not favoured by animal breeders.

TECHNIQUES WITH ANIMALS

TECHNIQUES WITH ANIMALS are less well advanced than those with plants because:

a animals have a longer generation time and few offspring

b more food will be made available from improved plants

c there are many ethical problems which limit genetic experiments with animals.

Two important animal techniques are:

Artificial insemination: allows sperm from a male with desirable characteristics to fertilize a number of female animals.

Embryo transplantation: allows the use of **surrogate mothers** (thus increasing number of offspring) and **cloning** (production of many identical animals with the desired characteristics).

POLYPLOIDY AND PLANT BREEDING

Polyploids contain **multiple sets of chromosomes** (chromosome multiplication can be induced by treatment with **colchicine** during mitosis – this inhibits spindle formation and prevents chromatid separation).

Autopolyploids (all chromosomes from the **same** species) e.g. all **bananas** are **triploid** – they are infertile and contain no seeds. Most **potatoes** are **tetraploid** – cells are bigger and tubers are larger. Cultivated **strawberries** are **octoploid.**

Allopolyploidy (sets of chromosomes from **different** species) is possible if the two species have a chromosome complement similar in number and shape. This might allow plant breeders to **combine the beneficial characteristics of more than one species.**

The evolution of **bread wheat** is an important example.

Wild wheat: has brittle ears which fall off on harvesting.

    SELECTIVE BREEDING

Einkorn wheat: non-brittle but low yielding.

    POLYPLOIDY with *Agropyron* grass

Emmer wheat: high yielding but difficult to separate seed during threshing.

    POLYPLOIDY with *Aegilops* grass

Bread wheat: high yielding with easily separated 'naked' seeds.

PROTOPLAST FUSION

PROTOPLAST FUSION is a modern method for production of hybrids in plants.

e.g. production of virus-resistant tobacco.

Reproductive isolation and speciation

Reproductive isolation is essential for speciation: **allopatric speciation** occurs when populations occupy different environments; **sympatric speciation** occurs when populations are reproductively isolated within the same environment.

Geographical isolation takes place when two populations occupy two different environments which are separated by some physical barrier, such as a mountain range, a river or even a road system;
e.g. eastern and western races of the golden-mantled rosella, an Australian parakeet.

Mechanical isolation takes place when the reproductive structures are physically incompatible.
e.g. a Great Dane will not mate with a Chihuahua, and some flower species cannot be entered by the same pollinating insect.

Temporal isolation takes place when two or more species or populations live within the same area but are reproductively active at different times,
e.g. American frog species.

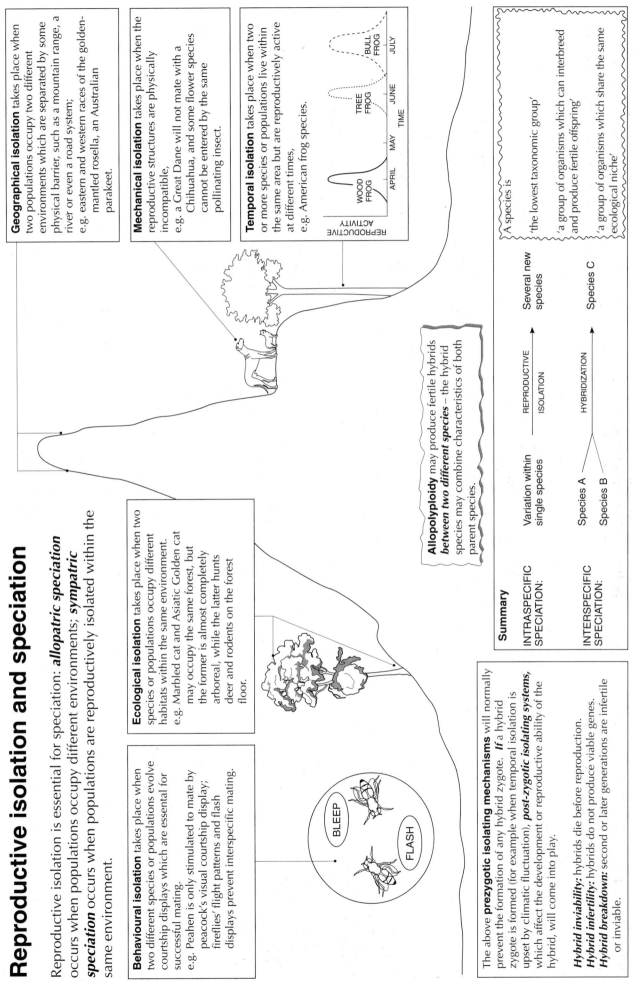

Behavioural isolation takes place when two different species or populations evolve courtship displays which are essential for successful mating.
e.g. Peahen is only stimulated to mate by peacock's visual courtship display; fireflies' flight patterns and flash displays prevent interspecific mating.

Ecological isolation takes place when two species or populations occupy different habitats within the same environment.
e.g. Marbled cat and Asiatic Golden cat may occupy the same forest, but the former is almost completely arboreal, while the latter hunts deer and rodents on the forest floor.

BLEEP

FLASH

The above **prezygotic isolating mechanisms** will normally prevent the formation of any hybrid zygote. *If* a hybrid zygote is formed (for example when temporal isolation is upset by climatic fluctuation), **post-zygotic isolating systems,** which affect the development or reproductive ability of the hybrid, will come into play.

Hybrid inviability: hybrids die before reproduction.
Hybrid infertility: hybrids do not produce viable genes.
Hybrid breakdown: second or later generations are infertile or inviable.

Allopolyploidy may produce fertile hybrids *between two different species* – the hybrid species may combine characteristics of both parent species.

A species is
'the lowest taxonomic group'
'a group of organisms which can interbreed and produce fertile offspring'
'a group of organisms which share the same ecological niche'

Summary

INTRASPECIFIC SPECIATION:

Variation within single species → REPRODUCTIVE ISOLATION → Several new species

INTERSPECIFIC SPECIATION:

Species A
Species B → HYBRIDIZATION → Species C

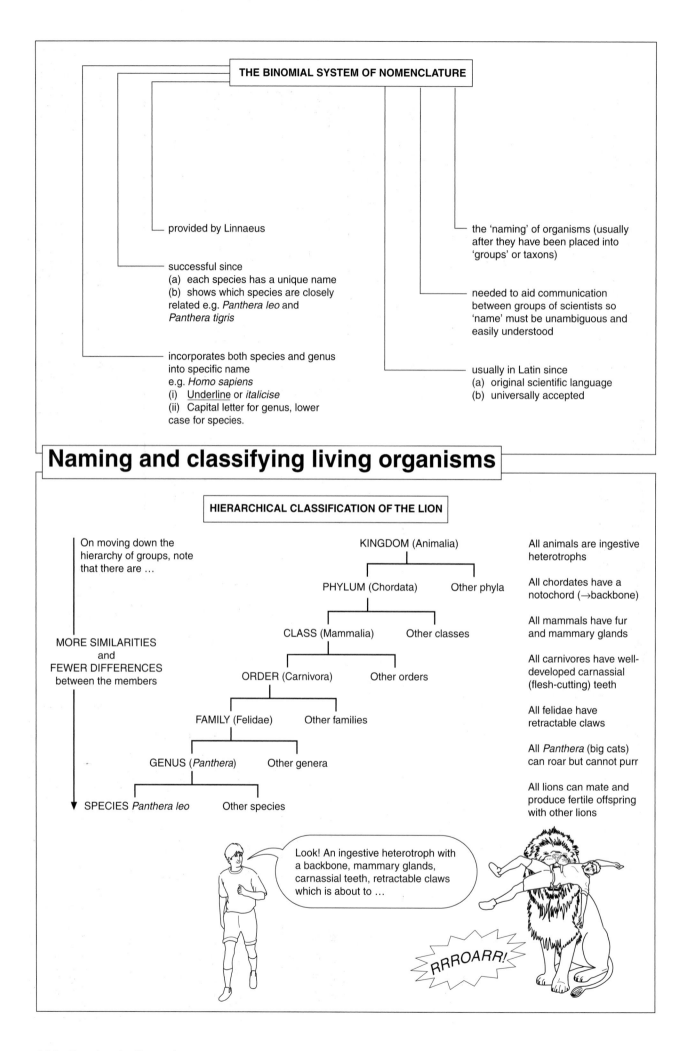

THE BINOMIAL SYSTEM OF NOMENCLATURE

— provided by Linnaeus

— successful since
(a) each species has a unique name
(b) shows which species are closely related e.g. *Panthera leo* and *Panthera tigris*

— incorporates both species and genus into specific name
e.g. *Homo sapiens*
(i) Underline or *italicise*
(ii) Capital letter for genus, lower case for species.

— the 'naming' of organisms (usually after they have been placed into 'groups' or taxons)

— needed to aid communication between groups of scientists so 'name' must be unambiguous and easily understood

— usually in Latin since
(a) original scientific language
(b) universally accepted

Naming and classifying living organisms

HIERARCHICAL CLASSIFICATION OF THE LION

On moving down the hierarchy of groups, note that there are …

MORE SIMILARITIES
and
FEWER DIFFERENCES
between the members

KINGDOM (Animalia)

PHYLUM (Chordata)   Other phyla

CLASS (Mammalia)   Other classes

ORDER (Carnivora)   Other orders

FAMILY (Felidae)   Other families

GENUS (*Panthera*)   Other genera

SPECIES *Panthera leo*   Other species

All animals are ingestive heterotrophs

All chordates have a notochord (→backbone)

All mammals have fur and mammary glands

All carnivores have well-developed carnassial (flesh-cutting) teeth

All felidae have retractable claws

All *Panthera* (big cats) can roar but cannot purr

All lions can mate and produce fertile offspring with other lions

Look! An ingestive heterotroph with a backbone, mammary glands, carnassial teeth, retractable claws which is about to …

RRROARR!

Keys and classification

* the FIVE KINGDOMS

A key enables identification of an organism by observation of its characteristics. Close observation allows series of questions (the branch points in this key) to be answered, eventually leading to the organism being studied.

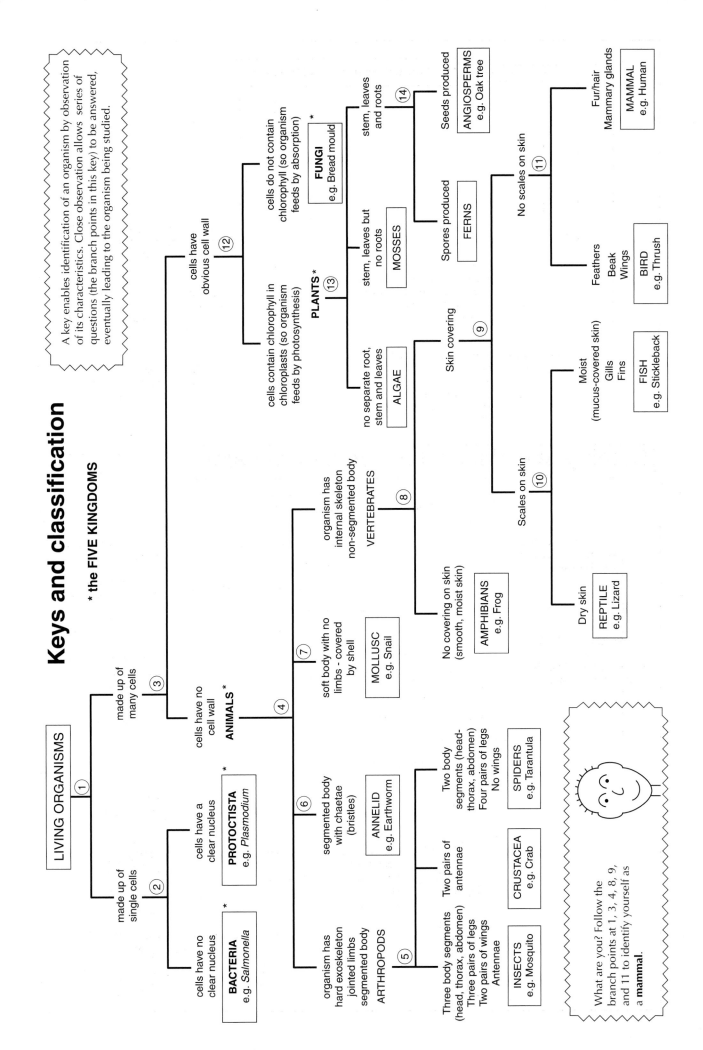

LIVING ORGANISMS ①

made up of single cells ②

  cells have no clear nucleus *
  BACTERIA e.g. *Salmonella*

  cells have a clear nucleus *
  PROTOCTISTA e.g. *Plasmodium*

made up of many cells ③

  cells have no cell wall
  ANIMALS * ④

  cells have obvious cell wall ⑫

    cells do not contain chlorophyll (so organism feeds by absorption) *
    FUNGI e.g. Bread mould

    cells contain chlorophyll in chloroplasts (so organism feeds by photosynthesis)
    PLANTS * ⑬

      no separate root, stem and leaves
      ALGAE

      stem, leaves but no roots
      MOSSES

      stem, leaves and roots ⑭

        Spores produced
        FERNS

        Seeds produced
        ANGIOSPERMS e.g. Oak tree

ANIMALS ④

  soft body with no limbs - covered by shell ⑦
  MOLLUSC e.g. Snail

  segmented body with chaetae (bristles) ⑥
  ANNELID e.g. Earthworm

  organism has hard exoskeleton jointed limbs segmented body
  ARTHROPODS ⑤

    Three body segments (head, thorax, abdomen) Three pairs of legs Two pairs of wings Antennae
    INSECTS e.g. Mosquito

    Two pairs of antennae
    CRUSTACEA e.g. Crab

    Two body segments (head-thorax, abdomen) Four pairs of legs No wings
    SPIDERS e.g. Tarantula

  organism has internal skeleton non-segmented body
  VERTEBRATES ⑧

    No covering on skin (smooth, moist skin)
    AMPHIBIANS e.g. Frog

    Skin covering ⑨

      Scales on skin ⑩

        Dry skin
        REPTILE e.g. Lizard

        Moist (mucus-covered skin) Gills Fins
        FISH e.g. Stickleback

      No scales on skin ⑪

        Feathers Beak Wings
        BIRD e.g. Thrush

        Fur/hair Mammary glands
        MAMMAL e.g. Human

What are you? Follow the branch points at 1, 3, 4, 8, 9, and 11 to identify yourself as a **mammal**.

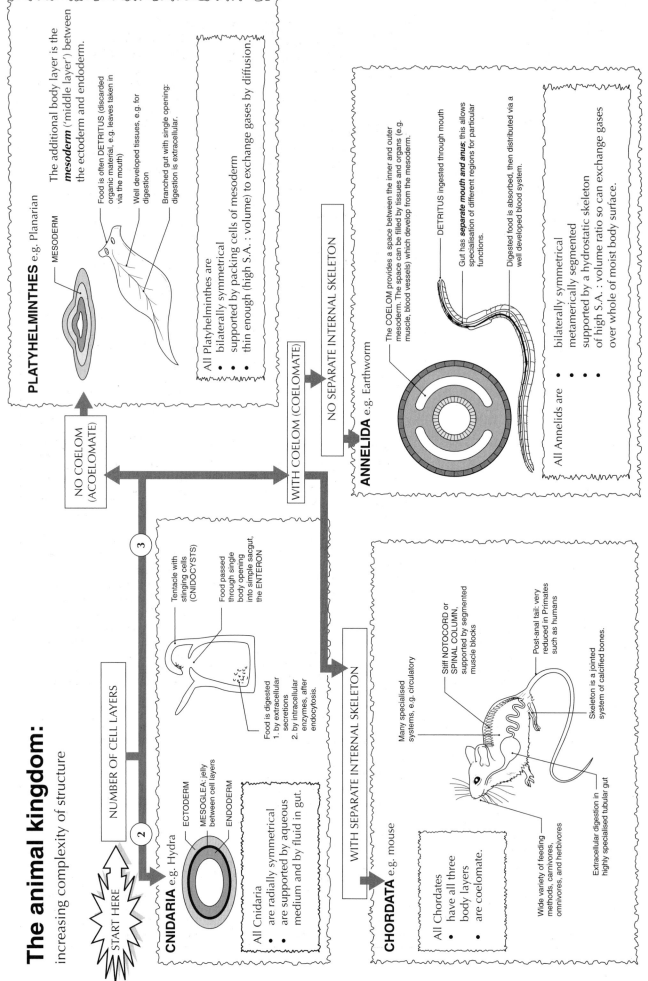

Section M Biotechnology

Genetic profiling: minisatellites and probes

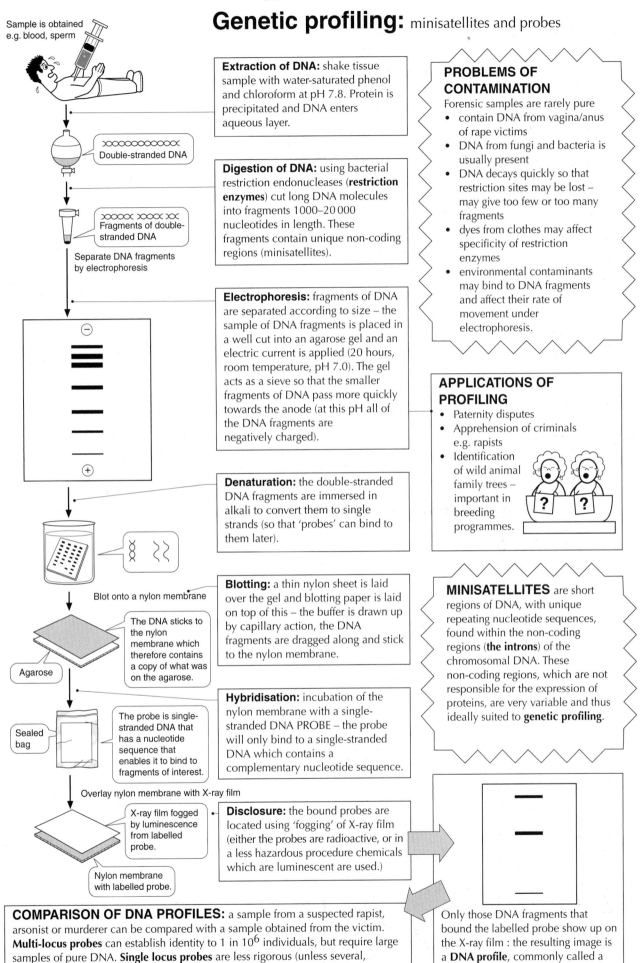

Sample is obtained e.g. blood, sperm

Double-stranded DNA

Fragments of double-stranded DNA

Separate DNA fragments by electrophoresis

Blot onto a nylon membrane

Agarose

The DNA sticks to the nylon membrane which therefore contains a copy of what was on the agarose.

Sealed bag

The probe is single-stranded DNA that has a nucleotide sequence that enables it to bind to fragments of interest.

Overlay nylon membrane with X-ray film

X-ray film fogged by luminescence from labelled probe.

Nylon membrane with labelled probe.

Extraction of DNA: shake tissue sample with water-saturated phenol and chloroform at pH 7.8. Protein is precipitated and DNA enters aqueous layer.

Digestion of DNA: using bacterial restriction endonucleases (**restriction enzymes**) cut long DNA molecules into fragments 1000–20 000 nucleotides in length. These fragments contain unique non-coding regions (minisatellites).

Electrophoresis: fragments of DNA are separated according to size – the sample of DNA fragments is placed in a well cut into an agarose gel and an electric current is applied (20 hours, room temperature, pH 7.0). The gel acts as a sieve so that the smaller fragments of DNA pass more quickly towards the anode (at this pH all of the DNA fragments are negatively charged).

Denaturation: the double-stranded DNA fragments are immersed in alkali to convert them to single strands (so that 'probes' can bind to them later).

Blotting: a thin nylon sheet is laid over the gel and blotting paper is laid on top of this – the buffer is drawn up by capillary action, the DNA fragments are dragged along and stick to the nylon membrane.

Hybridisation: incubation of the nylon membrane with a single-stranded DNA PROBE – the probe will only bind to a single-stranded DNA which contains a complementary nucleotide sequence.

Disclosure: the bound probes are located using 'fogging' of X-ray film (either the probes are radioactive, or in a less hazardous procedure chemicals which are luminescent are used.)

PROBLEMS OF CONTAMINATION

Forensic samples are rarely pure
- contain DNA from vagina/anus of rape victims
- DNA from fungi and bacteria is usually present
- DNA decays quickly so that restriction sites may be lost – may give too few or too many fragments
- dyes from clothes may affect specificity of restriction enzymes
- environmental contaminants may bind to DNA fragments and affect their rate of movement under electrophoresis.

APPLICATIONS OF PROFILING
- Paternity disputes
- Apprehension of criminals e.g. rapists
- Identification of wild animal family trees – important in breeding programmes.

MINISATELLITES are short regions of DNA, with unique repeating nucleotide sequences, found within the non-coding regions (**the introns**) of the chromosomal DNA. These non-coding regions, which are not responsible for the expression of proteins, are very variable and thus ideally suited to **genetic profiling**.

Only those DNA fragments that bound the labelled probe show up on the X-ray film : the resulting image is a **DNA profile**, commonly called a **DNA fingerprint**.

COMPARISON OF DNA PROFILES: a sample from a suspected rapist, arsonist or murderer can be compared with a sample obtained from the victim. **Multi-locus probes** can establish identity to 1 in 10^6 individuals, but require large samples of pure DNA. **Single locus probes** are less rigorous (unless several, typically 4 or 5, are used sequentially) but can be applied to tiny samples.

Gene cloning

involves *recombination,* *transformation* and *selection*.

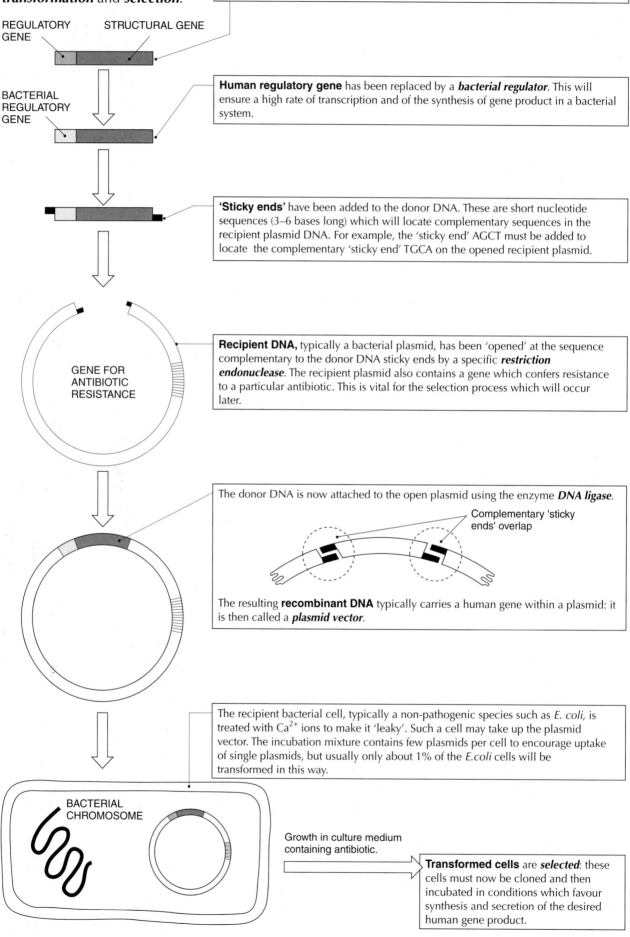

Donor DNA, e.g. a human gene. This may have been extracted from a *genome library*, been manufactured from messenger RNA using *reverse transcriptase* or, very rarely, synthesized in an *automatic polynucleotide synthesizer*.

REGULATORY GENE   STRUCTURAL GENE

Human regulatory gene has been replaced by a *bacterial regulator*. This will ensure a high rate of transcription and of the synthesis of gene product in a bacterial system.

BACTERIAL REGULATORY GENE

'Sticky ends' have been added to the donor DNA. These are short nucleotide sequences (3–6 bases long) which will locate complementary sequences in the recipient plasmid DNA. For example, the 'sticky end' AGCT must be added to locate the complementary 'sticky end' TGCA on the opened recipient plasmid.

GENE FOR ANTIBIOTIC RESISTANCE

Recipient DNA, typically a bacterial plasmid, has been 'opened' at the sequence complementary to the donor DNA sticky ends by a specific *restriction endonuclease*. The recipient plasmid also contains a gene which confers resistance to a particular antibiotic. This is vital for the selection process which will occur later.

The donor DNA is now attached to the open plasmid using the enzyme *DNA ligase*.

Complementary 'sticky ends' overlap

The resulting **recombinant DNA** typically carries a human gene within a plasmid: it is then called a *plasmid vector*.

The recipient bacterial cell, typically a non-pathogenic species such as *E. coli*, is treated with Ca^{2+} ions to make it 'leaky'. Such a cell may take up the plasmid vector. The incubation mixture contains few plasmids per cell to encourage uptake of single plasmids, but usually only about 1% of the *E.coli* cells will be transformed in this way.

BACTERIAL CHROMOSOME

Growth in culture medium containing antibiotic.

Transformed cells are *selected*: these cells must now be cloned and then incubated in conditions which favour synthesis and secretion of the desired human gene product.

Enzymes and genetic engineering

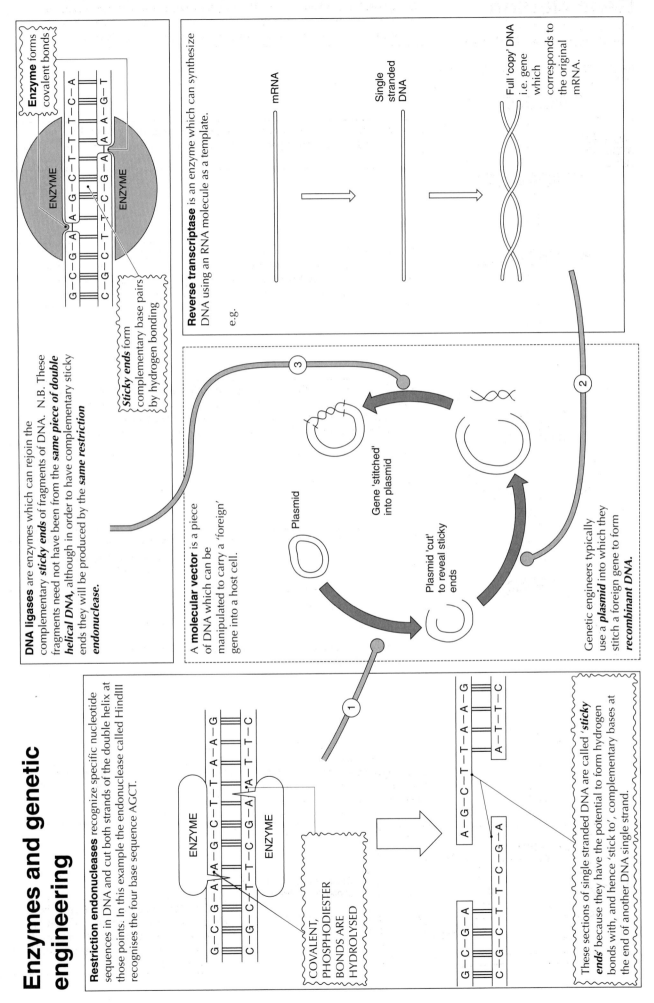

Restriction endonucleases recognize specific nucleotide sequences in DNA and cut both strands of the double helix at those points. In this example the endonuclease called HindIII recognises the four base sequence AGCT.

ENZYME

G – C – G – A │ A – G – C – T – T – A – A – G

C – G – C – T – T – C – G – A │ A – T – T – C

ENZYME

COVALENT, PHOSPHODIESTER BONDS ARE HYDROLYSED

G – C – G – A   A – G – C – T – T – A – A – G

C – G – C – T – T – C – G – A   A – T – T – C

These sections of single stranded DNA are called 'sticky ends' because they have the potential to form hydrogen bonds with, and hence 'stick' to', complementary bases at the end of another DNA single strand.

A **molecular vector** is a piece of DNA which can be manipulated to carry a 'foreign' gene into a host cell.

Plasmid

①

Plasmid 'cut' to reveal sticky ends

Gene 'stitched' into plasmid

② Genetic engineers typically use a *plasmid* into which they stitch a foreign gene to form *recombinant DNA.*

③

DNA ligases are enzymes which can rejoin the complementary *sticky ends* of fragments of DNA. N.B. These fragments need not have been from the *same piece of double helical DNA*, although in order to have complementary sticky ends they will be produced by the *same restriction endonuclease.*

Enzyme forms covalent bonds

ENZYME

G – C – G – A – A – G – C – T – T – T – C – A

C – G – C – T – T – C – G – A – A – A – G – T

ENZYME

Sticky ends form complementary base pairs by hydrogen bonding

Reverse transcriptase is an enzyme which can synthesize DNA using an RNA molecule as a template.

e.g.

mRNA

Single stranded DNA

Full 'copy' DNA i.e. gene which corresponds to the original mRNA.

Bioreactors/fermenters exploit microbes for commercial reasons.

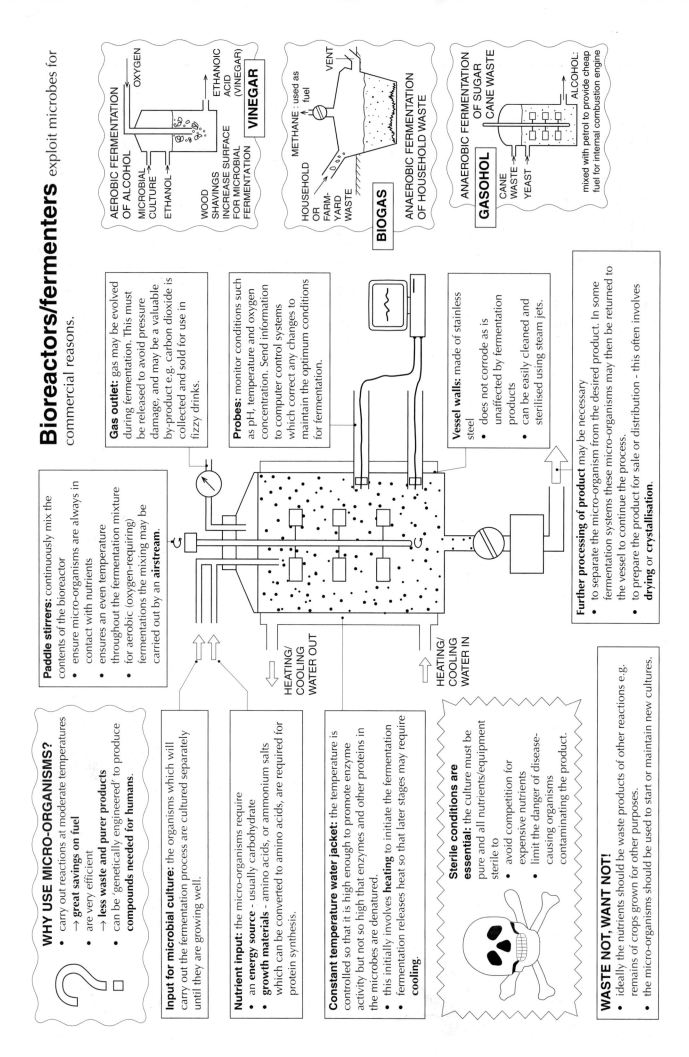

VINEGAR

AEROBIC FERMENTATION OF ALCOHOL — OXYGEN — ETHANOIC ACID (VINEGAR) — MICROBIAL CULTURE — ETHANOL — WOOD SHAVINGS INCREASE SURFACE FOR MICROBIAL FERMENTATION

BIOGAS

VENT — METHANE: used as fuel — ANAEROBIC FERMENTATION OF HOUSEHOLD WASTE — HOUSEHOLD OR FARM-YARD WASTE

GASOHOL

ANAEROBIC FERMENTATION OF SUGAR CANE WASTE — ALCOHOL: mixed with petrol to provide cheap fuel for internal combustion engine — CANE WASTE — YEAST

Gas outlet: gas may be evolved during fermentation. This must be released to avoid pressure damage, and may be a valuable by-product e.g. carbon dioxide is collected and sold for use in fizzy drinks.

Probes: monitor conditions such as pH, temperature and oxygen concentration. Send information to computer control systems which correct any changes to maintain the optimum conditions for fermentation.

Vessel walls: made of stainless steel
- does not corrode as is unaffected by fermentation products
- can be easily cleaned and sterilised using steam jets.

Further processing of product may be necessary
- to separate the micro-organism from the desired product. In some fermentation systems these micro-organisms may then be returned to the vessel to continue the process.
- to prepare the product for sale or distribution - this often involves **drying** or **crystallisation**.

Paddle stirrers: continuously mix the contents of the bioreactor
- ensure micro-organisms are always in contact with nutrients
- ensures an even temperature throughout the fermentation mixture
- for aerobic (oxygen-requiring) fermentations the mixing may be carried out by an **airstream**.

HEATING/COOLING WATER OUT

HEATING/COOLING WATER IN

WHY USE MICRO-ORGANISMS?
- carry out reactions at moderate temperatures → **great savings on fuel**
- are very efficient → **less waste and purer products**
- can be 'genetically engineered' to produce **compounds needed for humans.**

Input for microbial culture: the organisms which will carry out the fermentation process are cultured separately until they are growing well.

Nutrient input: the micro-organisms require
- an **energy source** - usually carbohydrate
- **growth materials** - amino acids, or ammonium salts which can be converted to amino acids, are required for protein synthesis.

Constant temperature water jacket: the temperature is controlled so that it is high enough to promote enzyme activity but not so high that enzymes and other proteins in the microbes are denatured.
- this initially involves **heating** to initiate the fermentation
- fermentation releases heat so that later stages may require **cooling.**

Sterile conditions are essential: the culture must be pure and all nutrients/equipment sterile to
- avoid competition for expensive nutrients
- limit the danger of disease-causing organisms contaminating the product.

WASTE NOT, WANT NOT!
- ideally the nutrients should be waste products of other reactions e.g. remains of crops grown for other purposes.
- the micro-organisms should be used to start or maintain new cultures.

Improved shelf life: many fruits are wasted because they deteriorate before they can be sold/eaten. A gene has been introduced into tomatoes which inhibits the enzymes causing deterioration – the new '**flavr-savr**' tomatoes last for several weeks.

TRANSFERRING GENES WITH *Agrobacterium tumefaciens*

Introduce desired gene into T1 plasmid.

Return plasmid to bacterium.

Bacterium infects plant – plant produces a tumour (crown gall). Each cell contains the plasmid with the desired gene.

Fragments of gall grow into identical plants each containing the desired gene.

Gene transfer
can promote desirable characteristics.

Nitrogen fixation: this process involves the reduction of nitrogen gas from the atmosphere into a form suitable for conversion into amino acids, nucleotides and other essential compounds.

$$N_2 \text{ from atmosphere} \longrightarrow NH_4^+ \text{ ammonium ions} \longrightarrow \text{amino acids and other compounds}$$

this key step is controlled by enzymes coded for by 12 genes called the Nif genes.

Most plants cannot fix nitrogen but it is hoped that gene transfer might either

- insert these genes directly into a plant
- make a plant more susceptible to the formation of root nodules with nitrogen-fixing bacteria of the genus *Rhizobium*.

This could

- produce cereal crops which also manufacture large amounts of protein;
- reduce the demand for nitrogenous fertilisers.

YUK!

RESISTANCE TO PESTS AND HERBICIDES

A gene is inserted into the plant which enables it to make **insecticidal crystal protein** (ICP) which affects the gut of the caterpillars so that they cannot feed and eventually die.

The crop plant has a gene transferred into it which makes it resistant to herbicides. The field of growing crop can then be sprayed with the herbicide which will selectively kill the 'weeds' since they do not possess the 'resistance' gene.

GENE TRANSFER IS ALSO IMPORTANT IN ANIMALS

TRANSGENIC ANIMALS

DNA containing desired gene can be introduced into nucleus using a fine pipette.

Animal releases protein made from desirable gene in its milk. Factor 8 (essential for blood clotting) is made in this way.

Cells are cultured, implanted into female animal which gives birth to transgenic animal.

GENE THERAPY may help in the treatment of cystic fibrosis by the replacement of a faulty CFTR gene.

The basic principle is

Adenovirus is an ideal vector since it can enter non-dividing cells such as epithelia in lungs.

Adenovirus coat – genetic material removed so no danger of infection

Healthy CFTR gene is inserted in bacterial plasmid

Adenovirus now carries healthy CFTR gene

Human epithelial cell

DNA is released and is transported to the nucleus

CFTR gene codes for protein synthesis

normal CFTR now present – chloride and water movement normal

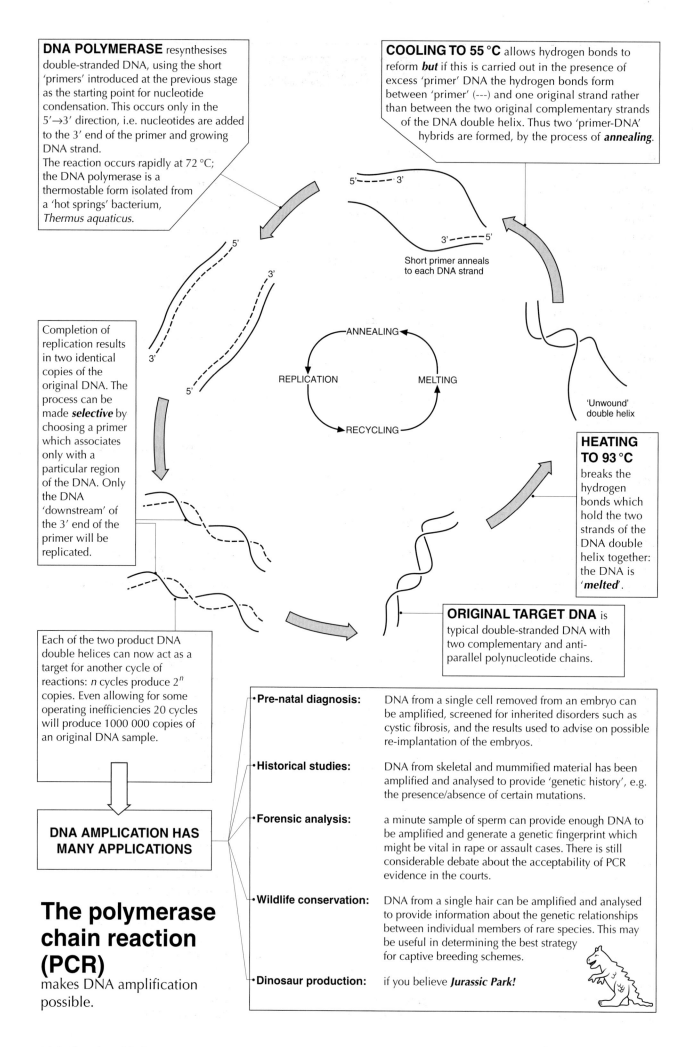

DNA POLYMERASE resynthesises double-stranded DNA, using the short 'primers' introduced at the previous stage as the starting point for nucleotide condensation. This occurs only in the $5' \rightarrow 3'$ direction, i.e. nucleotides are added to the 3' end of the primer and growing DNA strand.
The reaction occurs rapidly at 72 °C; the DNA polymerase is a thermostable form isolated from a 'hot springs' bacterium, *Thermus aquaticus*.

COOLING TO 55 °C allows hydrogen bonds to reform *but* if this is carried out in the presence of excess 'primer' DNA the hydrogen bonds form between 'primer' (---) and one original strand rather than between the two original complementary strands of the DNA double helix. Thus two 'primer-DNA' hybrids are formed, by the process of *annealing*.

5'- - - - -·3'

3'- - - - -·5'

Short primer anneals to each DNA strand

Completion of replication results in two identical copies of the original DNA. The process can be made *selective* by choosing a primer which associates only with a particular region of the DNA. Only the DNA 'downstream' of the 3' end of the primer will be replicated.

ANNEALING

REPLICATION   MELTING

RECYCLING

'Unwound' double helix

HEATING TO 93 °C breaks the hydrogen bonds which hold the two strands of the DNA double helix together: the DNA is *'melted'*.

Each of the two product DNA double helices can now act as a target for another cycle of reactions: n cycles produce 2^n copies. Even allowing for some operating inefficiencies 20 cycles will produce 1 000 000 copies of an original DNA sample.

ORIGINAL TARGET DNA is typical double-stranded DNA with two complementary and anti-parallel polynucleotide chains.

DNA AMPLICATION HAS MANY APPLICATIONS

The polymerase chain reaction (PCR)

makes DNA amplification possible.

• **Pre-natal diagnosis:** DNA from a single cell removed from an embryo can be amplified, screened for inherited disorders such as cystic fibrosis, and the results used to advise on possible re-implantation of the embryos.

• **Historical studies:** DNA from skeletal and mummified material has been amplified and analysed to provide 'genetic history', e.g. the presence/absence of certain mutations.

• **Forensic analysis:** a minute sample of sperm can provide enough DNA to be amplified and generate a genetic fingerprint which might be vital in rape or assault cases. There is still considerable debate about the acceptability of PCR evidence in the courts.

• **Wildlife conservation:** DNA from a single hair can be amplified and analysed to provide information about the genetic relationships between individual members of rare species. This may be useful in determining the best strategy for captive breeding schemes.

• **Dinosaur production:** if you believe *Jurassic Park!*

INDEX

chlorofluorocarbons (CFCs) 69, 71
chlorophyll 37, 39, 42
chloroplast 11, 39, 52
cholesterol 9, 122, 125
chordata 198
choroid 147
chromatid 181, 182
chromatin 10, 177
chromatography 39
chromosome 10, 177, 181, 182, 185, 190
chrysalis 105
chymosin 98
climax community 65
Clinistix 24, 127
cloning 142
Cnidaria 198
cocaine 159
cochlea 150
codominance 191
codon 33, 178, 179
coelom 198
coenzyme 22
cofactor 22
cohesion 5
cohesion–tension 49
collagen 21
collecting duct 129, 130
collenchyma 43
colloid 5
colon 85
colour blindness 192
community 59, 65
companion cell 46
competition, interspecific and intraspecific 62
complement 134
conditioned reflex 87
conditioning 154
cone cell 148, 149
consumer 63, 64
contraception 165, 170
corpus luteum 167
cortex 44, 129, 160
covalent bond 16
creatine phosphate 31
crop 73, 83
    rotation 80
CSF (cerebrospinal fluid) 156
cuticle 36
cystic fibrosis 187, 203
cytokinesis 181, 182
cytokinin 53
cytoplasm 10, 11
cytosine 32

D

Darwinism 193
DDT 78, 101, 145
deamination 101, 102
decomposer 63, 64, 65, 68
deforestation 72, 81, 82

dendrite 155
denitrification 68
deoxyribonucleic acid (DNA) see also DNA 32, 176, 177, 181, 182, 199, 200
depolarisation 157
desertification 81
diabetes 127
dialysis, kidney 131
diaphragm 85, 107, 110
diastole 121
diffusion, facilitated 9
digestion 85, 86, 89
dihybrid 189
disaccharide 18
disease 14, 134
disulphide bridge 20
DNA (deoxyribonucleic acid) 32, 176, 177, 181, 182, 199, 200
    fingerprinting 199
    polymerase 204
dormancy 53, 56
double fertilisation 56
double helix 32
Down's syndrome 182, 185
duodenum 85

E

ECG (electrocardiogram) 120, 121
ecology 59
ecosystem 59, 63, 64
ectotherm 133
effector 156
electrocardiogram (ECG) 120, 121
electrophoresis 199
embryo 56
emphysema 107, 108
emulsification 86
emulsion test 16
end product inhibition 23, 25
endocrine organ 151
endocytosis 138
endodermis 44, 49
endonuclease 200, 201
endoplasmic reticulum 10, 11
endosperm 56
endotherm 132
enzyme 22, 24, 25, 201
enzyme–substrate complex 22
epidermis 43, 44
epiglottis 85, 107
epistasis 191
epithelium 35
erythrocyte 106, 112, 117
ethanol 83, 99
ethene 53
etiolation 36
Eustachian tube 150
eutrophication 75
evolution 193
excretion 128, 129
exercise 31, 123, 164

exocytosis 10
expiration 108, 110
extensor 162
extinction 72
eye 147
eyepiece 1

F

fat 16, 75, 86, 89
fatty acid 16, 17
feedback inhibition 168
fermentation 83, 99, 104, 144, 202
fern 57
fertilisation 169
fibre 84
    muscle 163
fibrinogen 21, 112, 137
filament 54, 55
flagellum 12, 166
flexor 162
flower 58
flowering 53
follicle stimulating hormone (FSH) 168
food poisoning 95
fructose 18
FSH (follicle stimulating hormone) 168
fuel 82, 83, 202
fungicide 78, 80
fungus 103

G

gamete 182, 187
gasohol 202
gated channels 9, 157
gene 177, 178, 184, 200
    cloning 200
    therapy 203
generator potential 146
genetic engineering 127, 200, 201, 203
genetic modification (GM) 73, 203
genotype 187, 193
geotropism 53
germination 56
gibberellic acid 53, 56
gills 106
global warming 69
glomerulus 129
glucagon 127, 151
glucose 16, 18, 19, 24, 38, 101, 102, 127
glycocalyx 9
glycogen 19, 101, 102, 127
glycosidic bond 18
GM (genetic modification) 73, 203
goitre 90
Golgi apparatus 10, 11
Graafian follicle 167
granum 39
greenhouse effect 69, 71
grey matter 156

S

SA (sino-atrial) node 120
saliva 85, 87
salivary gland 87
saprobiontic nutrition 103
sarcomere 163
sclerenchyma 43
scrotum 165
secondary structure 20
seed 56, 58
semen 165
semi-conservative replication 176, 177
sensory neurone 155, 156
sepals 54, 55
sere 66
serial dilution 136
Sertoli cell 166
serum 112
sewage 74
sex linkage 192
sickle cell anaemia 184
sieve tube 46
sino-atrial (SA) node 120
sinusoid 101, 102
smoking 122, 125
sodium 130, 155
solute potential 47, 48, 113
solvent 5, 6, 114
sorghum 73
soy sauce 99
soya 73
speciation 194
species 72, 194, 196
specific heat capacity 5
sperm 165, 166
spermatocyte 166
spinal cord 153, 156
spirometer 111
spleen 122
spore 57
starch 16, 19, 86
statolith 53
stem 43
steroid 17, 151
stigma 54, 55
stimulus 126
stomach 85, 86
stomata 36, 52, 58
stroma 39, 167
suberin 11
succession 66
sucrase (invertase) 86
sulphur 15
    dioxide 70
summation 148
surface tension 5

sweating 132
sympathetic nervous system 147, 153
sympatric speciation 195
symplast 11, 49
synapse 159
synovial fluid 6, 161
systole 121

T

Taenia 103
tapeworm 103
taxis 154
TCA cycle 26, 29
telophase 181, 182
tertiary structure 20
testa 54
testis 151, 165
thermoregulation 6, 132, 133
threshold potential 146, 157
thylakoid 37, 39
thymine 32, 33
thyroid gland 151
thyroxine 132, 151
tidal volume 111
tissue 34, 35
    fluid 113
T-lymphocyte 139
tongue 87
trachea 107
tracheole 106
tracheophyte 57
transamination 29, 102
transcription 178, 186
transect 60
transfer RNA 33, 178, 179
transgenic organism 203
translation 178, 179, 186
translocation 45, 46, 183
transmembrane protein 9
transpiration 6, 49, 50
transplant, kidney 131
trisomy-21 185
trophic level 64, 65
troposphere 71
tubular reabsorption 129
Tullgren funnel 61
turgor 48, 52
tympanum 150

U

ultraviolet radiation 71
uracil 33
urea 27, 102, 128
ureter 129, 165

urethra 129
uric acid 128
urine 129, 130
uterus 165, 167

V

vaccination 140
vacuole 11
vagina 165, 167
vagus nerve 119
variation 182
variation
    continuous 183
    discontinuous 183
vas deferens 165
vasectomy 165, 170
vasoconstriction 132
vasomotor centre 123
vector 137, 145
vegetarian 91
vena cava 118, 129
ventilation 106
vesicle 10
viable count 136
villus 89, 171
virus 14, 141
vital capacity 111
vitamin 84, 102
    C 18
    D 17, 90

W

water 5, 6, 7, 84
water potential 7, 47, 48, 49
white matter 156

X

xanthophyll 39
xeromorph 51
xerophyte 51
xylem 36, 43, 44, 49, 58

Y

yeast 83, 99
yoghurt 98

Z

zygote 56, 169